Geology of Soils

Man's ingenuity and his highly developed technology have enabled him to modify his environment in many ways. Nevertheless, he remains entirely dependent upon plants for his existence, as do other animals. Plants, in turn, are dependent upon water. So too is man, and the availability of water in places remote from rivers depends upon the kind of ground: some kinds store water, and some do not. [Photograph by John Stacy, U.S.G.S.]

Geology of Soils

THEIR EVOLUTION, CLASSIFICATION, AND USES

Charles B. Hunt
THE JOHNS HOPKINS UNIVERSITY

W. H. Freeman and Company
SAN FRANCISCO

A Series of Books in Geology
EDITORS: James Gilluly, A. O. Woodford, and Thane H. McCulloh

Printed in the United States of America

Library of Congress Catalog Card Number: 71-158739

International Standard Book Number: 0-7167-0253-3

3   4   5   6   7   8   9

Preface

This book presents a discussion of soils in terms of three different disciplines—geology, soil science, and engineering. The presentation necessarily has a geological emphasis, for that has been my main field. It is hoped, however, that the approach taken here will lead to increased interdisciplinary study, and thus to a better understanding of the ground around us. For too long, the specialists—geologists, soil scientists, and engineers—have worked each in his own cell without the benefit of the knowledge that the others could provide.

My interest in surface deposits and soils was stimulated by two experiences during World War II. The first was when I became acquainted with valley fills in the Basin and Range Province, while studying deposits of bedded manganese. The first experience made me realize how little was known about unconsolidated deposits, because geologists had given so much emphasis to the hard rocks. The second came when I was Assistant Chief, and later Chief, of the Military Geology Unit of the United States Geological Survey, and had the opportunity to work with teams of soil scientists, plant geographers, hydrologists, and engineers, as well as geologists. My interest in mixing the disciplines became further stimulated after the war—while working in the Lake Bonneville Basin, where I developed an appreciation of the stratigraphic importance of the pre-Wisconsinan weathering and soils. This book is the result of those and other experiences.

The first several chapters of the book are given to the discussion of those factors responsible for the development and distribution of the various kinds of surface deposits and soils: geologic history (time), topography, weathering and erosion, climate, and organisms. Most of the remainder of the book discusses characteristics of the surface deposits and soils, including their geography, physical properties, mineralogy and geochemistry, fertility and conservation, and some engineering aspects.

To avoid filling the book with definitions, technical terms are used as little as possible, and simplified explanations are given for those that are used. Several hundred words from the various disciplines—terms not necessarily used in this book—are explained in a combined index and glossary.

January 1972   *Charles B. Hunt*

Contents

Geology of Soils

Barnard Glacier, Alaska. Glaciers once covered most of Canada and a large part of the United States, reaching as far south as Missouri and Illinois. That ground is mantled with various kinds of surface deposits of glacial origin, from bouldery till to fine loess. All of our modern soils, including the ones that developed on those surface deposits, are therefore younger than the last glaciation. [Photograph by Bradford Washburn.]

1 / *Perspective*

This book is an introduction to the ground around us—the ground on which we build roads, jetports, barns, and skyscrapers; the ground on which we grow crops and graze livestock; the ground in which burrowing animals make their home; the ground that includes exposures of the bedrock so prized by geologists. Derived from bedrock by weathering and erosion, and largely covering the bedrock, is a layer of loose stuff—dirt—which constitutes an important part of the terrestrial envelope of life. The loose surface deposits and soils that make up this thin, outermost layer of the earth are many and varied, but they all have one thing in common: they are *young*. A brief sketch of the part of earth history that is measured by geologic time will give a clear impression of just how young they are.

When the crust began solidifying, some four-and-a-half billion years ago, the surface of the primitive earth may have appeared much like the moon's surface today, and its composition may have been similar to that of stony meteorites. With the formation of the first crust of rock, geologic time began; earlier earth history is astronomical history. As the crust developed, so did the atmosphere; water vapor, nitrogen, carbon dioxide, and other gases gradually escaped from the earth's interior through vents in the slowly cooling crust. Water vapor may have formed clouds, but neither precipitation nor accumulation of water could have occurred until the surface had cooled to less than 100°C. The first eon of geologic time was a protoatmospheric stage ("proto": from the Greek *protos*, "primitive"). When the rains did come, water running off high ground collected to form the first streams and accumulated in depressions to form the first lakes and seas (Table 1.1).

TABLE 1.1
Rough chronology of the biosphere and of
the biochemical weathering that produces soil

Years before present	Event	Geologic formation
0	Soil changes caused by man (agriculture)—last 7,500 years, especially the last 100 years	Holocene
	Modern soils (last 15,000 years)	Late Pleistocene
	Modern fauna (last 60 million years)	Cenozoic
	Modern flora with forest, shrub, and grassland soils (last 125 million years)	Late Mesozoic
	Oldest known land plants; beginning of herbaceous soils	
	Oldest metazoan land animals; beginning of soil mixing and other soil changes caused by animals	Early Paleozoic
1 billion	Soils still primitive	
	Biosphere still primitive; oldest cells capable of sexual reproduction	Beck Spring Dolomite (California)
	Atmosphere becomes oxidizing	
2 billion	Oldest photosynthetic and nitrogen-fixing organisms; primitive atmosphere begins accumulating oxygen	Gunflint Formation (Ontario)
3 billion	Oldest known organisms; biochemical weathering begins	Figtree Formation (South Africa)
4 billion	Continued differentiation of earth's crust, mantle, and core; formation of ocean basins and nuclei of the continents	
	First surface water and protosoils	
	Protoatmosphere, no free oxygen	
4.6 billion	First rocks; beginning of the crust; beginning of geologic history	
	Formation of earth; astronomic history	

Source: Various.

Available evidence suggests that it was in the seas that life began, perhaps because only in deep water was there sufficient shielding to protect the earliest life forms from cosmic radiation. Not until photosynthesizing organisms evolved did oxygen become a constituent of the atmosphere and produce the envelope of ozone that now exists in the upper atmosphere and protects life from lethal cosmic rays.

The gradual evolution of an atmosphere spurred the evolution of life along paths that led toward increasing diversity and complexity. When life arrived on land, the biochemical processes of weathering and soil development could begin. Early soils (protosoils) must have differed markedly from those that have developed under the kinds of vegetation familiar to us.

The stages represented by the development of the protoatmosphere and protosoils may have lasted about half a billion years. In the course of the next three-and-a-half billion years of geologic time, slabs of the earth's crust thickened and broadened to form the nuclei of our several continents. The evolution of the higher forms of life and the development of the biochemical processes so important today in weathering and soil formation took place within the last half billion years. But our surface deposits and soils—"the pitiful remnants of the solid rocks"—developed much more recently. Most of the surface deposits and soils that are the subject of this book were formed within the past few million years, and man's influence upon these materials dates back less than ten thousand years. Man became a geological agent affecting the environment during the Holocene, when he began raising herds of domesticated livestock, practicing agriculture with irrigation, and building cities.

In professional geologic jargon, the loose deposits covering the bedrock are called "surficial deposits"; in this book they will be called *surface deposits*. They are composed largely of mineral matter rather than organic matter—that is, like the bedrock from which they were derived, they are largely inorganic. Their thickness, down to the bedrock, may be a few feet or a hundred or more feet. They are less dense than rock, and about twice as dense as water.

Most surface deposits are sediments weathered from bedrock in one area and transported by water, wind, or ice to another area. Thus not only are they much younger than the underlying bedrock; they are also unrelated to it (Fig. 1.1). Geologists classify such deposits according to the mode of transportation by which they were carried to the place of deposition. Some of the different kinds are:

Alluvium. Sediment transported by running water.

Dunes and loess. Sediment transported by wind. The difference between the two is primarily in the grain size of the material: dunes are composed of coarse grains of sand; loess is made up chiefly of silt particles.

Till, or glacial drift. Sediment transported by glaciers or their meltwaters.

Colluvium. Material moved downhill by gravity.

Some surface deposits are untransported, and are called *sedentary deposits*. Those formed *in situ* as a result of weathering (physical disintegration and chemical decomposition of bedrock) are called *residual deposits*. Still other sedentary deposits

4

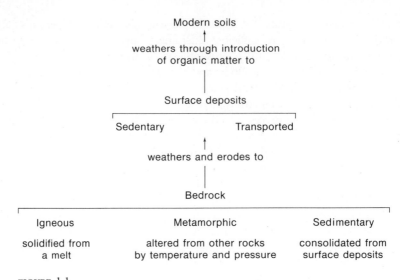

FIGURE 1.1
Diagram illustrating the relationships between igneous, metamorphic, and sedimentary rocks and how these become parent materials for surface deposits, which in turn become the parent materials of soils.

are formed by the accumulation of organic matter. Accumulations of plant material form *peat;* accumulations of seafowl excrement form *guano.*

Weathering of surface deposits causes the development of what appear to be layers, more or less parallel to the ground surface. These weathered near-surface layers constitute the *soil.* The soil layers produced by weathering should not be confused with the bedding or stratification in surface deposits that were laid down by wind or running water. The relationships between bedrock, surface deposits, and soils are shown in Figure 1.2.

Agriculturalists use the term "soil" to refer to the weathered uppermost layers of surface deposits, the layers in which plants anchor their roots and from which they derive the nutrients and water necessary for growth. At the surface is a layer of litter, or dead plant material. Below this is a layer containing older, decayed organic matter usually called *humus.* The soil below these organic layers is composed largely of weathered mineral matter. As rainwater seeps through the organic layers it reacts with the carbon-rich material to produce carbonic acid. It is this acid that enables the rainwater to dissolve some of the mineral matter in the upper layers of the soil, move it downward in solution (some also in suspension), and redeposit it deeper in the ground. These several weathered layers form the agriculturalists' soil, the zone in which plants root. These soil layers generally amount to a thickness of no more than 2 or 3 feet, although in some places roots may go much deeper. Such soil layers are still developing as present-day weathering continues. Below the evident soil layers, changes are slight, and for most practical purposes the deeper parts of the surface deposits may be regarded as unaltered. The agriculturalists' definition of soil, however, is not the only one. The term "soil" has other, quite different meanings, and is used ambiguously in many scientific reports.

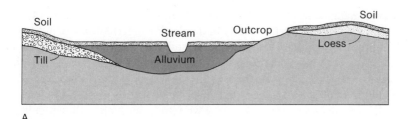

FIGURE 1.2
(A) Soils, generally about 2 to 3 feet thick, developed on transported surface deposits—glacial till, river alluvium, and wind-deposited loess. Below the surface deposit is bedrock, which crops out on the hillside. (B) Soil developed on a residual deposit is a product of recent weathering; the residual deposit itself (an ancient soil to geologists) is an untransported product of ancient weathering, and may be scores of feet deep. Near the top of the bedrock in (A) an ancient soil could lie buried beneath the transported deposits. Another buried soil could exist at the top of the till, under the alluvium. In such a section one might see three or more ages of weathering and soils.

Geologists use the term "soil" in a broader sense to refer to all materials produced by weathering in situ, regardless of their depth or utilization by plants. Geologists thus consider residual deposits as soils. Such soils may be buried by unweathered, younger surface deposits or they may be the deep part of a residual deposit produced by ancient weathering far below the present root zone. Thus, in the specialized sense of the agriculturalist, "soil" on a residual deposit may be only the surface part of an older, deeply weathered zone or "soil" in the geological sense of the term. That is, a geologist would recognize two soils—a young soil, which is still undergoing development, and an older and deeper soil that is parent material for the young one (Fig. 1.2). To geologists, the older soils are indicators of past weathering processes and climates.

Engineers have a still different usage, one that is highly empirical. In engineering, "soil" refers to ground that can be excavated by earth-moving equipment without blasting; the term has about the same meaning that "surface deposit" has to a geologist.

Two examples illustrate the ambiguity that develops when the terms *soil* and *surface deposit* are not clearly distinguished. An excellent and much-used textbook on soil mechanics contains a table entitled "Coefficient of permeability of common natural soil formations", but the table gives properties for unweathered surface deposits: river, glacial, wind deposits, as well as subaqueous lacustrine and marine offshore deposits.

In a second example, from a book on soil conservation, the author states that "Soil out of place is epitomized in the history of the Rio Puerco, riddled with huge gullies." But the author writes not about erosion of soil but about erosion of alluvium, a surface deposit that has undergone little or no weathering in situ. What, then, is *soil conservation?* In practice it refers to the protection of all surface deposits, not merely the near-surface, organic layers that are subject to present-day weathering.

Such different usages arise because those who study our subject, the agriculturalists, the geologists, the engineers, generally belong to separate organizations; whether they work in universities or in government agencies, they are separately housed. Still a fourth group are the chemists, who in their laboratories have contributed much that is known about clays and organic matter. With all these specialists so scattered, there has been inadequate interdisciplinary exchange of ideas and approaches. This book is an attempt to plow that interdisciplinary field, for it has lain fallow too long.

This book uses the term "soil" in two senses, that of the agriculturalist and that of the geologist. The two senses allow for the division of soils into two categories based on the age of the weathering that produced them. With rare exceptions, the weathering that produced the soils recognized by agriculturalists occurred during or since the last stages of the last glaciation, and most of them have developed quite recently and are still developing. Because they are the youngest soils, they will be referred to in this book as *modern soils,* or *soils of the agriculturalist.* The deeply weathered residual deposits of geological interest, and some buried soils, predate the last glaciation; many are much older. These products of ancient weathering will be referred to as *ancient soils.* Modern and ancient soils are both products of weathering in situ, weathering caused by acids produced chiefly by the decomposition of organic matter. The term "surface deposit" will be used in the same sense that engineers use the term "soil," to refer to any unconsolidated material at the surface. The term "ground" will be used in an inclusive sense to refer to soils, surface deposits, and the bare surfaces of the bedrock (Fig. 1.3). None of these terms is subject to exact definition; they are as inexact as the terms "hill" and "mountain", yet the distinctions need to be made for several reasons.

One must distinguish surface deposits from the modern soils developed on them in order to focus clearly upon problems pertaining to land use. Whether land is suitable for growing crops or raising livestock depends upon the kind and topographic position of the modern soil, its nutritive value, ease of working, slope, and susceptibility to flooding and erosion. On the other hand, problems pertaining to water supply, ground drainage, and irrigation, many erosion problems, and most ecological and engineering problems have to do with the surface deposit rather than the soil on it. Thus, kind of ground is a major factor that affects its suitability for a particular use.

Kind of ground is a major factor affecting its suitability for particular engineering uses. Some ground is easy to excavate, some difficult and costly; some ground provides a stable foundation for heavy structures, and other ground is unstable. Kind of ground, of course, is not the only factor that affects the worth of land and its suitability for various uses, mineral, scenic, recreational, or other.

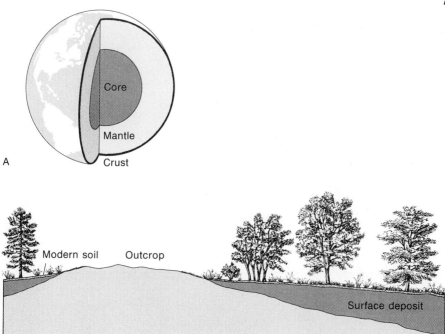

FIGURE 1.3
"Bedrock" is a generic term that embraces any of the solid rocks that make up the earth's crust. Where exposed at the surface, bedrock forms *outcrops*—samples of the crust. (A) Under the continents, the crust has an average thickness of twenty miles; under the ocean basins, the crust is about five miles thick. The continents are slabs of crust floating on the underlying, denser mantle. The density of the crust is about three times that of water, whereas the density of the mantle is about twice that of the crust. About a hundred times thicker than the crust, the mantle surrounds the earth's core, which has a diameter of about four thousand miles and a density about twelve times that of water. (B) Cross section illustrating four principal kinds of ground on the earth's surface: an outcrop of *bedrock; surface deposit* composed of gravel, sand, and clay such as might be deposited by a stream or glacier; a mat of *organic matter* (lichens, algae, and mosses) growing on the bedrock; and *soil*, the weathered layers (soil horizons) developed on the surface deposit and including the organic-rich surface layer.

In addition, social, economic, and psychological factors affect man's interest in particular tracts of ground. Accessibility is such a factor; farmers must be able to transport their products quickly and inexpensively to markets. On the other hand, from the point of view of a vacationist or hermit, inaccessibility may be a prime factor. A person considering whether to buy a lot may not think about the kind of ground, or the fertility of his garden, but will consider the cost of commuting. Moreover, factors that affect the value of land often counterbalance one another. The homesite with a view is more desirable than one without a view, but the owner of a homesite on a hill may have difficulty obtaining water.

We live in an era in which the uses of land are constantly changing. Increasing populations mean not only greater demands for products of the ground but greater

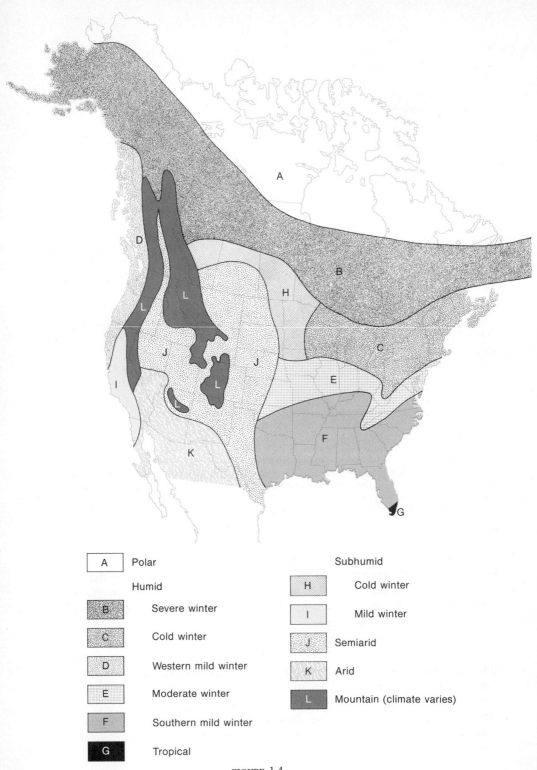

FIGURE 1.4
Climates of North America.

A	Polar	

Humid

B	Severe winter
C	Cold winter
D	Western mild winter
E	Moderate winter
F	Southern mild winter
G	Tropical

Subhumid

H	Cold winter
I	Mild winter
J	Semiarid
K	Arid
L	Mountain (climate varies)

demands for the ground itself; farm lands become suburbs, factory complexes, superhighways, and not infrequently become incorporated into cities.

The ground in the United States is much the most valuable on the planet. We are favored by topography, climate, forests, water supplies, minerals, scenery, and a great variety of fertile and productive soils.

Topographically, almost half of the United States consists of plains, a quarter of mountains, and another quarter of plateaus (Fig. 4.3). The plains, which have extensive smooth areas, favor agriculture and easy transportation. The plateaus are less usable; their smooth areas are less extensive because they are dissected by stream valleys. Only small areas within the mountains are suitable for agriculture, but mountains have compensating minerals, forests, recreational uses, and scenery.

Climatically the eastern half of the United States is humid (Fig. 1.4), with an average annual precipitation of more than 20 inches, enough to support good crops and a forest having a thousand or more species, as many, in fact, as any other forest in the world with the possible exception of that in eastern China. Climates in the western half of the United States, except in the mountain ranges and in the Pacific Northwest, vary from arid to semiarid. The southern half of the United States is warm and crowded with winter visitors; the northern half is cool to cold and crowded with summer visitors. Differences in precipitation are reflected in regional differences in runoff, water supply, patterns of drainage, number and size of rivers, flood frequency, rates of erosion, natural vegetation, and use and capacity of the land.

Lands in the conterminous states of the U.S. are used as follows (U.S. Department of Agriculture):

Use	Millions of acres	Percent
Cropland	465	24
Pasture and grazing, nonforested	633	35
Forestland, grazed	301	15
not grazed	314	16
Special purposes (cities, highways, parks, military reservations, etc.)	110	6
Rural land of little agricultural value	81	4
TOTAL	1,904	100

Man's use of land is not static, but is subject to change—by man and by nature. Farming that depletes mineral nutrients and causes soil erosion may make agricultural ground worthless; the deposition of salts from irrigation waters may leave soils unproductive. Conversely, new technologies may make poor land productive. Changes in economic conditions also can alter land use; a conspicuous example is our increasing urban sprawl, which changes farms, forests, and pastures into residential and industrial land, but too rarely into parks. Similarly, our bureaus of highways, pushed by our traffic needs (and by their bulldozing enthusiasm), have built super-highways through much of our open space, increasing their accessibility and inviting development. The pressures for developing and using *all* our land are increasing as our population increases.

Natural processes are constantly at work, changing the land all the time; most such processes, however, are slow. Nevertheless, some do produce abrupt changes. Unusual storms may cause streams to flood, or a stream may meander into a new course and isolate a field from its owner. Earthquakes may depress or raise shorelines, displace drainage and property lines, and cause landslides. Forest fires, including those started by lightning, denude mountain sides and hasten erosion. Man's activities can hasten the natural processes too, sometimes for the better, but more often for the worse. Stripped, bare ground at logged areas and at the sites of real estate and industrial developments erodes readily, causing streams and the nearby countryside to be muddied. Irrigated land may accumulate salts. Building on unstable ground can aggravate the instability.

Natural changes lead to consideration of longer spans of time: indeed, it took long periods of time for natural processes to produce the fertile glacial tills and loess, the alluvium of river deltas, the sands and gravels and clays of commerce. To understand the ground that is our land, we need to know its history; we need to appreciate *geologic time*.

BIBLIOGRAPHY

Cloud, P. 1968, Atmospheric and hydrospheric evolution on the primitive earth: Science, v. 160, pp. 729–736. Reprinted in Cloud, Preston (editor), 1971, Adventures in earth history: W. H. Freeman and Company, San Francisco.

———, and Gibor, A. 1970, The oxygen cycle: Scientific American, v. 223 (September). [Offprint 1192.]

Legget, R. F., 1967, Soil—its geology and use: Geol. Soc. America Bull., v. 78, pp. 1433–1460.

——— (editor), 1968, Soils in Canada—geological, pedological, and engineering studies: Univ. Toronto Press, 240 pp.

Platt, R. 1965, The great American forest: Prentice-Hall, Englewood Cliffs, N.J.

Sears, P. B., 1968, Lands beyond the forest: Prentice-Hall, Englewood Cliffs, N.J.

Although geologic processes are exceedingly slow, geologic time is vast, and the processes have literally all the time in the world in which to operate, and thereby cause immense changes. The layers of rock exposed in Grand Canyon record about a billion years of earth history. At the bottom of the canyon, in the inner gorge, are Precambrian rocks. The layers from there to the rim record most of the Paleozoic Era. In the butte on the skyline are Triassic rocks. A few of the surface deposits described in this book are as old as Grand Canyon (middle Tertiary?). The age of the canyon bears no relation to the ages of the rocks into which it is cut—except, of course, that it is younger. When the glacial deposits were laid down in the northern United States and in Canada, Grand Canyon was within 50 feet of its present depth. During the short time during which our modern soils have been developing, there have been but minor changes in the sides and bottom of the canyon. [Photograph by John S. Shelton.]

2 / Geologic Time

ITS VASTNESS, DIVISIONS

Paleontology, superposition; the Quaternary Period; Pleistocene, Holocene; Pleistocene-Holocene boundary, extermination of Pleistocene species; glaciations and interglaciations; Pliocene–Pleistocene boundary; section in Italy; glaciations in the Alps; Upper Pliocene and Lower Pleistocene in England; glaciations in North America; floral and faunal changes; paleogeography of the Baltic Sea; vegetation shifts in North America; archeology; estimating geologic time; radiometric methods, miscellaneous methods, fluorine, dendrochronology, varves; the estimates by Kay; time as a factor in erosion and weathering; bibliography.

Time is of the essence—a truism whether we keep track of it with a watch or with a calendar, by counting birthdays, or by taking the long view of historic and prehistoric episodes. The weathering and erosion of bedrock to form surface deposits and the weathering of these surface deposits to form soils involves time on a geologic scale. Geologic time is vast; geologic processes are slow, but they have literally all the time in the world in which to operate and thereby cause tremendous changes. Ever so slowly mountains of granite or other hard rock are uplifted; a foot of uplift in a thousand years is not earth shaking, but continual uplift at that rate for 10,000,000 years could build a mountain range 10,000 feet high. Slow and easy does it! Equally slowly, however, and equally persistently, those elevated rocks are weathered and eroded, resulting in the creation of surface deposits.

Geologic time is divided into *eras*, eras into *periods*, and periods into *epochs* (Tables 2.1, 2.2). Each of these divisions of geologic time has been based on the fossil record of species living then and considered diagnostic or characteristic; accordingly, their

TABLE 2.1
Geologic time

Millions of years ago	Eras	Periods
Today	Cenozoic (Recent life)	Quaternary
50		Tertiary
100	Mesozoic (Middle life)	Cretaceous
150		Jurassic
200		Triassic
250	Paleozoic (Ancient life)	Permian
300		Pennsylvanian
350		Mississippian
400		Devonian
450		Silurian
500		Ordovician
550		Cambrian
600		
Precambrian		
4,700		

Duration of Quaternary exaggerated relative to that of the Cenozoic Era. See Table 2.2 for beginning of Tertiary.

TABLE 2.2
Cenozoic time. Most surface deposits are Quaternary in age,
although some are as old as Tertiary. Few soils in the agricultural sense
of the term are older than late Wisconsinan; most are Holocene

Period	Epoch	Glaciation	Interglaciation	Years ago (estimated)
Quaternary	Holocene			Today
				11,000
	Pleistocene	Wisconsinan		70,000
			Sangamon	
		Illinoian		
			Yarmouthian	
		Kansan		
			Aftonian	750,000
		Nebraskan		
			Blancan ?	1,000,000?
				3,000,000
Tertiary	Pliocene			10,000,000
	Miocene			
	Oligocene			30,000,000
	Eocene			
	Paleocene			60,000,000

names are time terms defined on the basis of *paleontology*, the record of the fossils.
Names of the eras, except the earliest, refer to the general stage in the evolution
of living forms. Type areas are selected for defining the periods and smaller units,
and the location names commonly are used for the period, epoch, or other names.
Thus, Devonian refers to Devon, England.

Some of the time units are defined on the basis of the first appearance of species
regarded as diagnostic of the period. For example, the beginning of the Cambrian
is taken at the first appearance of trilobites and shell fish. Other time units are defined
on the basis of the last appearance of species regarded as diagnostic of the period.
The end of the Cretaceous is taken as the last appearance of the big reptiles, the
dinosaurs. Still other time units are defined on the basis of the greatest changes in
fauna.

The chronology of geologic deposits is based on the principle of superposition
of strata, the logical premise that younger deposits overlie older ones, and on correlation, the lateral tracing of individual deposits from one locality to another. Gravel
may be deposited in one place while muds are being deposited in another; the two
contemporaneous deposits will be found to grade into each other when traced

laterally. The paleogeographic conditions, including the paleoclimates, are recorded by the fossil faunas and floras and by the internal physical and mineralogical features of the deposits.

THE QUATERNARY PERIOD

In considering the ground, we are concerned primarily with the deposits and episodes of the Quaternary Period, which includes the Pleistocene and Holocene epochs. Some surface deposits are as old as Tertiary, and some as young as Holocene, but most are Pleistocene in age. Some soils in the agricultural sense are as old as late Pleistocene, but most are Holocene.

The time units "Pleistocene" and "Holocene" are an outgrowth of Sir Charles Lyell's efforts more than a hundred years ago to establish divisions of the Cenozoic Era on the basis of the fossil record of invertebrate animals that still survive. Of the marine molluscan species known in his time, Lyell found that about three percent of the Eocene species, about fifteen percent of the Miocene species, about thirty-five percent of the Pliocene species, and about ninety-five percent of the Pleistocene species survived into the present. As knowledge has increased, the percentages have changed, but the principle remains as the basis for the subdivision of the Cenozoic Era.

Pleistocene-Holocene Boundary

Of the Holocene Epoch (formerly known as "Recent"), Lyell wrote in 1863 that "In the Recent we may comprehend those deposits in which not only the shells but all the fossil mammalia are of living species." This stratigraphic boundary therefore also has a paleontological basis, specifically, the extermination of many of the vertebrate animals that lived during the Pleistocene. This places the Pleistocene-Holocene boundary at the stratigraphic horizon of greatest change in fauna; it emphasizes major faunal change rather than range of particular species. The elephant, camel, horse, and other animals disappeared from North America at the end of the Pleistocene. Why they became exterminated here is not known, but the stratigraphic fact is that their remains are common in Pleistocene deposits but are absent in Holocene deposits. A few individuals from the Pleistocene may have survived in some places into the earliest Holocene; after all, not all the animals died at once. But even in the central and eastern parts of the United States and Canada, where glaciers persisted long into the Holocene, very few of the typically Pleistocene animals ranged north of a line marked by glacial deposits that have been correlated with the Valders Moraine, in Wisconsin (Fig. 2.1). These deposits mark the position of the ice front at the end of the Pleistocene Epoch, as defined paleontologically. Abundant archeological and geological evidence indicates that few individuals or species survived that time, and that none survived for long.

FIGURE 2.1

The dots show occurrences of *Elephas columbi*, a Pleistocene elephant, in the northern United States and Canada. In the central and eastern parts of the United States, the northern limit of common occurrence is shown by the hachured line, which approximately coincides with a glacial deposit known as the Valders Moraine. This line marks the approximate southern limit of the ice at the beginning of the Holocene Epoch (see Fig. 2.2). North of the broken line, the surface deposits and the soils on them are no more than about 11,000 years old. [After Hay, 1923, 1924, 1927.]

Pleistocene Glaciations and Interglaciations

The Tertiary Period, which preceded the Quaternary, was a comparatively warm time; the Pleistocene Epoch of the Quaternary Period was alternately cold and warm, with glaciations occurring during the cold periods. At least four times ice collected in Canada in amounts sufficient to form masses of continental dimensions, causing ice sheets to move southward into the United States (Fig. 2.2). Similar and apparently contemporaneous ice sheets formed in northern Europe and spread southward. In Europe, however, at least five (some say six) major glaciations are recognized, the first of which may have occurred before the earliest recognized North American glaciation. In addition to the massive ice sheets, there were glaciers in high mountain valleys in the western United States and in the European Alps; remnants of *alpine glaciers* persist in many places in the United States, such as on Mount Rainier and in Glacier National Park. During the Holocene, regarded as post-glacial, the great ice sheets have been melting and ice fronts retreating, with only minor reversals. Remnants of the great ice sheets still persist in Greenland and Antarctica.

The extensive deposits left during the Pleistocene glaciations form an important part of the ground around us. Of course not all of them are glacial. In the nonglaciated, southerly latitudes the climate during the glaciations was humid and rainy—a climate called *pluvial*. The ground there includes lake and other deposits characteristic of wet climates. In general, the Pleistocene deposits are mostly

1. Glacial till: the gravel, sand, and mud laid down by the ice as it melted.

2. Fluvial-glacial deposits: the gravel, sand, and mud laid down by streams draining from the ice as it melted.

3. Lake deposits.

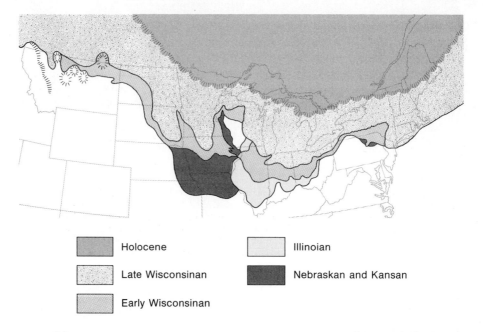

Holocene

Late Wisconsinan

Early Wisconsinan

Illinoian

Nebraskan and Kansan

FIGURE 2.2

Drift borders at the time of the glacial maxima and at the beginning of the Holocene. The surface deposits are progressively older southward. Ancient soils are extensive south of the Wisconsinan drift. The modern soils developed on them and on the Wisconsinan drift are mostly late Wisconsinan and Holocene in age. [After Alden (1953); Horberg and Anderson (1956); Denny (1956b).]

4. Wind-blown silt (loess) deposited by dust storms that were especially intense at that time.

5. Stream deposits, on flood plains and on alluvial fans.

Pleistocene—Pliocene Boundary

The type area of the boundary between the Pliocene and Pleistocene is in Italy; there, changes in the fossil marine fauna can be correlated with equivalent terrestrial fossil mammalian faunas. In this type area the lowermost Pleistocene includes as its basal members the marine Calabrian Formation with its lateral terrestrial equivalent, the Villafranchian Formation (Fig. 2.3). The Calabrian, which grades landward into the Villafranchian, contains foraminifera and molluscs characteristic of colder water than is indicated by the species found in the uppermost formation of the Pliocene. In terms of change, more than ten percent of the late Pliocene foraminiferal species are now extinct, while less than five percent of the Calabrian foraminifera are extinct. Of the molluscs, more than thirty-five percent of the late Pliocene species are extinct, while only about ten percent of the Calabrian species are extinct. Similar paleonto-logical changes are recorded in the late Pliocene and early Pleistocene formations

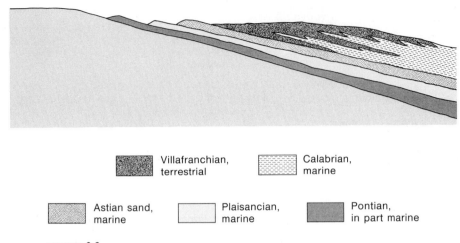

Villafranchian, terrestrial

Calabrian, marine

Astian sand, marine

Plaisancian, marine

Pontian, in part marine

FIGURE 2.3
Stratigraphy of the boundary between the Pliocene and Pleistocene in Italy. The boundary is defined as the top of the Astian Sand and base of the Villafranchian and Calabrian deposits.

in eastern England (Table 2.3). Comparably complete sections of intertonguing continental and marine deposits are not exposed around the shores of North America.

The Villafranchian contains species of *Equus* (the first modern horse), rhinoceros, and a mastodon (*Archidiskodon meridionalis*); the mastodon, commonly is taken as diagnostic of the earliest Pleistocene. Unfortunately, terrestrial equivalents of the late Pliocene are not found beneath the Villafranchian, and the change in vertebrate fauna at the Pliocene-Pleistocene boundary is not as well known as the change in the marine fauna. Table 2.4 lists the glaciations and interglaciations in the European Alps and their probable correlation with prehistoric human types and European culture periods.

The lower boundary of the Pleistocene Series in Italy is much older than the Nebraskan Glaciation (Table 2.2) in the United States if, as has generally been assumed, the Nebraskan correlates with the Günz Glaciation in Europe. Assuming the correlation to be correct, half or more of the Pleistocene Epoch in the United States is pre-Nebraskan. Because the Pleistocene includes the major glaciations (5 or 6 recognized in Europe, 4 in the Central United States, and 3 in the western United States), it is frequently referred to as the "glacial period." But the terms "Pleistocene" and "glacial period" are not synonymous, nor are climatic events a basis for establishment of Pleistocene boundaries.

The glacial deposits in the United States record a progressive eastward overlap of younger drifts onto older ones (Fig. 2.2). The Kansan overlaps (extends onto) the east side of the Nebraskan, the Illinoian overlaps the east side of the Kansan, the early Wisconsinan overlaps the east side of the Illinoian, and the late Wisconsinan overlaps the east side of the early Wisconsinan. Such distribution of maximum extent of the deposits suggests an eastward shifting of the ice center, or centers, dur-

TABLE 2.3
Paleontological changes in upper Pliocene and lower
Pleistocene formations in eastern England

	Stage	Marine or other deposit	Nature of fauna
Pleistocene	Cromer (=Mindel) Glaciation	Cromer fresh-water deposits	Cold-water invertebrates
	Interglaciation		
	Icenian (=Günz) Glaciation	Weybourne Crag*; marine sandy clay	20% arctic forms; 89% living species
	Interglaciation	Chillesford Beds,	
		Norwich Crag	
	Donau Glaciation	Norfolk Bone Bed,	
		Butley Crag; marine	Mostly cold-water forms; 87% living species
		Newbourne Crag; marine	Southern forms scarce; 68% living species
	Preglacial	Walton Crag; marine	20% southern and 5% northern species; 64% living species
		Suffolk Bone Bed	
Pliocene		Coralline Crag; marine	25% southern and 1% northern species; 62% living species

Sources: Wright, 1914; Baden-Powell, *in* Oakley, 1950.

*A "crag" is a sandy bed containing fragments of shells.

ing Pleistocene time. The ice maxima in North America are generally thought to have been contemporaneous with those elsewhere in the northern hemisphere; there is considerable doubt about the contemporaneity of ice maxima in the northern and southern hemispheres.

Floral and Faunal Changes

As environments change, animals must adapt to the change, migrate, or face extinction. During the Quaternary, changes in climate caused marked changes in environments; in response to these changes, animals must have migrated from areas where living conditions were deteriorating to areas where conditions were more favorable. Changes in environment are reflected in differences in the composition or kind of surface deposits, and the changes are further reflected by differences in fossil content of the deposits, each deposit having its characteristic faunal or floral record (Fig. 2.4). For example, in the Great Plains, fossil remains of a Pleistocene elephant are found mostly in the floodplain deposits along the streams, whereas fossil remains of a camel that lived at the same time as the mammoth are found mostly in the upland deposits.

TABLE 2.4
Glaciations and interglaciations in the European Alps
and their probable correlation with prehistoric human types
and European cultural episodes

	Stage	Human types	European cultures
Holocene		Modern man	Iron
			Bronze
			Neolithic
			transition { Maglemosean / Tardenoisian / Azilian
Pleistocene	Fini-glacial pause, end of Pleistocene	Cro-Magnon	Magdalenian
			Solutrean
	Würm Glaciation		Aurignacian
			late Mousterian
	R/W Interglaciation	Neanderthal	early Mousterian
	Riss Glaciation		Acheulian
	M/R Interglaciation		
	Mindel Glaciation		Chellean
	G/M Interglaciation		
	Günz Glaciation ?	*Pithecanthropus* in East Indies	Pre-Chellean
	D/G Interglaciation		
	Donau Glaciation ?	*Zinjanthropus* in Africa	
	Preglacial		

Source: Generalized after Oakley (1964); LeGros Clark (1966).

The map reproduced in Figure 2.1 is just one of a whole set of maps compiled in three volumes by Hay (1923, 1924, 1927) to show the distributions of species of Pleistocene animals in North America. From that map one can readily see a paleontological basis for choosing a glacial deposit, the Valders Moraine, as a physical marker of the end of the Pleistocene and the beginning of the Holocene. Subsequent studies have revealed that correlations exist between certain stages of the Pleistocene and certain changes in the vertebrate fauna. Table 2.5 shows known stratigraphic ranges for some of the 75 vertebrate species that once existed in the Great Plains and are now extinct.

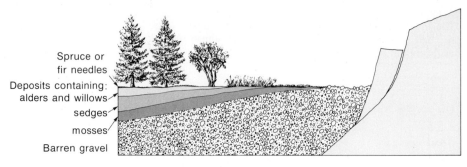

Spruce or
fir needles

Deposits containing:
alders and willows
sedges
mosses
Barren gravel

FIGURE 2.4

Paleontological record formed by vegetation zones in front of a retreating mass of ice. As the ice retreats, the mosses move onto the bare ground. With continued retreat of the ice, the mosses become buried by the advance of the sedges, and these in turn become buried by the advance of alders and willows and by the spruce-fir forest. Beneath the forest is a series of layers containing, in succession downward, remnants of alders and willows, sedges, mosses, and, at the bottom, barren gravel.

Invertebrate fossils, both fresh-water and terrestrial, have also proved to be useful indicators of environmental change during the Quaternary. Leonard (1953) has tabulated data on nearly 100 species of invertebrate shells found in the central United States. According to his data, about fifteen percent did not survive beyond the Nebraskan Glaciation, and about twelve percent of those that appeared during the Nebraskan or Kansan glaciations did not survive beyond the Illinoian.

The climatic changes of the Quaternary Period caused major shifts in stands of vegetation (Table 2.6). The fossil record abounds with evidence of those changes. For example, the retreat of ice masses is marked by the paleontological record of vegetation zones that formed in front of them (Fig. 2.5). The changes in climate are also widely recorded by pollens in sediments that accumulated in lakes and bogs.

An excellent example of the use of pollen chronology to show the succession of forest types is the detailed stratigraphic study that made possible the reconstruction of the paleogeographic changes that led to the formation of the present Baltic Sea:

1. An ice-bound lake (Baltic Ice Lake) became drained as the ice melted, and a seaway (Yoldia Sea) connected the Arctic Ocean and North Sea across the south front of the remaining ice.

2. The land rose as the load of ice was removed by melting, causing Finland to emerge, Denmark to become joined with Sweden, and a fresh-water lake (Ancylus Lake) to form.

3. Continued rise of sea level again connected the North Sea with the Baltic, forming a sea characterized by a shell fish, *Littorina;* temperatures in the sea were higher then than at present.

4. Finally, the present Baltic Sea was formed.

The vegetation has shifted its position by reoccupying ground that had been covered by ice or by glacial lakes, and the weathering profiles of the modern soils

TABLE 2.5
Stratigraphic range of some of the Pleistocene vertebrate animals on the
Great Plains

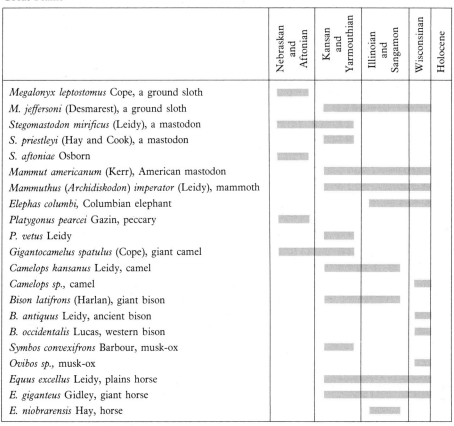

	Nebraskan and Aftonian	Kansan and Yarmouthian	Illinoian and Sangamon	Wisconsinan	Holocene
Megalonyx leptostomus Cope, a ground sloth	▬				
M. jeffersoni (Desmarest), a ground sloth		▬▬▬▬▬▬▬▬			
Stegomastodon mirificus (Leidy), a mastodon	▬▬▬▬				
S. priestleyi (Hay and Cook), a mastodon		▬			
S. aftoniae Osborn	▬				
Mammut americanum (Kerr), American mastodon		▬▬▬▬▬▬▬			
Mammuthus (*Archidiskodon*) *imperator* (Leidy), mammoth		▬▬▬▬▬▬▬			
Elephas columbi, Columbian elephant			▬▬▬		
Platygonus pearcei Gazin, peccary	▬				
P. vetus Leidy		▬			
Gigantocamelus spatulus (Cope), giant camel	▬▬▬				
Camelops kansanus Leidy, camel		▬▬▬▬▬			
Camelops sp., camel				▬	
Bison latifrons (Harlan), giant bison		▬▬▬▬▬			
B. antiquus Leidy, ancient bison				▬	
B. occidentalis Lucas, western bison				▬	
Symbos convexifrons Barbour, musk-ox		▬			
Ovibos sp., musk-ox				▬	
Equus excellus Leidy, plains horse		▬▬▬▬▬▬			
E. giganteus Gidley, giant horse		▬▬▬▬▬▬			
E. niobrarensis Hay, horse			▬▬		

Source: Chiefly from Hay (1923, 1924, 1927); Hibbard, *in* Flint (1957); Schultz and others (1951).

there seem to reflect the present vegetation. Perhaps this means that the ground was revegetated quickly when freed of ice.

In North America, the southward advance of ice overwhelmed and destroyed the northern forests. Tundra, Taiga, and conifer forests probably were crowded together near the ice front. After the ice had reached its maximum, the southerly part of it may have been blanketed with loess and could have been forested, as is the front of the Malaspina Glacier in Alaska. As the ice melted back, forests and other vegetation moved northward onto new ground. The shift in vegetation zones may still be continuing. According to some ecologists, beech and sugar maple are moving northward into pine and spruce areas; in Alaska, spruce has been reported moving into tundra; trees of the eastern forests are said to be spreading westward in Ne-

TABLE 2.6
Chronology of late Pleistocene and Holocene vegetation changes in the United States

Probable age	Climatic interpretation	Washington coast	Oregon coast	Columbia Plateau	Northern Rocky Mountains	Northern Middle West	Eastern United States
Holocene	Cooler, moister than before	Hemlock	Douglas-fir	Yellow pine predominant	White pine predominant	Local return of spruce	Oak-chestnut
Holocene	Maximum warmth and dryness	White pine minimum; hemlock increase; Douglas-fir decline	Oak maximum; Douglas-fir static	Chenopod-composite maximum	— — Grass maximum — —	Pine decline	Oak-hickory; pine decline
Pleistocene and Holocene	Increasing warmth and dryness	Douglas-fir predominant	Douglas-fir increases; oak absent	Grass maximum		Pine-oak-elm Pine	Oak-hemlock-poplar Maple-white pine
Late Pleistocene	Cool, moist	Lodgepole pine predominant	Lodgepole pine predominant	Lodgepole pine predominant	Lodgepole pine maximum	Spruce-fir	Spruce-fir
Late Pleistocene	Cold, wet	No tundra		?	?	Tundra	Tundra

Source: Data from Hansen (1947); Deevey (1949).

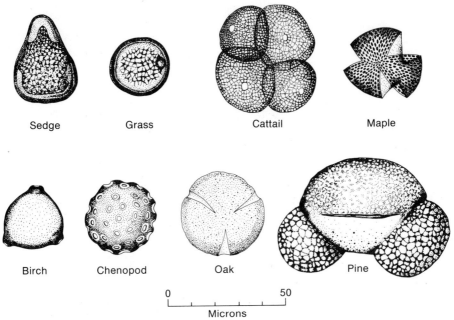

Sedge   Grass   Cattail   Maple

Birch   Chenopod   Oak   Pine

0   50

Microns

FIGURE 2.5
Examples of some differences in shapes and sizes of pollen grains.

braska, and yellow pine like that in the Rocky Mountains may be spreading east-ward on the Great Plains.

ARCHEOLOGY

Archeology contributes much to the study of Quaternary stratigraphy. In Europe the history of man dates far back into early Pleistocene, and in Africa and southeastern Asia remains of early man date back to earliest Pleistocene (Table 2.4). In North America, archeological stratigraphy does not begin until late Pleistocene time; man arrived in North America while the Pleistocene animals still were extant, and his artifacts have been found associated with the remains of such animals as mammoth, camel, and the long-horned Pleistocene bison. The associated tools at such sites indicate a hunting culture; projectile points were for spears or atlatls.

Later, during the early and middle Holocene, the Paleo-Indians of western North America became gatherers as well as hunters (Fig. 2.6). During the middle Holocene in the southeastern United States and in the Mississippi and Ohio River valleys, the occupants became sedentary; they lived in villages, practiced agriculture, used the bow and arrow, and made pottery. In the western United States, the trend toward a sedentary culture did not begin until about A.D. 1. The succession of archeological changes is about the same in the eastern and western parts of the United States, but lateral tracing of the deposits indicates that changes occurred earlier in the east than in the west. Archeological history and its application to the stratigraphy of

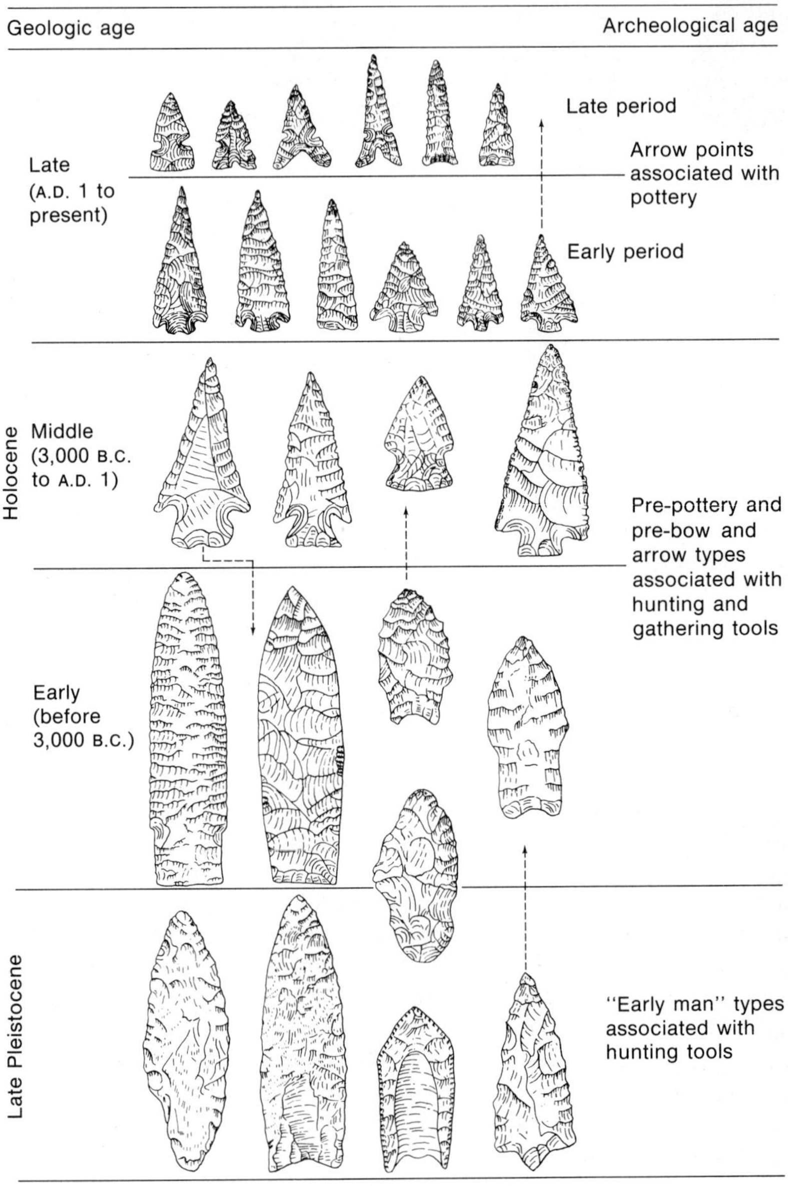

Geologic age | Archeological age

Late (A.D. 1 to present) — Late period — Arrow points associated with pottery — Early period

Holocene

Middle (3,000 B.C. to A.D. 1)

Pre-pottery and pre-bow and arrow types associated with hunting and gathering tools

Early (before 3,000 B.C.)

Late Pleistocene

"Early man" types associated with hunting tools

FIGURE 2.6
Chronology of projectile points in the western United States.
[From U.S. Geol. Survey Prof. Paper 424, p. B 195.]

surface deposits involves the same principles as paleontological history and its application. An isolated artifact, like an isolated fossil, has limited significance even though found in place; it may have been reworked into a younger deposit from an older one by human or animal or by water or some other agency of transportation. An entire fauna, however, or an articulated skeleton of a vertebrate animal, could not have been so moved and is regarded as stratigraphically significant.

The duration of archeological history in North America is particularly important in studies of the ground because the period of that history practically coincides with the time during which soils, in the agricultural sense, were formed—the late Pleistocene and Holocene. In Europe, modern soils have formed since the time of Neanderthal Man, and most of them have formed since the time of the Paleolithic cultures.

ESTIMATING GEOLOGIC TIME

Radiometric Methods

Stratigraphy and paleontology provide the basis for determining the chronological sequence of geologic events, but in order to estimate the duration of episodes it is necessary to be able to date geologic events—that is, to measure the age of a geologic event in years. The most satisfactory methods for estimating the age of rocks and surface deposits are radiometric. These methods are based on the decay, or transformation, of radioactive elements in naturally occurring minerals. The underlying principle is disarmingly simple. A radioactive element decays—that is, undergoes a transformation from parent to daughter element—at a constant and known rate. The rate of decay can be expressed in terms of the half-life of an element, the time required for half the parent element to decay to the daughter element. No two radioactive elements have the same half-life. Knowing the decay rate of a radioactive element, one need only measure the proportions of parent and daughter to determine the age of a sample. The major uncertainty in applying the methods to rocks and minerals or to surface deposits and soils is in the assumption that there has been neither depletion nor enrichment of either parent or daughter elements, a difficult condition to find in nature.

The methods for analyzing radioactive substances (by mass spectroscopy) have reached a high degree of refinement. But it is well to keep in mind that an age determination may be inaccurate despite analytical precision in the laboratory because there may have been depletion or enrichment of either the parent or daughter element caused by the natural conditions that have operated during the long history of burial (or exposure) of the sample. Samples preserved under conditions simulating those of a sealed test tube would give accurate dates, but such samples are not easily obtained in natural environments.

A satisfactory check on particular dates can be obtained by determining the ages of two or three radioactive elements having different rates of decay. If two or more pairs of parent and daughter elements having different rates of decay give identical ages, the probability is very great that a correct age has been determined. This is far more satisfactory than making three measurements using only a single pair of parent and daughter elements. Consistent results from three samples of a single pair may merely reflect a consistent, built-in error. That this condition is common is indicated by the frequency with which different pairs of parent and daughter elements give significantly discordant dates.

The radiometric methods that have been most used thus far for measuring geologic time are those based on uranium-lead, thorium-lead, potassium-argon, and radio-carbon. In the uranium-lead and thorium-lead methods, measurements are made on three pairs of elements that occur together and have different decay rates:

Uranium 238 $\longrightarrow$ lead 206, half-life $4\frac{1}{2}$ billion years.

Uranium 235 $\longrightarrow$ lead 207, half-life 700 million years.

Thorium 232 $\longrightarrow$ lead 208, half-life 14 billion years.

Although the method has the advantage that three quite different decay ratios are measured, it is not suitable for dating surface deposits or soils, because the minerals in which uranium and thorium occur do not originate in such deposits. For dating surface deposits and soils other methods are used.

One such method used to date young deposits is the potassium-argon method. A great advantage of this method is the abundance of potassium-bearing minerals in the crust and in surface deposits. Potassium-40 undergoes a branching decay to argon-40 and calcium-40. The half-life of potassium-40 to argon-40 is 11.8 billion years, but the method is being applied to Pleistocene deposits no more than a hundred thousand years old, and the age determinations are being correlated with the paleon-tologic and stratigraphic evidence.

Another useful method is based on radiocarbon. Carbon-14 decays to nitrogen-14 with a half-life of about 5,000 years. With this short age range the method is use-ful for dating events of the last 50,000 years, those of the Holocene and those of the latter half of the Wisconsinan Glaciation. This makes the method of special interest in connection with studies of modern soils.

The method involves measuring the ratio of the two forms of carbon that exist in nature, carbon-12, which does not decay, and carbon-14, which decays to nitro-gen-14. The method came into extensive use when samples of wood from Egyptian tombs, already dated by historic records, were dated by the carbon-14 method and the results were in agreement with the known record. The method provides very satisfactory dates for samples that have been preserved under conditions that protected the sample against alteration, but extravagant pronouncements have been made about dates obtained from samples taken from exposed environments.

Applying the method to the organic fraction of soils poses difficulties because decomposition of organic matter is biochemical. The decay of organic matter in soil is the result of attack by microorganisms, and to the degree that they derive any of their carbon from soil moisture or soil air, they introduce younger carbon into the system. Moreover, the quantities of carbon-14 compared to that of carbon-12 are so small that a little contamination with carbon younger than that in the sample introduces enough carbon-14 to affect the apparent age significantly, whereas intro-duction of considerable "dead" carbon (C-12) older than that in the sample makes little difference in the total carbon-12 content and little difference in the apparent age. The proportions are such that a sample contaminated with about 20 percent

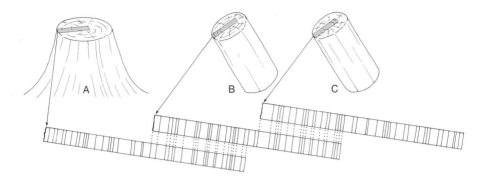

FIGURE 2.7
The principle of the tree-ring method of dating. (A) Section from living tree. (B) Section from dead tree, such as a beam in a ruin, that overlaps A in age. (C) Section of older specimen that overlaps B in age. A tree-ring chronology has been extended back to approximately A.D. 1. [After Arizona Univ. Staff (1955).]

present-day carbon will not give an age greater than 11,000 years no matter how old it really is. Contamination by about 5 percent of such young carbon in a sample 30,000 years old would give an apparent age one-third lower.

If the radiocarbon dates for a particular geologic event are tabulated, most of the dates are bunched together near the middle number. Above are a few smaller numbers (apparent dates that are younger), and below are a few larger numbers (dates that are older). It is common practice to take the average of the bunched dates as the age for the event, and discard the others as "aberrant." This, however, is not even good arithmetic unless the error is random, and probably the error is not random but is systematic. The average of the numbers probably errs in giving an apparent date that is too young, unless the samples have been preserved in environments that protected them against the introduction of younger carbon.

Miscellaneous Methods

Geochemical changes that take place in buried objects can be used to determine the relative ages of similar materials from similar environments. For example, bone that has lain in the ground will have incorporated fluorine into its structure in the form of fluorapatite. The rate at which fluorine enters the bone, however, is not exclusively a function of time; it is chiefly a function of the concentration of fluorine in the particular piece of ground or in the waters passing through it. For these reasons, the fluorine method is not without serious limitations. Nevertheless, it was applied successfully in proving the famous Piltdown skull to be compounded of both recent and fossil bones (see Strauss, in *Science*, Feb. 26, 1954).

Among the better of the nonradiometric methods is *dendrochronology*, the measurement of age via annual tree rings (Fig. 2.7). The method is useful primarily in arid and semiarid regions, where tree growth is highly sensitive to changes in annual or seasonal rainfall. Although the chief application of dendrochronology is in the direct dating of archeological ruins in which timbers are found, it serves indirectly for dating many late Holocene deposits that contain pottery or other artifacts that can be correlated with timbers in the ruins.

Another method of age determination consists in counting *varves* in glacial lake deposits. Many such deposits, now dry, are layered with extraordinary regularity. Layers of coarse silt or sand alternate with layers of fine clay. A pair of such layers is called a varve, and these are regarded as annual deposits that indicate seasonal differences in sedimentation. During the spring thaw, when much sediment discharges from the ice into the lake, sand and silt are deposited. Late in the year when freezing halts discharge into the lake, the fine clay suspended in the lake waters settles to the bottom. The thickness of the paired layers varies, depending on local and seasonal circumstances, but generally ranges from a fraction of an inch to a few inches. Measuring the thicknesses of varves in sequence allows for correlation of deposits between neighboring lakes on the basis of orderly repetitions of relative thicknesses of the varves. Conditions that would produce a thick or thin layer in one lake would probably produce a similar layer in neighboring lakes.

Counting the varves provides a measure of the duration of an individual lake; correlation from one lake deposit to another gives a measure of the rate of ice retreat. This method was first applied to the study of glacial retreat in Scandinavia and Germany. The results showed that the last ice sheet began its retreat in Germany about 25,000 years ago; by 13,700 years ago it had retreated to southern Sweden; by 10,200 years ago it had retreated to the latitude of Oslo and the Gulf of Finland; and by 8,700 years ago the ice cap remaining in the Scandinavian highlands was breached at Ragunda (Table 2.6). A similar study in New England and southeastern Canada by Antevs (1928, 1953) unfortunately had some gaps, but suggested an interval of about 28,000 years for retreat of the ice from its maximum southern reach on Long Island to a position near Cochrane, Ontario, about 600 miles north. Radiocarbon dates suggest a somewhat shorter interval.

EARLY ESTIMATES OF GEOLOGIC TIME

In the past we have had to rely on estimates of age that have been based on rates of sedimentation, rates of erosion, rate of cooling of the earth, rate of influx of salt into the oceans, rates of leaching in soils, and rates of recession of the galaxies. These estimates were by no means exact, but they have been of great use in suggesting magnitudes. The following list indicates the many bases of early estimates of age.

1. Volumes of till, lake deposits, alluvium, or other sediments.

2. Amount of plant growth and humus accumulation.

3. Amount of erosion.

4. Cutting of gorges and retreat of falls.

5. Degree of development of beach features.

6. Weathering of pebbles and cobbles.

7. Degree and extent of oxidation; depth of leaching of calcium carbonate.

8. Rates of migration of plants and animals.

9. Rates of evolution of some animals.

10. Accumulation of salts in desert lakes.

It is not surprising that estimates based on such criteria vary widely. One such estimate, however, has proven satisfactory. Kay (1931), basing time estimates on the depth of removal by solution of carbonates in certain midwestern soils and on assumed average rates of advance and retreat of ice fronts, reached the following conclusions for the major stages of the Quaternary:

Stage	Duration	Years ago
Post-late Wisconsinan time in Iowa	25,000 years	25,000
Late Wisconsinan ice in Iowa	3,000	28,000
Mid-Wisconsinan time	55,000	83,000
Early Wisconsinan time	3,000	86,000
Sangamon time	120,000	206,000
Illinoian time	9,000	215,000
Yarmouth time	300,000	515,000
Kansan time	7,500	522,500
Aftonian time	200,000	722,500
Nebraskan time	7,500	730,000

These estimates were regarded by Kay as minima: he concluded that the total probably was a million years and perhaps two million years. Radiometric dating, it should be noted, has not changed Kay's estimates greatly (see Table 2.2). If the Nebraskan correlates with the Gunz in Europe, as has been generally assumed, the North American glacial deposits represent only the second half of the Pleistocene Epoch.

Kay's estimates indicate that the glaciations were brief episodes in contrast with the interglaciations. According to his table, the Nebraskan, Kansan, Illinoian, and Wisconsinan glaciations together lasted only about 5 percent as long as the interglaciations.

In brief, we emphasize again that geologic time, because of its vastness, is of the essence for understanding the changes that have occurred on and in the ground. When geologic time is considered, the variables multiply greatly. The physical stratigraphy and paleontology of the Pleistocene and Holocene record major changes in the climate and other environmental conditions, which in turn have greatly affected the processes of erosion, weathering, and soil development. An understanding of geologic history, therefore, is essential to our understanding of the changes that have occurred on and in the ground.

On the other hand, although time is one of the important factors in the erosion and weathering of rocks and surface deposits as well as in the development of soil profiles, doubling the time does not necessarily double the effect. This is because the processes of erosion, weathering, and soil development do not progress at uniform rates. With the passage of time, some processes accelerate, others are slowed. For example, a rock breaks up rapidly once it has started to crack, as can be seen at tombstones or other monuments. But because erosion on steep mountain slopes is more rapid than on plains, other factors being equal, it follows that erosion becomes slower as mountains become reduced. The next chapter considers some of the processes of weathering and erosion.

BIBLIOGRAPHY

Antevs, E., 1928, The last glaciation, with reference to the ice retreat in northeastern North America: Am. Geog. Res. Ser. 17, 292 pp.

Arizona Univ., staff, 1955, Geochronology: Arizona Univ. Phys. Sci. Bull. 2, 195 pp.

Charlesworth, J. K., 1957, The Quaternary Era: E. Arnold, London, 2 vols.

Flint, R. F., 1957, Glacial and Pleistocene geology: Wiley, New York, 553 pp.

Frye, J. C., and others, 1968, Definition of Wisconsinan Stage: U.S. Geol. Survey Bull. 1274-E, 22 pp.

Hay, O. P., 1923, 1924, 1927, The Pleistocene of North America and its vertebrate animals (3 vols.): Carnegie Institute Washington Publ. 322, 322A, 322B.

Hester, J., 1960, Late Pleistocene extinctions and radiocarbon dating: American Antiquity, v. 26, pp. 58-71.

Hunt, Chas. B., 1953, Pleistocene-Recent boundary in the Rocky Mountain region: U.S. Geol. Survey Bull. 996-A, pp. 1-24.

Jennings, J. D., 1968, Prehistory of North America: McGraw-Hill, New York, 391 pp.

Kay, G. F., 1931, Classification and duration of the Pleistocene Period: Geol. Soc. America Bull. v. 42, pp. 425-466.

Krieger, A. D., 1950, A suggested general sequence in North American projectile points: Anthropological Papers, Univ. Utah, no. 11, pp. 117-124.

Kurten, B., 1968, Pleistocene mammals of Europe: Aldine Publ. Co., Chicago, 317 pp.

Leonard, A. B., 1953, The relation of fossil mollusks to Quaternary stratigraphy in the central United States of America: Actes du IV Cong. de l'Assoc. Internat. pour l'Etude du Quaternaire, Rome.

Lyell, C., 1863, The geologic evidences of the antiquity of man, with remarks on the theories of the origin of species by variation: Philadelphia, Childs, 518 pp.

Morrison, R. B., and others, 1957, In behalf of the Recent: Am. Jour. Sci. v. 255, pp. 385-393.

Oakley, K. P., and others, 1950, The Pliocene-Pleistocene boundary: Report 18th Internat. Geol. Cong., Pt. 9, Great Britain, 1948, 130 pp.

Oakley, K. P., 1964, Framework for dating fossil man: Weidenfeld and Nicolson, London, 355 pp.

Rankama, K. (ed), 1965, The Quaternary (Vol. 1, Denmark, Norway, Sweden, Finland): Wiley-Interscience, New York, 300 pp.

———, 1967, The Quaternary (Vol. 2, British Isles, France, Germany, Netherlands): Wiley-Interscience, New York, 477 pp.

Schultz, C. B., and others, 1951, A graphic resume of the Pleistocene of Nebraska: Nebraska State Mus. Bull. v. 3, no. 6, 41 pp.

Wormington, H. M., 1957, Ancient man in North America (4th ed.): Denver Mus. Nat. Hist., Pop. Ser. 4, 322 pp.

Wright, W. B., 1914, The Quaternary ice age: MacMillan, New York, 464 pp.

Zeuner, F. E., 1959, The Pleistocene Period: Hutchinson, London, 447 pp.

Erosion can be considered in terms of different scales, which are related to duration of process. The big scale, illustrated earlier by the picture of Grand Canyon, represents the progress of erosion during tens of millions of years—so much time that both the kind and rate of erosion were affected by earth movements, volcanic eruptions, and climatic change. At the other extreme, the small scale, is the erosion one may witness in his backyard—the kind commonly called "soil erosion." An intermediate stage is represented here by this badlands scene near the Henry Mountains, Utah, where a fantastic network of gullies and ridges has developed as a result of the kind of ground and the stage of erosion.

3 / *Weathering and Erosion*

THE PHYSICAL PROCESSES

Concept of structure, process, and stage applied to weathering; structure of the rocks; structure in surface deposits; processes; importance of water; stage, rates of weathering; erosion—a selective process; wind erosion; erodibility of different kinds of ground; bibliography.

Surface deposits are products of weathering and erosion of rocks and minerals, the result of their physical disintegration and chemical decay. Except for the residual deposits, which are ancient deeply weathered soils, surface deposits have been transported from their parent source rock by gravity, running water, wind, or ice and later deposited as sediments (Fig. 1.1). These sediments include fragments of the original rock. Some of the minerals of which the source rocks were composed may remain unchanged; others, however, may be altered to new minerals by weathering. Modern soils have been formed, and are still forming, by the weathering of these surface deposits.

Weathering and soil formation can be considered in terms of the *structure* of the parent materials, the *process* or processes that have operated on them or that still are operating on them, and the *stage*, or geologic history, of the deposit. Structure, as used in this sense, pertains to those properties of rocks that resist or facilitate alterations:

1. Structure of the rocks, whether flat-lying or folded; their cementation, hardness, and texture (especially permeability); and their bedding, foliation, and jointing.

2. Structure of the minerals; their molecular structures, including chemical composition and solubility, density, hardness, cleavage, and crystal size.

This chapter considers some aspects of the structure of rocks; structure of the minerals is considered in chapter 11.

Three categories of processes contribute to the development of surface deposits and soils:

1. Mechanical processes that (a) cause disintegration of rock at the place of origin, such as the splitting action of frost, roots, and fire, and (b) cause comminution and abrasion of rock and mineral fragments during transport as sediment.

2. Chemical processes, such as oxidation, reduction, and hydration.

3. Biological processes, especially those that contribute organic acids, which aid in weathering the deposit.

This chapter considers some of the mechanical processes; chemical and biological processes are considered in chapter 11.

Stage refers not only to the duration of a particular process but also to the succession of different processes that may have accompanied a sequence of environmental changes. Stage, or geologic history, is of particular importance in the study of weathering and development of ancient soils, the residual deposits: considerable changes in climate took place during the Cenozoic, and these resulted in a great variety of physical and chemical environments under which the superimposed weathering profiles developed. The weathered products contained in a surface deposit include:

1. Fragments of rock, which may be either altered or fresh.

2. Fragments of minerals from the rock, which also may be either altered or fresh.

3. Precipitates, which include new minerals formed in the surface deposit or soil.

4. Organic matter, aqueous suspensions, and solutions.

STRUCTURE OF THE ROCKS

Bedrock, the parent material of surface deposits, and ultimately of soils, is classified according to mode of genesis: igneous, sedimentary, and metamorphic. There are many varieties of each kind, depending on mineralogy and texture. Igneous rocks are those that once were molten like lava. Some examples are illustrated in Figure 3.1. Porphyries consist of large crystals (phenocrysts) in a fine-grained groundmass. The big crystals formed early and were floating in a melt that chilled quickly to form the fine-grained groundmass. Volcanic ash, or tuff, is rock that was chilled quickly from the molten state to become glass; there was no time for minerals to crystallize. Ash beds may contain angular fragments of glass, called shards, and angular fragments of rocks and/or minerals. A tuff that contains abundant mineral fragments is known as a *crystal tuff;* one that contains abundant rock fragments is a *lithic tuff.* Lavas, like porphyries, may consist of large crystals in a glassy groundmass; the elongated minerals commonly are oriented parallel, showing a flow structure. Many lavas are vesicular due to vugs formed by gases expanding in the molten rock. Granitic rocks are coarsely crystalline and *massive*—that is, without structure.

Sedimentary rocks (Fig. 3.1) once were unconsolidated sediments and subsequently

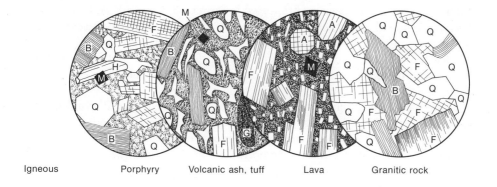

Igneous   Porphyry   Volcanic ash, tuff   Lava   Granitic rock

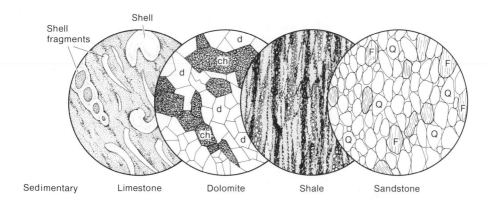

Sedimentary   Limestone   Dolomite   Shale   Sandstone

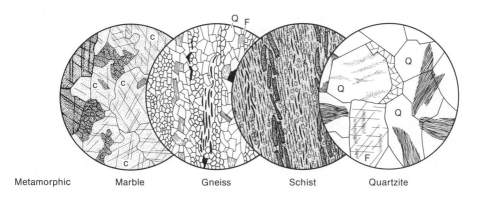

Metamorphic   Marble   Gneiss   Schist   Quartzite

FIGURE 3.1
Examples of textures in various kinds of igneous, sedimentary, and metamorphic rocks. A, augite; B, biotite; C, calcite; ch, chert; d, dolomite. F, feldspar; G, volcanic glass; H, hornblende; M, magnetite; Q, quartz. Diameter of fields, 2.5 mm.

became compacted and cemented (*lithified*) to form rock. Some are *chemical precipitates,* like limestone and dolomite. The *clastic* sediments, such as sandstone and shale,
were carried in suspension by running water or rolled along the bottom. Limestone
is calcium carbonate; dolomite is calcium-magnesium carbonate. Both may be either
fine-grained or coarse-grained. Commonly they contain irregular masses of amorphous silica known as chert.

Shale, the lithified equivalent of mud, generally is well bedded. Sandstone is
composed of rounded grains of quartz accompanied by grains of other minerals,
especially feldspar. The grains may be cemented with carbonate, silica, or clay, or
the space between the grains may be open. The kinds of sedimentary rocks grade
into each other, giving shaly limestone, sandy shale, limey sandstone, etc.

Shale is likely to be thinly bedded and break into thin flakes or plates. Sandstone,
limestone, and dolomite tend to be more thickly bedded; some are massive. Sedimentary deposits laid down in marine or fresh-water environments may contain
fossil shells; deposits formed on land may contain remains of terrestrial animals or
plants, including petrified wood, leaf impressions, or coal beds.

Metamorphic rocks are those that were once igneous or sedimentary rocks, but
that were subsequently altered or recrystallized by heat and/or pressure at great depth
in the earth's crust. Marble, the metamorphic equivalent of limestone or dolomite,
may be either coarse- or fine-grained and commonly is streaked. Gneiss, which is
metamorphosed igneous or sedimentary rock, consists of layers of coarse grains,
usually quartz or feldspar, alternating with layers of fine-grained minerals. Schist,
metamorphosed shale, consists of micaceous layers. Quartzite, metamorphosed sandstone, may be coarse or fine-grained; the recrystallized quartz forms a tight cement
between the grains.

Permeability—the degree to which water can enter and change rock—is a
structure-related property of rock that is important because of its effect on weatherability. Some sandstones are highly permeable; specimens immersed in a glass of
water for a few hours may absorb an amount of water equal to 10 percent of the
dry weight of the specimen. Specimens of coarse-grained igneous rock may absorb
an amount equal to one or two percent of their dry weight. A block of coarse granite
in New York had to be coated with paraffin to protect it against disintegration; its
220 square yards of surface absorbed nearly 70 pounds of paraffin. Water absorbed
by rock surfaces causes both physical and chemical changes.

Permeability of rock is affected not only by size and spacing of minerals but also
by the kinds of minerals of which it is composed. Some minerals weather readily;
others weather but slowly, as will be brought out in chapter 11 (Mineralogy and
Geochemistry). Rock disintegration, though, can be favored by the presence of a
mineral that is more readily weathered than those around it. For example, granitic
rocks in many places disintegrate to form sandy ground because the biotite (Fig.
3.1) weathers readily and causes the other minerals to break apart. Sandstone may
disintegrate readily to sand, particularly if the cement in the rock is the easily soluble
calcium carbonate.

Fractures are another structural characteristic of importance to the weatherability
of rock, because they allow water to enter rock and alter it. Fractures in massive
rocks, such as granite, are called *fissures* or *joints* (Fig. 3.2, A). Some were produced

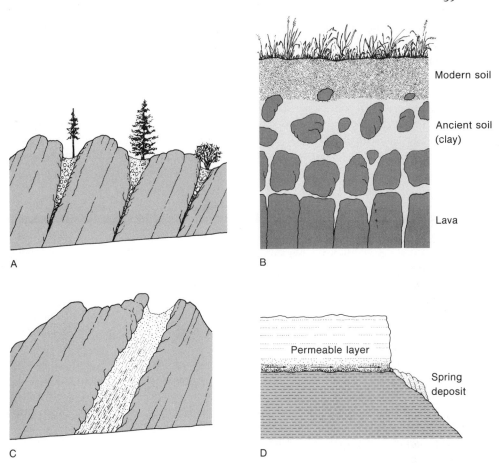

FIGURE 3.2

Examples of how rock structures affect weathering. (A) Widely spaced, steeply dipping joint planes in granite provide entryways for water to weather the rock. The amount of weathering increases upward; weathered rock and surficial materials collect in the openings, forming a substrate in which plants can grow. Two sets of joints at about right angles to each other lead to the development of rounded granite masses isolated from one another. (B) Columnar jointed lavas weather to clay in areas where there are deep residual deposits. The vertical columnar joints are widely spaced at the bottom of the cut. As weathering progresses, rounded masses of the lava become isolated in the clayey rock rot, and the residual rounded masses become smaller upward. (C) Where formations are steeply dipping, permeable beds enable water to filter downward, and these permeable beds weather more deeply than the impermeable ones. (D) Where gently dipping permeable beds overlie impermeable ones, water can seep along the base of the permeable bed and cause it to weather.

by strains induced in the rock during cooling; others were caused by the relief of stresses produced by earth movements during the rock's history. In any given area fractures commonly have preferred directions, reflecting kind and direction of the stress that produced them, and the fractures may be widely or closely spaced. Water carrying dissolved acids seeps into such cracks and reacts with the fissure walls, causing chemical alteration of the minerals. The more fractured the rock, the more susceptible it is to weathering. The dissolved acids are of two kinds: organic acids that are generated by the decay of organic matter at the surface, and mineral acids generated by alteration of sulfide, chloride, or carbonate minerals in the rock.

Basalt is an igneous rock that is extruded as a lava and that commonly forms widespread sheet-like masses 20 to 40 feet thick. It is a fine-grained massive rock. As the solidified mass cools, shrinkage cracks develop that are oriented perpendicular to the principal cooling surfaces; that is, the cracks are vertical. Like mud cracks, they tend to form closely spaced six-sided columns that generally are no more than 2 or 3 feet in diameter (Fig. 3.2, B). Such rock has many entrances for water, and the extent of surfaces subject to weathering is very much greater than in rocks that are broken only by widely spaced joints. Moreover, basaltic rocks contain a high proportion of minerals that are readily broken down by weathering.

Few sedimentary or metamorphic rocks are massive; most are divided by closely spaced *partings.* In sedimentary rocks these are the bedding planes that separate successive *strata,* as where a sand has been deposited on mud, or vice versa. Water can seep along bedding planes, especially where a permeable stratum, such as sand or gravel, overlies a relatively impermeable one, such as clay or shale (Fig. 3.2, D). In metamorphic rocks the partings are generally controlled by parallelism, or *foliation,* of the metamorphic minerals, but these too provide access for moisture (Fig. 3.1). In addition to stratification and foliation, the sedimentary and metamorphic rocks have open fractures, or joints, like those in the igneous rocks; the jointing is generally at a high angle to the bedding or foliation.

The combination of closely set bedding or foliation in sedimentary and metamorphic rocks produces flat, tabular fragments (*channers*). Flagstone and slate are familiar examples. Massive rocks produce blocky fragments that are more nearly equidimensional. The flat, tabular rock fragments tend to break down faster than the blocky ones because their thin edges break easily, and more surface area per unit volume is exposed to weathering.

STRUCTURE IN SURFACE DEPOSITS

In structure and composition, the surface deposits exhibit about the same extreme variety as the sedimentary rocks. Some are massive, others thinly bedded; some are readily permeable, others less so. Deposits of volcanic ash have the composition of the parent magma; they range in composition from about that of basalt to about that of granite, and their average composition would be that of the average igneous rock.

Even more extreme are the composition ranges of such surface deposits as caliche (mostly calcium carbonate), sand (mostly silicon dioxide) in dunes or beaches, and clays (mostly hydrous aluminum silicates) in lake or stream beds. Most surface deposits, however, are intergradational mixtures of these three kinds, making for even greater variety of composition. In general, surface deposits differ in composition from average bedrock in having lower amounts of the alkalis (sodium, potassium) and alkaline earths (calcium, magnesium), since those elements are easily leached, and differ also in having relatively more of the less soluble silica and alumina. Also, surface deposits contain more carbon, especially as carbonates, because of the addition of organic matter in the layers at the ground surface.

Weatherability of a surface deposit also depends on particle size; the finer the particles, the greater the surface that is exposed per unit volume. One cubic inch of rock has 6 sides totalling 6 square inches of surface exposed to weathering. Dividing this into cubes 0.1 inch on a side gives 1,000 small cubes having surfaces that aggregate 60 square inches $(1,000 \times 0.1 \times 0.1 \times 6)$. Further dividing these into cubes 0.01 inch on a side gives 1,000,000 tiny cubes having 600 square inches of surface exposed to weathering $(1,000,000 \times 0.01 \times 0.01 \times 6)$.

In brief, the structures of both surface deposits and rock formations—their textures, fractures, and bedding planes—allow water to enter the deposit and weather the toughest kinds of rock or mineral.

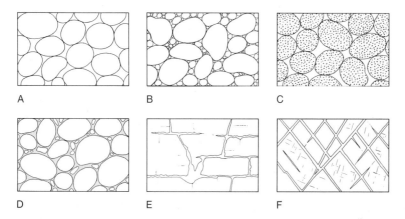

FIGURE 3.3
Diagram showing several types of rock interstices and the relation of rock texture to porosity. (A) Well-sorted deposit with high porosity. (B) Poorly sorted deposit with low porosity. (C) Well-sorted deposit consisting of pebbles that are themselves porous, making the deposit highly porous. (D) Well-sorted deposit whose porosity has been diminished by the deposition of mineral matter in the interstices. (E) Rock rendered porous by solution. (F) Rock rendered porous by fracturing. [From Meinzer, 1923a.] (Note that permeability is not to be confused with porosity. Two kinds of ground may have the same porosity but very different permeabilities, as illustrated in Fig. 10.1.)

PROCESSES

The processes that cause weathering, whether mechanical, chemical, or biological, depend on the entrance of water into joints or partings in rock or into the pore spaces between mineral grains (Fig. 3.3). If the water freezes, it can break the rock as effectively as freezing can burst a household waterpipe or the engine block of an automobile. Slabs and blocks of rock break off to form rubble or colluvium (Fig. 3.4). As joints become wider, they collect fine particles and material suitable for plant growth. Added to the wedging action of ice is the wedging action of roots, the effects of which are illustrated by the disrupted slabs of concrete along many city sidewalks. The crystallization of salts introduced into the pore spaces in rock surfaces produces an effect similar to that of frost-wedging.

FIGURE 3.4
Example of rock being split by alternate freeze and thaw. The part shown is about 12 feet high; with each freeze the central joint is widened, and the stone holding the walls apart slips downward a little and pushes them farther apart. [After Hunt, 1953.]

Water absorbed into the pores of bare rock surfaces, causes sheets or shell-like slabs of rock to expand and break away from the surface—a process known as *exfoliation* (Fig. 3.5). The process seems to have four chief causes: (1) frost-wedging; (2) crystallization of salts that enter the pores; (3) chemical combination of water with certain minerals, causing them to swell (*hydration*); and (4) unloading of over-burden by erosion, causing a fracturing analogous to rock bursts in mines and quarries. (A rock burst is the explosive expansion of part of a rock face from which the surrounding, confining rock has been removed.) The rounding of magnificent Half Dome in Yosemite National Park and of some of the bald rock surfaces in northern New England and in the southern Blue Ridge have been attributed to exfoliation.

Expansion and contraction of rock as a result of heating and cooling have been thought to contribute to exfoliation, but laboratory tests indicate that such changes, *in the absence of water,* are not effective. Furthermore, the exfoliation shells are just as well developed on shaded exposures, where temperature changes are minimal, as they are on sunny ones.

Freeze and thaw of water in the ground and the hydration of minerals (especially the clays) are major factors in mass wasting and in the development of landslides and creep. Freezing and thawing also produce patterned ground (10.7), and can even destroy the layering in a soil profile by mixing the different layers, particularly in arctic soils. Other mechanical effects of water include the lubrication of the ground, which causes solifluction or other forms of slump on hillsides; pelting of the surface by raindrops; erosion by floods; abrasion of rock and mineral particles by wave action,

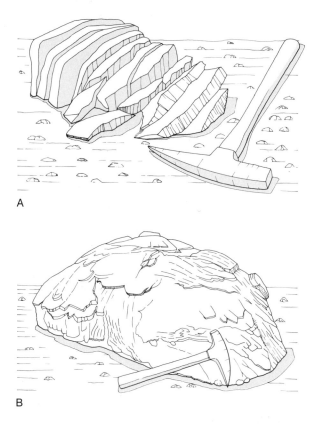

A

B

FIGURE 3.5
On old gravel fans, boulders of massive rocks like chert and
quartzite tend to break into slabs along transverse fractures (A).
Porphyritic igneous rocks tend to fragment by exfoliation (B).
[After Hunt, 1966a.]

grinding effect of ice, and stream transportation; and the ground effects of frost and
hail.

The distinction between acid and alkaline soils, discussed in Chapter 8, is due
chiefly to differences in water supply. The weathering that produced the modern
soils took place mostly in aerated ground above the water table, where oxidation
and leaching can occur. Where the water table is shallow, the ground is deficient
in oxygen, organic matter causes a reducing chemical environment, and gley soils
are produced. Under such anaerobic conditions, organic matter putrefies and accumulates rather than being oxidized and destroyed. In humid regions such ground
becomes strongly acid.

The processes of mineral alteration and soil development are affected also by the
geologic and topographic situation. High water tables (Figs. 3.6, 3.7) cause waterlogging, whether in wet bottom lands, on hillsides at seeps, or on hilltops where
ground water is perched on impervious strata. Hillsides tend to be better drained
than flat hilltops or valley bottoms. Deposits underlain by sand or gravel are better
drained than those underlain by less permeable beds. The geologic and topographic

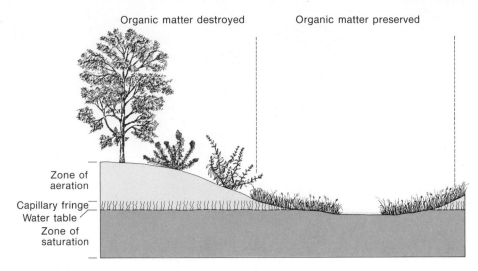

Organic matter destroyed   Organic matter preserved

FIGURE 3.6
The water table and weathering. Position of the water table greatly affects the kind and rate of weathering. In humid regions water percolates through the zone of aeration to the water table. Organic matter
is destroyed by oxidation in the aerated zone, but is preserved where the saturated zone is near the
surface.

situation, therefore, affects the availability of water, the frequency of wetting and
drying. The kind of ground, and its setting, affect its cooling and heating and thus
its susceptibility to freezing and thawing, and all these factors in turn affect the climate
and biota both on and in the ground.

STAGE; RATES OF WEATHERING

Stage, as already noted, refers to the duration of the processes that cause weathering
and to the succession of different processes that result from climatic or other changes
in the environment. The factor of *time* in weathering and soil development is important chiefly because the history of a particular soil or piece of ground may extend
back into Pleistocene and earlier times, when the climates, and consequently the
processes, were quite different from those we know today. For example, whatever
the weather statistics, the effective climate of western Utah was very different when,
during the Pleistocene, the area that is now desert was submerged under the thousand feet of Lake Bonneville water (p. 135).

The time required for weathering of rocks has long been a subject of much interest.
Experimental work done almost 100 years ago led one scholar to conclude "that
the disintegration of Jurassic limestone to the depth of 1 metre would occur at the
end of 728,000 years of atmospheric action; while the same depth of syenite would
be similarly affected after a lapse of 731,400 years" (Stockbridge, 1888). The reli-
ability or precision of such estimates is not of course to be taken seriously, but much
of the present fad for quantitative determination and classification of the natural world
probably will appear equally naive a hundred years from now.

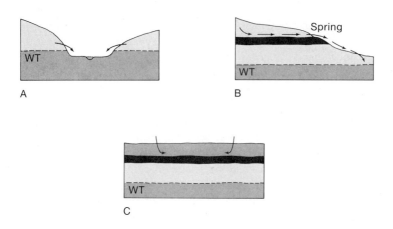

FIGURE 3.7
Variations in the position of the water table affect weathering. (A) Water tables higher than the streams discharge to them. (B) Water tables may be deep, but poorly permeable beds (black) perch ground water, which discharges in springs. (C) Water tables may be deep, but shallow permeable layers may be waterlogged in wet seasons by ground water perched on poorly permeable beds.

The kind and rate of weathering depend on the intensity of the process, which is largely controlled by climate, and on the susceptibility of the material to alteration, which is largely controlled by its structure and composition. In cool temperate regions, exposed rock surfaces planed and grooved by glacial ice in late Pleistocene time are still little weathered; some even retain a glacial polish. Striated cobbles and pebbles contained in glacial deposits also retain their glacial markings. Such occurrences indicate that in those environments, weathering has had little effect for the past 10,000 to 20,000 years. In arid regions like the Nile Valley and our own Southwest, well-preserved prehistoric inscriptions on rock surfaces indicate a slow rate of weathering throughout the past few thousand years.

An example of rapid weathering is provided by changes observed in young beds of volcanic ash, such as the alteration of the ash ejected when Krakatau, a volcanic island in Sunda Strait between Java and Sumatra, erupted in 1883. Fifty years later there was measurable loss of silica (about 5 percent) from the upper layers and residual enrichment in clay. Such weathering, however, is exceptional because volcanic glass is notably unstable, especially when it is both fine grained and subjected to the intense acid leaching of a tropical environment. A related phenomenon is the leaching of silica from pottery sherds at archeological sites only a few hundred years old. The sherds, made of baked clay, are porous and rather like volcanic ash. Many are coated with a thin film of silica.

Caliche may be deposited quickly too, particularly at moist sites in arid regions. It commonly coats pottery sherds and other Indian artifacts no more than a few hundred years old. Most stones embedded in the ground in semiarid and arid regions have a coating of caliche. Given enough moisture, of course, caliche may be deposited quickly enough to clog a drain pipe. Desert varnish may form quickly, although most desert varnish in the arid and semiarid west is middle Holocene or older (p. 155). Examples of rapid chemical changes are described in Chapter 11.

Despite such examples of rapid weathering, the geologic record is clear that during the late Pleistocene and Holocene the environments in most of the world have not been such as to cause much weathering of rocks or minerals. This is notably true in North America and Europe, where the record is best documented.

Rates of development of soil profiles, on the other hand, are not quite comparable to rates of weathering of the rocks and minerals, even though the layering in a soil profile is a result of weathering processes. For example, the conspicuous layering of acid forest soils is due mainly to the downward transport of soluble original constituents; other constituents may be unaltered or only slightly altered. In most agricultural soils the layering is quite young.

Organic matter in soils is largely the product of the present biota, and is of no great antiquity; for practical purposes most of it can be considered as late Holocene. But even the entire Holocene has not been long enough, at least in the temperate, arctic, and arid regions, for mineral matter in surface deposits and soils to have been greatly affected. Where there has been much rock or mineral alteration, the characteristics of the mineral fraction are largely inherited from the processes that operated during the Pleistocene or during earlier epochs, when the climate and biota were very different from what they are now.

The deep and advanced weathering that characterizes ancient soils, and that has reduced hard rocks to clay, commonly is attributed to the antiquity of the soils and the long duration of the processes have operated. But this is only half the truth; the other half, and perhaps the more important half, concerns the environment and process that prevailed during the Pleistocene, when climates were different. The important corollary of this is the usefulness of the clayey alteration in soil stratigraphy, as described in Chapter 9, Ancient Soils.

EROSION— A SELECTIVE PROCESS

The present rate of erosion in the world is estimated to average about 1 inch per thousand years (see Leopold, Wolman, and Miller, 1954, p. 27). Assuming this rate also is about average for all of the Holocene, we may infer that the land surface has been lowered on the average about 1 foot since Pleistocene time. The estimates seem reasonable, but these are of course averages. Geographically, the rate of erosion has been quite variable. Areas with much resistant rock contributed little to the sediment being transported by streams from the continents to the oceans. That this is true, is indicated by the preservation of glacial striae in resistant rock formations that were overridden by Pleistocene glaciers. Sandstone formations coated with desert varnish have eroded but little in the past 2,000 years. And there are photographic records of monoliths showing no detectable erosion in 100 years.

Erosion is selective. In the eastern United States, surface deposits generally are thick over bedrock formations and are principal sources of the sediment being eroded in that part of the country. Easily eroded formations (for example, Cretaceous and Tertiary shales) occur in the Atlantic and Gulf Coastal Plains, but those areas are

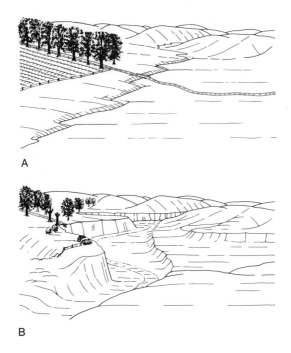

A

B

FIGURE 3.8
Modern arroyo-cutting on the Colorado Plateau. Pleasant Creek at
Notom, Utah, about 1890, before the arroyo-cutting (A) and today
(B). The arroyo is about 20 feet deep. [After Hunt, 1953.]

at low altitudes and are nearly flat; erosion there, however, is comparatively slight.

In the arid and semiarid western states, erosion is highly selective. It is mainly eroding the surface deposits, especially colluvium on hillsides and alluvium on valley floors. In areas of easily eroded shale, like the Cretaceous shale, erosion is mostly in the badlands, which have steep hillsides mantled with a fluffy aggregate of shale flakes that wash readily. Shale flats, or pediments, that border the badlands are compact surfaces and are eroding slowly; photographic records made on some in Utah show that tiny rills have undergone very little change over a 30-year period, and that there has been no noticeable change in level at small wood stakes after 30 years of washing. These pediments act as surfaces for transporting the sediment eroded from the badlands.

Arroyo-cutting in the floodplains (Fig. 3.8), which has been a major problem in much of the west, is another example of selective erosion. This erosion has been attributed to land use, especially overgrazing, and it is a fact that much of the erosion occurred soon after the western valleys became occupied. But arroyo-cutting on an even greater scale occurred in the past too (Fig. 7.10), and in general the stratigraphic record seems to indicate alluviation during wet periods and arroyo-cutting during dry ones.

Wind has been the major factor in the formation of certain important surface deposits, notably the blankets of loess and sand dunes. The extensive deposits of sand and silt left on broad flood plains after the meltwaters of the Pleistocene glaciers

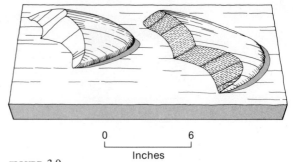

0   6

Inches

FIGURE 3.9

Wind-faceted cobbles. Argillite (left) is smoothly faceted; limestone (right) has rillen on the facets. Both specimens oriented as
in the field. [After Hunt, 1966a.]

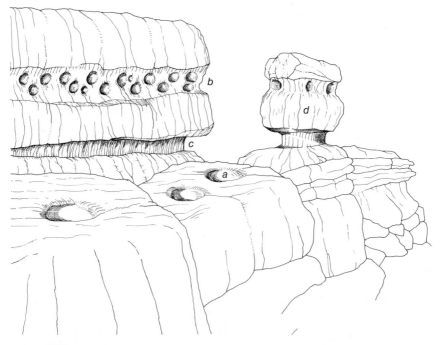

FIGURE 3.10

Desert features partly attributable to wind erosion: a, natural depressions, waterpockets or tanks;
b, niches; c, alcoves; d, pedestal rocks. Rains and running water, even though infrequent, can dissolve the cement matrix of some desert sandstones; the sand grains become loosened and are removed
by the wind. Water standing on the surface seeps into the sandstone, dissolves the cement, and the
wind scours a depression (a), forming a waterpocket or tank. Some of these are large and hold water
for long periods following rains. Sandstone beds in which the cement is spottily distributed may develop irregular small pockets or niches (b). Where bedding planes favor movement of perched water
(c), there may be large alcoves, some big enough to contain apartment houses of the prehistoric cliff
dwellers. Where a caprock overlies an easily eroded bed, pedestal rocks (d) form monoliths isolated
by erosion.

subsided were eroded by wind—the process of *deflation*. The wind blowing across the alluvium moved the fine particles (silt size) onto the uplands, and these deposits (loess) became parent materials for some of our most productive soils.

Wind on semiarid plains and in deserts erodes deflation hollows, which collect runoff and provide a source of water for range animals. In the southern plains they are known as buffalo wallows. Wind erosion in deserts is commonly overrated, even though it can cause sandblasting (Fig. 3.9). It is chiefly a transporting agent rather than an eroding agent. Water issuing from sandstone formations or rains beating against sandstone walls dissolve the cement holding the grains together, and the loosened sand grains can be blown away (Fig. 3.10). Winds also distribute pollen given off by plants, volcanic ash erupted by volcanoes, and radioactive fallout produced by man.

ERODIBILITY OF DIFFERENT KINDS OF GROUND

Present-day erosion in the United States is illustrated in Figure 3.11. In the eastern United States the severely eroded areas are on ancient, deep residual soils on which red and yellow soils are developed. In the central United States the most severe erosion takes place where there are loessial deposits. In the western United States erosion is most severe in areas where Mesozoic and Cenozoic shale formations are exposed. The map (Fig. 3.11) might have been made by showing the easily erodible formations rather than the areas of actual erosion, which is what are shown in Figure 3.11.

Erodibility of surface deposits is a function of several variables, including resistance to erosion (structure), vigor of the processes, and degree of protection, mainly by a cover of vegetation. Some kinds of deposits, such as gravel, resist erosion; loesses and residual soils erode readily. Sun-dried mud (adobe) bricks are widely used for construction in arid regions; in humid regions they would wash away. In humid climates limy deposits, gypsum, and rock salt are readily removed by solution; in arid climates these resist erosion. In some deserts rock salt may survive on the ground for 2,000 years. In humid regions, erosion by solution is accelerated not only because there is more moisture, but also because the water in the ground generally is acid.

Erosion, which is controlled largely by kind of ground and by kind of protective vegetative cover, is greatly accelerated by disturbance of that vegetative cover and topsoil. Muddiness is conspicuous at most construction projects, whether for highways, factories, or homes. Farming accelerates erosion too.

Erosion of bare ground is not necessarily accompanied by the development of rills. The ground may be eroded layer by layer, a process called sheet erosion. More severe erosion usually creates rills, and where erosion is most severe the rills are deepened and widened to form gullies. In homogeneous ground, gullies are likely to be V-shaped, but where the capping layer is tougher than the underlying ones, gully walls are steeper and bottoms broader and flatter, giving a U-shaped cross section.

Rain splash is insignificant on flat ground, but on hillsides it may move much sediment downslope. Moreover, rain may cause puddling where the ground is composed of clods that collapse when they become wet, and this can increase the runoff and amount of sediment even though the ground beneath remains dry. Any

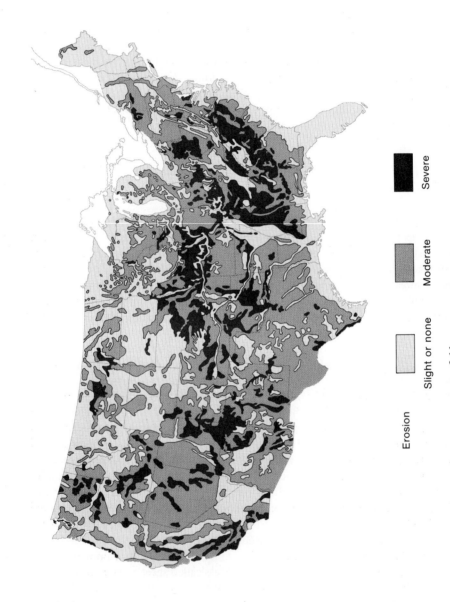

Erosion

Slight or none

Moderate

Severe

FIGURE 3.11
Present-day erosion in the United States. [After
U.S.D.A.]

of these forms of erosion can lead to selective removal of the fine-grained materials, leaving behind a lag concentrate of coarse materials.

Erosion rates are roughly proportional to slope; other conditions being equal, doubling the angle of slope increases the loss of sediment by about two and a half times. Some examples are given in these measurements by the Soil Conservation Service:

> Texas. Sandy loam in cotton with slope of 8 percent lost 20 tons per acre per year; same with slope of 16 percent lost 60 tons.
>
> Missouri. Loam in corn with slope of 4 percent lost 20 tons per acre per year; same with slope of 8 percent lost 50 tons.
>
> Ohio. Loam in corn with zero slope lost 2 tons per acre per year; same with slope of 2 percent lost 7 tons.

Doubling the length of slopes increases the loss about one and a half times.

The protection offered by plant cover affects the rates of erosion. Desert shrubs are widely spaced; in most stands more than half the ground is bare, and many stands comprise no more than a few shrubs per acre. Lands with bunchgrass may have bare areas between the bunches where runoff becomes channeled; other grassland has nearly continuous sod. Most eastern forests, whether coniferous or deciduous, have a nearly continuous mat of litter composed of fallen leaves and twigs. Although rainfall increases from desert shrub to grassland and from there to the eastern forests, the loss of sediment decreases.

In grassland areas the rate of erosion has been greatly increased by farming. The amount of soil eroded from crop lands is 50 to 100 times as much as is eroded from grassland (Leopold, Wolman, and Miller, 1964). Wind also contributes to the erosion of soil from farmed lands, especially where the ground is dry enough to provide granular missiles for blasting loose other particles.

Probably at least half, and more likely three-quarters, of the material being eroded in the United States is from the surface deposits, and it seems doubtful that these are being replaced at the rate they are being eroded. It would seem, then, that the continent is destined to become bare rock stripped of its few feet of overburden. At present rates of erosion and weathering, the material that remains certainly would last a few tens of thousands of years, so the problem is not exactly acute. Nevertheless, the gradual removal of this thin film of loose material could be one of the mechanisms that, in the geologic past, has contributed to increased biological competition and extermination of species.

To summarize, the weathering and erosion of bedrock to produce surface deposits are controlled by: (1) the kind and structure of the parent bedrock; (2) processes that weathered and eroded the bedrock and that formed the surface deposit; and (3) stage, that is, length of time the processes operated. Weathering to produce soil (in the agricultural sense) also is controlled by structure (kind of parent material in the surface deposit), and by process and stage. The processes of soil formation in turn are controlled largely by climate, organisms on and in the ground, and topography.

The importance of time has already been discussed. Our next chapter considers the effects of topography and landforms on kind of ground; after that, we consider climates (Chapter 5) and organisms (Chapter 6) on and in the ground. The second half of the book looks more closely at the surface deposits and soils that form the ground.

BIBLIOGRAPHY

Blackwelder, E., 1925, Exfoliation as a phase of rock weathering: Jour. Geol., v. XXXIII, pp. 793-806.
———, 1933, The insolation hypothesis of rock weathering: Am. Jour. Sci., v. XXVI, pp. 97-113.
Chandler, R. F., Jr., 1942, The time required for podzol profile formation as evidenced by the Mendenhall glacial deposits near Juneau, Alaska: Proc. Soil Sci. Soc. America, v. 7, pp. 454-459.
Chepil, W. S., and Woodruff, N. P., 1963, The physics of wind erosion and its control: Advances in Agronomy, v. 15, pp. 211-302.
Denny, C. S., 1951, Pleistocene frost action near the border of the Wisconsin drift in Pennsylvania: Ohio Jour. Sci., v. 51, pp. 116-125.
Flint, R. F., 1949, Leaching of carbonates in glacial drift and loess as a basis for age correlation: Jour. Geol., v. 57, pp. 297-303.
Frederickson, A. F., 1951, Mechanisms of weathering: Geol. Soc. America Bull., v. 62, pp. 221-232.
Gilbert, G. K., 1914, The transportation of debris by running water: U.S. Geol. Survey Prof. Paper 86, 160 pp.
———, 1917, Hydraulic-mining debris in the Sierra Nevada: U.S. Geol. Survey Prof. Paper 105, 154 pp.
Grawe, O. R., 1936, Ice as an agent of rock weathering: Jour. Geol., v. 44, no. 2, pp. 173-182.
Griggs, D. T., 1936, The factor of fatigue in rock exfoliation: Jour. Geol., v. 44, no. 7, pp. 783-796.
Hill, D. E., and Tedrow, J. C. F., 1961, Weathering and soil formation in the arctic environment: Am. Jour. Sci., v. 259, pp. 84-101.
Hough, G. J., Gile, P. L., and Foster, Z. S., 1951, Rock weathering and soil profile development in the Hawaiian Islands: U.S. Dept. Agriculture Tech. Bull. 752, 43 pp.
Judson, S., and Ritter, D. F., 1964, Rates of regional denudation in the United States: Jour. Geophys. Res., v. 69, pp. 3395-3401.
Leverett, F., 1909, Weathering and erosion as time measures: Am. Jour. Sci., 4th ser., v. 27, pp. 349-368.

MacClintock, P., 1954, Leaching of Wisconsin glacial gravels in eastern North America: Bull. Geol. Soc. America, v. 65, no. 5, pp. 369-383.

Nikiforoff, C. C., 1935, Weathering and soil formation, Trans. 3rd Internat. Cong. Soil Sci., Pt. I, pp. 324-326.

Nikiforoff, C. C., 1949, Weathering and soil evolution, Soil Sci., v. 67, pp. 219-230.

Polynov, B. B., 1937, The cycle of weathering (trans. by A. Muir): Murby, London.

Reiche, P., 1950, A survey of weathering processes and products: Univ. New Mexico Press Publ. in Geology no. 3, 95 pp.

Stauffer, R. S., 1935, Influence of parent material on soil character in a humid temperate climate: Jour. Am. Soc. Agronomy, v. 27, pp. 885-894.

Storer, F. H., 1883, Rock disintegration in hot, moist climates: Science, v. 1, p. 39.

Taber, S., 1929, Frost heaving: Jour. Geol., v. 37, pp. 428.

Tedrow, J. C. F., and Wilkerson, A. A., 1953, Weathering of glacial soil material: Soil Sci., v. 75, pp. 345-353.

Troll, C., 1958, Structure, soils, solifluction, and frost climates of the earth: U.S. Army Snow Ice and Permafrost Research Establishment, Willmette, Ill., Trans. 43, 121 pp.

The earth's surface is not static, for the earth's crust is not stable. A particular hill is geologically ephemeral, but hills in general are everlasting in the sense that we will always have them. While erosion destroys hills, earth movements create new ones. The shapes of hills and mountains are controlled by their geologic structure, and by the processes and stage of erosion, as illustrated in this view of Mt. Hillers, in the Henry Mountains, Utah. The mountain was formed by an igneous intrusion (dark, central mass of rock) that was forcibly injected upward through the sedimentary strata, the eroded edges of which are exposed and steeply upturned at the base of the intrusion (the light-colored, bedded rocks). Some 5,000 to 10,000 feet of sedimentary rocks have been eroded from above this mountain since it was formed, during the middle Tertiary. (Photograph by Fairchild Aerial Surveys for United States Geological Survey.)

4 / Landforms

Topography, geomorphology defined; scales of landforms; bedrock geology—structural control of landforms; plateaus, plains, mountains, and drainage patterns; physiographic regions and their landforms— Coastal Plain, Appalachian Highlands, central United States, Rocky Mountains, Colorado Plateau, Basin and Range Province, Columbia Plateau and Cascade Range, Sierra Nevada and Coast Ranges; some effects of topography on kinds of ground; bibliography.

The term "topography" refers to the configuration—the relief and contours—of the features that give variety to our landscape: our plains, plateaus, valleys, mountains, and other landforms. The study of landforms is the science of *geomorphology.* Any landform records two quite different chapters of earth history. The first reveals the manner in which the bedrock was formed and how its structure developed—whether by uplift, folding, faulting, or some combination of the three. The second chapter reveals the processes of weathering and erosion that sculptured the bedrock and produced the landform. These two chapters of earth history may be separated by hundreds or millions of years. Marine fossils are found at high elevations in many mountains, yet the sea never reached such levels. The rocks in which such fossils are found were once sediments that accumulated below sea level. Over vast periods of time they became *lithified* by pressure or heat caused by burial beneath younger sediments, and were subsequently raised to their present heights by movements of the earth's crust.

Landforms can be considered on at least four scales: (1) the continents, (2) the physiographic regions, (3) the plains, plateaus, and mountains, and (4) in *microrelief,* such features as mudcracks, ripple marks, and tussock mounds on sand dunes. In studying the ground we gain an understanding of how topography affects the kind and thickness of the weathered materials that cover the bedrock, and how surface deposits can build or modify the landforms.

BEDROCK GEOLOGY AND THE STRUCTURE OF LARGE-SCALE LANDFORMS

All continents are divisible, on the basis of structure, into clearly recognizable physiographic regions, each having distinctive landforms that reflect their structures. In the anatomy of a continent, these structural members are the bones to which the flesh (the weathered materials that cover the bedrock) and the circulatory systems (the drainage—streams and rivers) are attached.

The distribution of the mountains, plateaus, and plains in the United States reflects the paleogeography of the Paleozoic and early Mesozoic and to a lesser extent that of the Cretaceous. Many of our present-day mountain systems formed on or near the axes of former longitudinal downwarps that were flooded. The plains, those in the interior and along the east coast, are the sites of ancient shallow seas (shelf areas) in which thin deposits accumulated. In a sense, therefore, the landforms attributed to structure are the product of a series of structures, each dependent upon something that preceded.

Plateaus

The geologic structure of a plateau is easy to visualize, for the nearly horizontal strata can be seen along canyon walls. The surfaces of plateaus have been elevated high above sea level, and in the most highly elevated ones the streams are deeply incised, as in the Grand Canyon of the Colorado River and in Hell's Canyon of the Snake River. The development of plateaus requires not only the uplift of nearly horizontal formations but the presence of resistant formations to maintain the upland surface. In areas where all the formations are easily eroded, the result is badland topography.

In the east, the plateaus are old and deeply weathered; their ground surfaces have evolved over a long span of time under humid climates. In the west, the plateaus are younger, and the ground surfaces have evolved under semiarid climates. Some of the western plateaus are lava rather than sedimentary formations.

A flat-topped butte, known in the Southwest by the Spanish term "mesa," is a form of plateau. A typical mesa has a resistant caprock—a bed of hard sandstone or a lava—and sloping sides formed of more easily eroded shale. Mesas may be a few hundred or even a few thousand feet high, and they may cover a few acres or many square miles. When sufficiently extensive, mesas become plateaus.

Plains

Structurally, plains are like plateaus, for their formations are horizontal, or nearly so, but they have not been as deeply dissected by the streams and rivers. Interstream areas on plains are but little higher than the drainage courses, and the topographic relief is low. There are different kinds of plains, each kind reflecting differences in the geologic history of the ground.

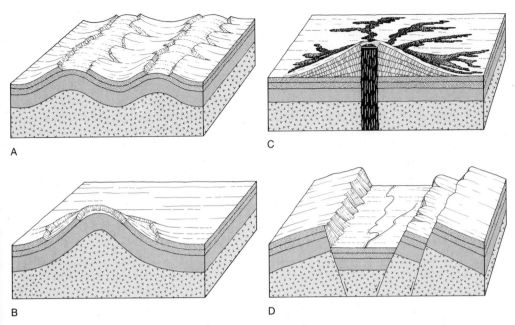

FIGURE 4.1
Four different kinds of mountains. (A) Fold mountains.
(B) Dome mountain. (C) Volcano. (D) Block mountain.

The flattest plains are the beds of dry lakes (playas), like the Bonneville Flats west of Great Salt Lake and the floor of Death Valley. Others that are nearly as flat as playas, are formed by river floodplains, like the Central Valley of California and the floodplains along the Mississippi River. Such plains have a distinct though slight slope. Some plains are formed by resistant formations that are horizontal or nearly so, like the plains in much of central United States. If uplifted and dissected such plains would become plateaus for their surfaces are like those capping plateaus.

The continental shelf off the Atlantic coast is a submerged plain, and parts of it that have been uplifted since Cretaceous time form the Coastal Plain that borders the Atlantic Ocean and the Gulf of Mexico.

Finally, some plains can be formed by erosion that smoothly bevels hard or soft rocks with complex structures.

Mountains

The most complex group of large-scale landforms are the mountains, of which there are many kinds, each reflecting different geologic structures or different kinds of rocks.

Some mountains are due to erosion of layered sedimentary rocks that have been folded, like the folds produced when opposite sides of a rug or piece of paper are pushed together (Fig. 4.1, A). These are fold mountains. Upfolds are *anticlines,* and the erosion of these forms anticlinal mountains; downfolds are *synclines,* and erosion of these produces synclinal mountains. Fold mountains are linear parallel to the axis and flanks of the fold. Dome mountains (Fig. 4.1, B) are actually another form of fold mountains in which the upfolding is more equidimensional than linear. Volcanoes may produce mountains that are rather like some dome mountains, but volcanoes are built by the accumulation of lavas and ash erupted from a vent and accumulated in a heap around it (Fig. 4.1, C). Still other mountains are due to faulting (Fig. 4.1, D), and are referred to as block mountains. These tend to be asymmetrical, with a steep front on the side that is faulted and a long gentle slope on the opposite side.

Where mountains have developed from uplifted masses of granite, the landforms are likely to be more rounded and more irregular than those on other kinds of mountains.

Drainage Patterns

The bedrock geology and large-scale landforms are reflected also in drainage patterns on the land (Fig. 4.2). Drainage lines tend to be parallel where the ground has a general slope in one direction—as, for example, in the southeast Atlantic coastal states, the Gulf coast, the plains from Oklahoma to the Dakotas, and the west side of the Sierra Nevada in California. Where the topography is more irregular, the drainage may develop patterns like the branching of a tree, referred to as *dendritic* drainage. The drainage patterns of the Snake, Colorado, and Ohio rivers are examples. In fold mountains the drainage lines tend to be parallel to the axes of the folds and offset by right angle bends both around the ends of the folds and where drainage cuts through folds. Such a drainage system forms a *trellis pattern,* which is well illustrated by the headwaters of the Potomac River (Fig. 4.2, B). Drainage off volcanoes and dome mountains is radial, like that in the Adirondacks (Fig. 4.2, A). Drainage courses may be changed and valleys abandoned because of damming by glaciers, lava flows, or earth movements. For example, in the Dakotas all the tributaries of the Missouri River are from the west. Formerly these streams turned northeastward and discharged to Hudson Bay, but the Pleistocene glaciers blocked that drainage and turned the Missouri River southward; the abandoned valleys are filled with glacial deposits.

THE PHYSIOGRAPHIC REGIONS

Figure 4.3 shows the boundaries of the physiographic provinces of the United States (those of Hawaii and Alaska are not shown). For the most part, the boundaries between the provinces are sharp, reflecting the fact that the chief differences are structural. An example of a sharply defined boundary is that between the Atlantic Coastal Plain and the more elevated Piedmont Province. The boundary there is marked by a line of falls. Some provinces, however, are not sharply bounded but grade into one another over wide areas, even though they are structurally distinct. Perhaps the best examples are the boundaries of the Middle Rocky Mountains and the Basin and Range Province, for these provinces grade into the mountains, plateaus, and basins of neighboring provinces.

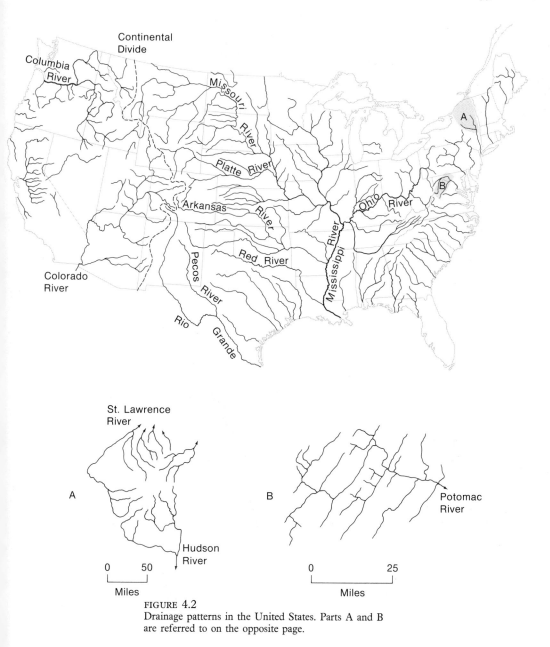

FIGURE 4.2
Drainage patterns in the United States. Parts A and B
are referred to on the opposite page.

Coastal Plain

A broad plain extends along the coast from Long Island southward to Florida and
westward around the Gulf of Mexico. In places this plain is 300 miles wide, and
almost all of it is below 500 feet in altitude. Most of the ground is level or nearly
so; much of it is poorly drained and even swampy. This coastal plain is crossed by
many large but sluggish rivers in broad shallow valleys.

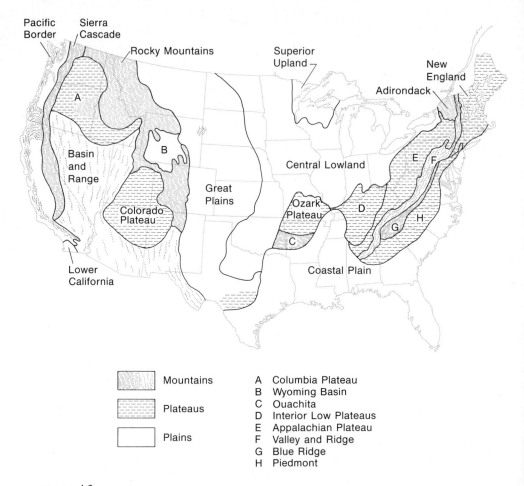

Pacific Border — Sierra Cascade — Rocky Mountains — Superior Upland — New England — Adirondack

A — Basin and Range — B — Central Lowland — E — F

Great Plains — Colorado Plateau — Ozark Plateau — D — H

Lower California — C — G — Coastal Plain

	Mountains	A	Columbia Plateau
	Plateaus	B	Wyoming Basin
		C	Ouachita
		D	Interior Low Plateaus
	Plains	E	Appalachian Plateau
		F	Valley and Ridge
		G	Blue Ridge
		H	Piedmont

FIGURE 4.3
Physiographic regions of the United States and their dominant landforms. About one-quarter of the land is mountains, one-quarter plateaus, and about half plains.

The Coastal Plain is an elevated part of the continental shelf, which continues seaward under the Atlantic Ocean for another 100 to 200 miles. It ends at a depth of about 600 feet, at the rim of a steep slope that forms the edge of the continent.

Both the shelf and the Coastal Plain are formed of rather poorly consolidated sandstone, shale, and limestone formations of Cretaceous, Tertiary, and Quaternary age. The formations slope gently toward the sea, as does the ground surface, and consequently the rivers are parallel. The formations crop out in belts of valleys and ridges parallel to the coast and to the inner edge of the province and at right angles to the drainage system. Erosion has formed valleys in the weakly consolidated formations, and the resistant ones crop out as *cuestas*, which are asymmetrical ridges (Fig. 4.4). The cuestas have long, smooth seaward slopes and short, steep escarpments

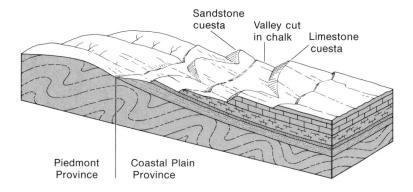

FIGURE 4.4
Landforms characteristic of the Coastal Plain. Cuestas are asymmetrical ridges formed where gently dipping resistant formations crop out at the surface; they have long smooth slopes down the dip and short steep escarpments facing updip. Between the cuestas in this block diagram is a valley eroded in soft Cretaceous chalk. This example is a north-south section across part of Alabama, where the Cretaceous formations extend onto the folded Paleozoic formations at the inner edge of the Coastal Plain. The valley between the cuestas is the Black Belt, named for the black grassland soil developed on the chalk.

facing inland. At the inner edge of the coastal plain, Cretaceous formations extend onto (*overlap*) older rocks of the Piedmont Province. In Cretaceous time the shore was at the inner edge of these Cretaceous formations; the lands of the coastal plain have been added to the continent since that time.

Despite the simplicity of the geologic structure of the Coastal Plain formations, the shorelines are very different along the different sections of the coast, chiefly because of crustal warping. The continent has been tilted northeastward, and the coastal plain passes below sea level east of Long Island. Nantucket and Martha's Vineyard are simply the tops of coastal plain hills protruding above the sea. From Long Island southward to Cape Hatteras, the coastal plain is low enough so that the sea has embayed the river valleys, forming New York, Delaware, and Chesapeake bays. Along the drowned valleys, tidewater extends all the way across the plain.

From Cape Hatteras to Cape Romain, the shoreline is cuspate with four prominent capes—Hatteras, Lookout, Fear, and Romain—separated by three broadly arcuate indentations. This section of the coast, referred to as the Cape Fear Arch, has been uplifted. The coast exposed to ocean waves from here to Long Island is characterized by sandy barrier beaches (Fig. 4.5).

The coast of South Carolina and Georgia has been downwarped and is characterized by numerous sandy islands. The advance of the sea landward in this section is recent enough so that the barrier beaches have been breached to form islands; oak trees back of them have been killed by the salt water.

In the southern part of the Coastal Plain, characteristic surface features include the many lakes in Florida, the alluvial floodplain of the Mississippi River, the bayous at its delta, and barrier beaches and related shore features east and west of the delta.

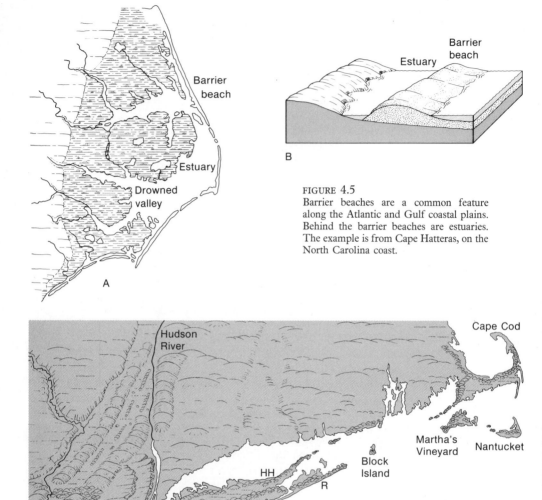

Barrier
beach

Estuary

Drowned
valley

A

B

Estuary

Barrier
beach

FIGURE 4.5
Barrier beaches are a common feature along the Atlantic and Gulf coastal plains. Behind the barrier beaches are estuaries. The example is from Cape Hatteras, on the North Carolina coast.

Hudson
River

Cape Cod

Martha's
Vineyard

Nantucket

Block
Island

HH

R

Long Island

Delaware
River

FIGURE 4.6
End moraines on Long Island. HH, Harbor Hill Moraine; R, Ronkonkoma Moraine.

Much of Florida is composed of limestone formations, and large areas of the peninsula are cavernous; lakes have formed where the caverns have collapsed. Marshes and swamps are extensive near the coast in parts of Florida and on the delta of the Mississippi River.

The northern part of the Coastal Plain was glaciated in the Pleistocene, and the topography and ground there are formed by deposits left when the glaciers retreated. Two ridges of glacially deposited gravels extend the length of Long Island and form the two fishtail peninsulas at the east end of the island (Fig. 4.6). Such deposits, known as *end moraines*, formed at glacial fronts during stillstands. The Narrows in

New York Harbor between Long Island and Staten Island is a cleft in the moraines. The moraines on Long Island are composed of boulders and cobbles of granite and other varieties of crystalline rocks found farther north in New England. Thus there is no doubt about the source of those deposits or the direction of ice movement. The moraines rest on a cuesta of Cretaceous sedimentary rock.

One of the effects of the Pleistocene glaciation was the lowering of sea level, the result of withdrawing vast amounts of water to form the continental ice sheets. Enough water was stored in the glaciers to lower the sea level by hundreds of feet; estimates vary, but a 500-foot lowering of sea level seems to be a reasonable estimate. As the ice sheets melted, sea level rose again, and it is rising still. The average rise along the Atlantic coast, based on 50-year records, is about one foot per hundred years. If this rate is becoming slower, which seems probable, one can estimate by extrapolation that a rise in sea level of more than 20 feet must have taken place since the birth of Christ. Fluctuations in sea level are referred to as *eustatic* changes.

Appalachian Highlands

Inland from the Coastal Plain are the Appalachian Highlands, a major physiographic division that includes several quite different provinces (Fig. 4.3): the Piedmont Plateau, Blue Ridge, Valley and Ridge, Appalachian Plateaus, Adirondack, and New England provinces. Northwest of the New England Province, along the St. Lawrence River, is a lowland.

The Piedmont Plateau is exceptional among our plateaus. Most plateaus are flat because their underlying structure consists of resistant, horizontal formations, but the flatness of the Piedmont Plateau is due to erosion that has bevelled intensely deformed rocks. The erosion is old—Tertiary or even Cretaceous. The rock formations are Precambrian and early Paleozoic; the erosion has obscured their structure. The rocks have been deeply weathered, and much of the ground is mantled with decomposed bedrock.

The surface of the Plateau consists of rolling hills at altitudes between 500 and 1,000 feet in the south and between 100 and 500 feet in the north. River valleys crossing the plateau are a few hundred feet deep. Streams cross the resistant formations in rapids; those stretches of streams that cross easily eroded formations are sluggish. Where the rivers tumble from the Piedmont Plateau to the lower Coastal Plain a line of rapids or falls marks the boundary between the two provinces—a boundary known as the Fall Line. Here the hard rocks of the Piedmont Plateau pass eastward beneath the Cretaceous formations of the Coastal Plain. The Fall Line is the head of tidewater and the reason why so many cities are located there, such as Philadelphia, Wilmington, Baltimore, Washington, Richmond.

Parts of the Piedmont Plateau are characterized by cuestas that are topographically like those of the Coastal Plain but of very different age and bedrock. The Piedmont cuestas are composed of sheet-like masses of once-molten rock (diabase) interlayered with bright red sedimentary formations (red beds) of Triassic age: the Palisades along the Hudson are a good example of the diabase ridges; others are the Watchung Ridges near Newark, New Jersey, Cemetery Ridge at Gettysburg, and the ridges along the Connecticut River valley (Fig. 4.7).

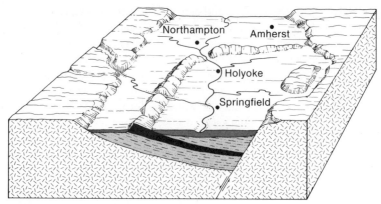

FIGURE 4.7

Cuestas on the Piedmont Plateau and in New England are formed by sheets of igneous rock (diabase, shown in black in the section), which is more resistant than the bright red sedimentary formations above and below. The diagram shows the Holyoke Ridges in the Connecticut River valley. These Triassic formations are downfaulted into the older gneisses and schists; the valley floor is mantled with Quaternary glacial deposits.

Why is the structure of the Triassic rocks so clearly reflected in the topography whereas the structure of the older rocks on the Piedmont Plateau is not? Some of the older rocks (known as gabbro) are mineralogically similar to the diabase; if the one has resisted weathering and erosion, why not the other? The reasons are not at all clear, and discussion of the problem is deferred to Chapters 8 and 9.

The Piedmont Plateau ends westward at a belt of mountains 50 to 75 miles wide that consist of strongly folded Precambrian and Paleozoic volcanic and sedimentary formations; erosion of these formations has formed a series of parallel valleys and ridges—the Blue Ridge and Valley and Ridge regions (Fig. 4.8). The rivers form a trellis pattern (Fig. 4.2) and cut across the mountain ridges in watergaps.

The intensity of the folding diminishes westward. In the western half of the Appalachian Highlands are the Appalachian Plateaus, where the formations are almost horizontal. The plateaus have an average altitude between 2,500 and 3,000 feet. Much of it is deeply dissected, and the terrain is mountainous. Here again, process dominates structure; the rock formations form nearly flat, broad uplands, but erosion has cut the plateau into mountains.

A structure quite different from that of the rest of the Appalachian Highlands is found at the Adirondack Mountains. These mountains are a domal uplift—that is, a structure formed by an upward movement of the underlying igneous and metamorphic Precambrian bedrock. The eroded edges of the overlying Paleozoic formations, raised by the uplift, form cuestas with steep escarpments facing the center of the dome. The drainage pattern is radial (Fig. 4.2).

The crystalline rocks and the complex structures of the Piedmont Plateau extend northward into New England, but elevations are greater there and the land is much hillier and even mountainous. The higher peaks rise more than 5,000 feet in altitude. This part of the Appalachian Highlands, together with the Adirondacks and the northern part of the Appalachian Plateaus, was glaciated; in those areas topography is modified by erosional and depositional landforms caused by the glaciation (Fig.

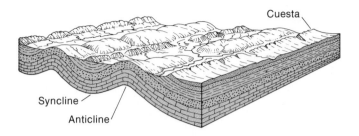

FIGURE 4.8
The Valley and Ridge region is composed of Paleozoic formations folded by horizontal compression. Erosion of the folds has formed parallel ridges where the resistant formations crop out, and parallel valleys on the easily eroded formations. [After U.S.G.S.]

4.9). The valleys were deepened by the ice—some enough so that they now contain lakes. The Finger Lakes in New York are a familiar example. Ground that was glaciated in late Pleistocene time commonly has a characteristic, hummocky topography built of glacial deposits, some of which were laid down by the ice itself and others deposited by its meltwaters. Drainage systems are haphazard and interrupted by numerous undrained depressions containing marshes, ponds, or lakes.

In places the southern limit of the ice advance is marked by end moraines, like the two on Long Island (Fig. 4.6). Farther to the north, other end moraines mark places where the ice temporarily stood still during its retreat.

Other glacial features include *kettle holes,* or depressions, eskers (Fig. 4.9), and *drumlins,* elliptical hills of gravel and sand, with the long axes parallel to the direction of ice movement. These hills appear to have been deposited as moraines and subsequently overridden by readvance of the ice. Drumlins generally occur in clusters. There is a group in Boston Basin (Bunker Hill is a drumlin), another in northern New York, and another in Wisconsin. They also are extensive in northern Canada west of Hudson Bay. In the hilly parts of the Appalachian Highlands, the different glacial landforms may be intimately mixed.

Depositional landforms laid down by glacial streams rather than by the glacier itself include:

1. *Outwash plains.* The deposits left by glacial meltwaters in front of the ice.

2. *Ice-contact features.* Deposits formed behind the ice front, where one or more side walls of ice contained glacial debris; the edges of these deposits collapsed when the ice melted. Among them are:

   kames, isolated hills of sand and gravel where debris accumulated in pits in the ice;

   kame terraces, formed by streams in open spaces between ice and valley side;

   kame deltas, formed by deposits in lakes at the edge of the ice;

   kame plains, debris that accumulated in extensive open areas between stagnating blocks of ice.

3. *Eskers.* Deposits that formed in stream channels on or under the ice; these ridges of gravel resemble meandering railroad beds, with branches where tributaries joined the main glacial stream.

66

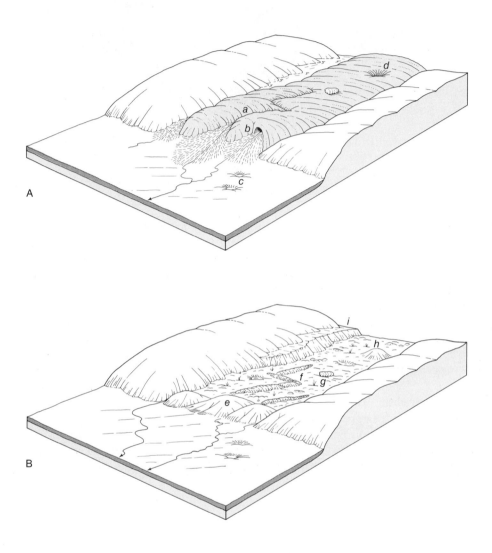

FIGURE 4.9
Nomenclature and origin of some glacial features. (A) Glacial ice in a valley has streams draining from tunnels (*b*) and crevasses (*a*) in the ice and from the base of the ice sheet itself; gravel fans are deposited where these discharge at the ice front. Beyond the ice is a plain of stream-deposited outwash where blocks of ice washed from the glacier melt and leave depressions, *kettle holes* (*c*), in the plain. In and on the ice are boulders from upvalley. Shafts (*d*) with funnel-shaped openings collect gravel and sand. Lakes ponded against the valley walls contain small deltas. (B) When the ice melts, the gravel fans in front of the ice collapse and form an *end moraine* (*e*); the stream channels on and under the ice are marked by ridges of gravel, *eskers* (*f*). The valley floor is hummocky ground moraine on which boulders rest as *glacial erratics* (*g*). The gravel and sand deposited in the shafts form hills, *kames* (*h*). Collapsed deltas in the marginal lakes form *kame terraces* (*i*).

Central United States

The Central United States embraces several physiographic provinces, each having its distinctive landforms. In two, the Central Lowland and the Superior Upland (Fig. 4.3), the landforms chiefly reflect the continental glaciation that extended southward to the Ohio and Missouri rivers. Two of the provinces are plateaus, one east of the Mississippi River valley, the Interior Low Plateaus, and one to the west, the Ozark Plateau. South of the Ozark Plateau is the Ouachita Province, formed of fold mountains of Paleozoic formations like those of the Valley and Ridge Province in the Appalachian Highlands. At the west is the Great Plains Province, where Paleozoic and Mesozoic formations were warped by folding and the folded formations were planed by erosion in early and middle Tertiary time. In late Tertiary time the eroded surface became covered with gravels washed eastward from the Rocky Mountains. The landforms in the Central United States, therefore, are determined partly by the geologic structure and partly by the processes of erosion and sedimentation that have operated there.

The Central Lowlands descend from the west foot of the Appalachian Plateaus, where the altitude is about 1,000 feet, to less than 500 feet along the Mississippi River, where it is joined by the Ohio and Missouri rivers. From there the lowland rises again westward about 1,500 feet in 500 miles to the edge of the Great Plains at an altitude of about 2,000 feet near the 100th meridian; the Great Plains rise westward for another 500 miles to about 5,000 feet at the foot of the Rocky Mountains. Although this is one vast plain, the forms of the stream valleys on it vary greatly from one place to another (Fig. 4.10).

Lakes characterize the glaciated regions. The largest existing ones occupy depressions scoured into bedrock, as at the Great Lakes and the Finger Lakes. Some were scoured below sea level. The bottoms of Lake Superior and Lake Ontario are more than 500 feet below sea level. The bottom of Seneca Lake, deepest of the Finger Lakes in New York, reaches a depth of 174 feet below sea level. How the beds of these big lakes became so deepened is not clear.

The greatest of the lakes no longer exist. They occupied valleys that once drained northward to Hudson Bay but were obstructed by the continental ice sheets. During the late Pleistocene and early Holocene, ice dammed the valley of the St. Lawrence River, and the Great Lakes at that time were much more extensive than they are now. They overflowed first to the Mississippi River, then to the Susquehanna and Hudson, and finally to the St. Lawrence. At one stage the lakes were joined as one. The now-dry lake bottoms are tremendous flats.

Greatest of the lakes was Lake Agassiz (Fig. 4.11), which was mostly in Manitoba and Ontario but extended southward along the Red River valley between North Dakota and Minnesota. Rainy Lake, Lake of the Woods, Lake Winnipeg, and Manitoba Lake are remnants of Lake Agassiz. The bed of the old lake is a vast plain of lake-deposited clays, and richly productive of wheat, both in the United States and Canada. The lake grew as the ice dam retreated northward; it spilled southward into the head of the Minnesota River and there cut a valley a mile wide and as much as 200 feet deep. So small a stream as the Minnesota River in so big a valley is referred to by geomorphologists as an underfit stream. Many, or most, rivers in the glaciated regions are underfit.

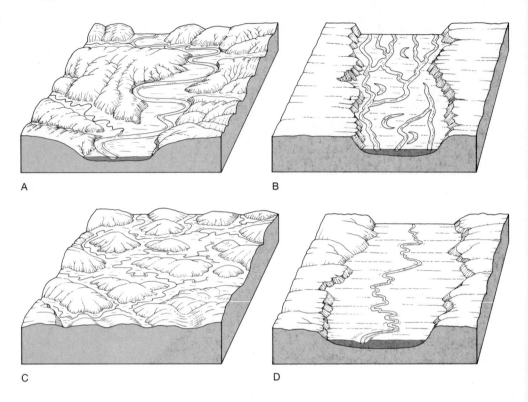

A   B

C   D

FIGURE 4.10
Floodplain and other features in the central United States. (A) Some streams entrenched below flat uplands have broad floodplains and have undercut the valley walls at the outside of meanders. (B) The Mississippi River meanders lazily in a braided pattern in a broad floodplain characterized by numerous sloughs and oxbow lakes. (C) In the north, where there is young, hummocky morainal topography, the drainage is disintegrated. (D) Some streams meander wildly within their floodplain without reaching the valley sides.

    Smaller ice-marginal lakes formed where the Missouri River and its tributaries in Montana were dammed by ice (Fig. 4.12). One was along the Yellowstone River and another along the Musselshell. A third lake, of considerable size, was situated along the upper part of the Missouri. The preglacial course of the Missouri River was around the north side of the Bearpaw Mountains (broken line in Fig. 4.12) and then east along Milk River, but when this course was blocked by ice the river was turned eastward and cut a valley (Shonkin Sag) across the north foot of the Highwood Mountains. Thereafter the river maintained its course around the south side of the Bearpaw and Little Rocky mountains, although it shifted northward from Shonkin Sag.

    The irregular shoreline of Lake Superior is structurally controlled. Ridges trending northeast form peninsulas that extend into the lake; they are the structural remains of ancient Precambrian mountains. The altitudes of the ridges range between 1,000 and 2,000 feet.

    The other Great Lakes have smoothly rounded shorelines that are controlled by process—specifically, by glaciation. Parallel end moraines of Wisconsinan age, de-

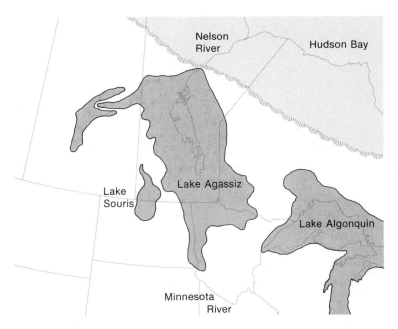

FIGURE 4.11
Lake Agassiz, one of the largest of the glacial lakes, formed when the ice front stood across the Nelson River and blocked the drainage, which overflowed via the Minnesota River. The lake bed is a vast plain.

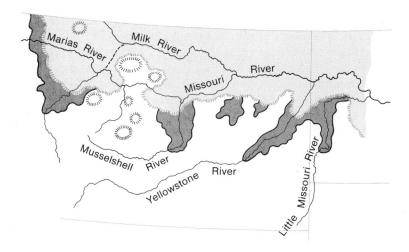

FIGURE 4.12
Glacial lakes in Montana, where the Missouri River and its tributaries were dammed by the Pleistocene ice sheets. The broken line shows the former course of the Missouri River (northward to Havre) on the west side of the Bearpaw Mountains. [After Lemke, Laird, Tipton, and Lindvall, 1965.]

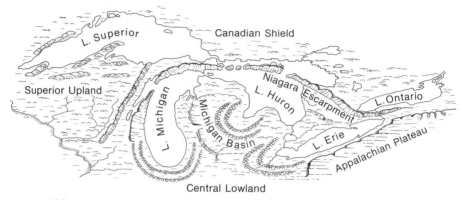

FIGURE 4.13

Diagrammatic birdseye view north across the Great Lakes region. Lake Huron and Lake Michigan are in easily eroded Devonian formations, which dip into the Michigan Basin, south of the Niagara Escarpment. The Niagara Escarpment, named for Niagara Falls, at the east end of Lake Erie, extends from Rochester to Milwaukee; it is a persistent 800-mile-long cuesta formed of Silurian formations that dip away from the Canadian Shield in Ontario and the Superior Upland. End moraines deposited by the Pleistocene glaciers form arcuate ridges at the west end of Lake Erie, the head of Saginaw Bay, and the south end of Lake Michigan. Width of the view, about 800 miles.

posited as the ice receded, form the arcuate ridges around the west end of Lake Erie, the south end of Saginaw Bay, and the south end of Lake Michigan (Fig. 4.13).

Lake Erie drains into Lake Ontario by overflowing a cuesta at Niagara Falls (Fig. 4.13). This cuesta, one of the most persistent topographic features in the country, swings in an arc around the northeast side of Lake Huron and the northeast side of Lake Michigan. As in all cuestas, the steep escarpment faces the structural uplift, and the moderately sloping opposite side faces a structural basin.

The moraines bordering the Great Lakes extend around the north end of the driftless areas in Wisconsin and then turn southward into Iowa (Fig. 2.2). Across South Dakota and North Dakota the morainal ridge, called the Coteau des Missouri, is paralleled by the Missouri River. The river there is deeply entrenched in a position originally determined by the ice front. Before the glaciations, the Missouri River drained to Hudson Bay, as did its tributaries from the west in North and South Dakota (Fig. 4.14).

North of the end moraines much of the central United States has a hummocky, knob and kettle topography, which is young—late Wisconsinan in age. Lakes and ponds abound, and no through-flowing streams have developed there. This section is the southern end of a vast area that extends to the arctic and is characterized by myriads of lakes. During the Pleistocene Epoch lakes larger than the present Great Lakes formed in front of the ice sheet as it retreated. The ground once covered by these lakes is now fertile plains.

From the limit of the Wisconsinan glaciation southward to the Ohio and Missouri rivers, knob and kettle topography is absent. The ground there is older than that to the north; it is formed of earlier Pleistocene glacial deposits that have been dissected by a well-integrated drainage system. The ground south of the rivers was not glaciated. South of the Ohio River are the Interior Low Plateaus, which are mostly less than 2,000 feet in altitude and are formed of gently warped Paleozoic formations

FIGURE 4.14
Capture pattern of the Missouri River in North and South Dakota.
Preglacial courses are shown by broken lines.

that crop out in cuestas. Altitudes in the Ozark Plateau and Ouachita Province also are mostly less than 2,000 feet. The generally smooth surface of the Great Plains in the north is interrupted by dome mountains that rise a few thousand feet higher than the surrounding plains. Best known of these are the Black Hills (Fig. 4.15).

Rocky Mountains

West of the Great Plains is the Rocky Mountain System, which extends from New Mexico northward into Canada. Within the United States the system is divided into four major parts: the Southern Rocky Mountains, mostly in Colorado and south; the Wyoming Basin; the Middle Rocky Mountains, which almost encircle the basin in Wyoming; and the Northern Rocky Mountains in Montana and Idaho, which are the southern end of the Canadian Rockies.

The Rocky Mountains are not the highest mountains in the country, but they do form the continental divide (Fig. 4.2), from which rivers flow to opposite sides of the continent. The altitude at the base of the mountains is about 5,000 feet, about the elevation of the highest summits in the Appalachians. Many peaks in the Rockies rise above 14,000 feet.

As the name implies, these mountains have extensive areas of bare rock, many more than the Appalachians. The reason for the difference is the slower rate of weathering, a result of less precipitation in the Rockies. Furthermore, because precipitation is low, fewer streams drain the Rocky Mountains; runoff in the Appalachians is five to ten times as much as in comparable areas of the Rocky Mountains. Rivers are fewer and smaller in the Rocky Mountains.

FIGURE 4.15
The Black Hills are a domal uplift of Precambrian rocks flanked by cuestas of Paleozoic, Cretaceous, and Tertiary formations.

The height and rugged topography of the Rocky Mountains are in large part expressions of their youth. The ranges were first uplifted in very late Cretaceous and early Tertiary time, but most of their height was attained in late Tertiary time. Structurally the mountains are of several kinds, some examples of which are illustrated in Figure 4.16. Not much of this mountainous ground is older than Quaternary.

How are such mighty mountains formed? We can see this for ourselves, because the uplifting and faulting are continuing. In 1959, at Hebgen Lake, Montana, just west of Yellowstone Park, earth movement tipped the lake bottom 20 feet toward the northeast, and half that displacement was taken up by a fault that displaced roads, fences, and buildings by ten feet (Fig. 4.17). The earthquake and damage it caused provide a classic example of how such mountains grow—by small increments of movement repeated over long periods of time.

The mountains were built by uplift of the Precambrian rocks that had previously formed the basement under the Paleozoic and younger formations. Cretaceous and

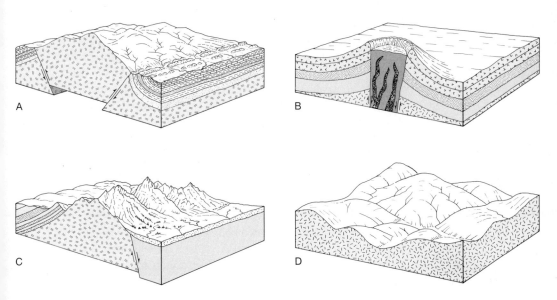

FIGURE 4.16
Some of the different kinds of mountains in the Rocky Mountains. (A) Anticlinal uplifts of Precambrian rocks flanked by the upturned edges of Paleozoic and Mesozoic formations (Front Range, Colorado; Uinta Mts., Utah; Wind River and Bighorn Mts., Wyoming). [After U.S.G.S.] (B) Faulted volcanic pile on top of domal uplift involving Precambrian, Paleozoic, and Mesozoic formations (San Juan Mountains, Colorado). (C) Block mountains, faulted on one side and having steep escarpment at the fault; the block tilts away from the fault (Teton Range, Wyoming; Wastach Range, Utah, and numerous ranges in southwestern Montana). [After U.S.G.S.] (D) Granite Mountains; general aspect much less precipitous than in the other kinds because rocks are homogeneous. (Salmon and other mountains in central Idaho). [After U.S.G.S.]

FIGURE 4.17
The earth movements at Hebgen Lake, Montana, during the 1969 earthquake illustrate how the Rocky Mountains formed. The ground was tilted 20 feet to the northeast (right), flooding houses and highways on the northeast shore and stranding docks on the southwest shore. Half the displacement was taken up by movement along a fault. Mountains are built by repeated small movements such as this.

older formations are turned up steeply along the foot of the mountains; the upturned edges of red Paleozoic formations form the Garden of the Gods near Colorado Springs and Red Rocks Theater west of Denver. When uplift started, that land was nearly at sea level. As a result of subsequent folding, faulting, and uplift, the entire region was elevated until the mountain base reached its present-day elevation of a mile above sea level.

The Wyoming Basin is a structural as well as topographic basin. When the individual ranges of the Rocky Mountains were uplifted, the large segment of land that became the Wyoming Basin sagged behind. Actually, it is not one but several coalescing basins that developed gradually during the Tertiary (and may still be

developing) while individual ranges like the Bighorn, Wind River, Tetons, and Uinta
Mountains were being uplifted. The Wyoming Basin, however, participated in the
regional uplift during middle and late Tertiary time that raised the base of the Rockies
a mile above sea level. The Basin has about the same altitude as the Great Plains.

Although the principal topographic features of the Rocky Mountains reflect the
geologic structure, the summits and upper valleys have been greatly modified by
glaciation. During the Pleistocene, snow collected on the mountains in depths sufficient to form alpine glaciers that moved down the valleys. The original V-shaped,
stream-cut valleys became U-shaped, steep-walled basins called *cirques* were eroded
in bedrock at the heads of the glaciers. Terminal moraines were deposited at the
fronts of these glaciers, and lateral moraines were deposited along the valley sides,
where debris from the mountains collected at the edges of the ice. In the Southern
Rocky Mountains the glaciers did not extend far down the valleys, only about half
way to the base, but their meltwaters deposited thick fills of bouldery gravel that
now form gravel terraces along the rivers where they leave the mountains. In the
Northern Rocky Mountains the glaciers descended eastward to the Great Plains,
where they joined the continental ice sheet moving southwestward from its source
in central Canada.

Colorado Plateau

West of the Southern Rocky Mountains and south of the Middle Rockies and
Wyoming Basin is the Colorado Plateau, named for the Colorado River that crosses
it. It is our highest plateau, averaging about 5,000 feet above sea level. Along its
western edge are the High Plateaus of Utah, 10,000 feet or more above sea level;
they are among the highest plateaus in the world. They cast a rain shadow to the
east across the Canyonlands Section of the Colorado River, where precipitation
averages less than 6 inches annually.

Thus, the canyonlands are desert, much of it gorgeously colored bare rock. Some
areas are sand desert with dunes. Some parts are shale desert with badlands "badder"
than those at Badlands National Monument—so bad, some would say, that they are
unfit to be a monument. Twenty-five of our national parks and monuments are on
the plateau, a measure of the grandness of the scenery.

Total relief is about the same as in the Rockies, with altitudes ranging from about
2,500 feet along the Colorado River in Grand Canyon to more than 10,000 feet on
the High Plateaus and more than 12,000 feet on some of the isolated mountains.

The plateau is built mostly of flat-lying Mesozoic formations. Precambrian rocks
are exposed in the bottom of Grand Canyon. Paleozoic rocks are exposed along the
southwest edge of the plateau, in Grand Canyon, and in the cores of some broad
asymmetrical anticlines (Circle Cliffs Uplift, San Rafael Swell, and Monument
Upwarp). Tertiary formations form the surface in the Uinta Basin along the north
and northeast edge of the plateau. Generally the plateau is a flat surface formed of
gently dipping formations, but the flat surface is interrupted in several ways.

Along the steep flanks of the asymmetrical anticlines, especially in Utah, the
formations are turned up in hogbacks forming colorful rock ridges 200 to 1,500 feet
high. One of the better known and longer ones is the Waterpocket Fold. A similar

Transverse valleys   Longitudinal valleys

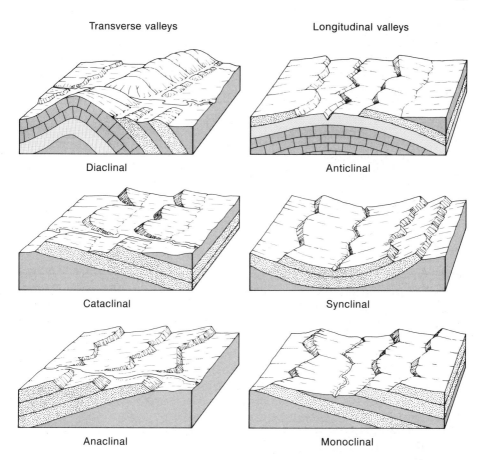

Diaclinal   Anticlinal

Cataclinal   Synclinal

Anaclinal   Monoclinal

FIGURE 4.18
Powell's classification of transverse and longitudinal valleys. On the basis of morphologic relation-ships of streams to great folds, Powell distinguished two principal orders of valleys, each with three variants. Those in his first order, *transverse valleys,* have directions at right angles to the strike of the structures. There are three varieties: *diaclinal valleys* pass through folds; *cataclinal valleys* flow down the dip; *anaclinal valleys* flow against the dip. By these definitions a diaclinal valley has two parts, one cataclinal and the other anaclinal. Those in his second order, *longitudinal valleys,* have directions parallel to the strike of the structures. Again, there are three varieties: *anticlinal valleys* follow anticlinal axes; *synclinal valleys* follow synclinal axes; *monoclinal valleys* are cut along the flanks of folds and are parallel to the strike of the formations.

hogback, crossed by Interstate 70, is along the east side of the San Rafael Swell. A third, Comb Ridge, at the east side of the Monument Upwarp, extends across the San Juan River at Monument Valley. The relationships of rivers to great folds in rocks are well illustrated on the Colorado Plateau (Fig. 4.18). These relationships were pointed out first by John Wesley Powell, and the principles he reported are applicable generally.

The flat surface of the Colorado Plateau also is interrupted by isolated mountains 5,000 to 7,000 feet higher than the plateau surface. Some of these are Tertiary volcanic cones, notably Mount Taylor, New Mexico, and San Francisco Mountains, Arizona.

These volcanic fields also have many necks that stand 500 to 1,500 feet high, and mesas capped by lavas. In the central part of the plateau are similarly isolated but less conical mountains formed by igneous intrusions (laccoliths) that domed the Paleozoic and Mesozoic formations. Best known of these are the Henry Mountains, Utah, a classic area in geology.

The greatest interruption to the plateau surface, however, is provided by the canyons, 1,000 to 5,000 feet deep. The best known, of course, is Grand Canyon, but canyons extend almost the whole length of the Colorado River and its principal tributaries. These canyons have been referred to as "mountains inside out."

The topographic features of the Colorado Plateau are mostly due to earth movements, igneous activity, and erosion during the middle and late Tertiary.

Basin and Range Province

West and south of the Colorado Plateau is the Basin and Range Province, characterized by block mountains and desert basins; Death Valley is a representative example (Fig. 4.19). The ridges and valleys in this province differ from those in the other provinces because the valleys are structural, not erosional. Some of the valleys are below sea level, like Death Valley and Salton Sea in Imperial Valley. Other valleys are at altitudes of 5,000 feet. Some of the mountains reach altitudes above 10,000 feet.

The western part of the Basin and Range Province is in the rain shadow of the Sierra Nevada, and is even more of a desert than the Colorado Plateau. The southern part is continuous with the Sonoran Desert of Mexico.

The mountains are asymmetrical fault blocks, having been uplifted by faulting along one side and tilted toward the other. Usually a steep escarpment marks the faulted front whereas the other side has a long gentle slope. The block faulting and tilting of the blocks occurred during the middle and late Tertiary and in many places is continuing. Tiltmeters on fault blocks in Death Valley show active earth movement there; Lake Mead has tilted measurably southwestward in the last 30 years. At many places in the Basin and Range Province, there has been historic faulting like that at Hebgen Lake, Montana, and prehistoric faulting recent enough to leave fresh fault scarps. Belts of active earthquake centers are located along the western valleys near the Sierra Nevada, along the foot of the Middle Rockies from Salt Lake City to Montana, and along the Basin and Range stretches of the Rio Grande. The topography of the Basin and Range Province is still developing structurally.

The ground in the Basin and Range Region is quite varied. The mountains that flank the valley floors are mostly bare rock. Sloping from the foot of the mountains to the valley floors are gravel fans composed of debris eroded from the bedrock. Depending on the geologic history of the mountain front, these fans have several different forms (Fig. 4.20) and different kinds of ground. Some have their apex at the mouth of a canyon, others extend into canyons, and still others extend so far into the mountains that they partly engulf them. Some fans are bouldery and some fine-grained, depending on the kinds of source rocks in the mountains. Some valley floors are dry lake beds, called *playas;* others are natural salt pans—that is, playas encrusted with salts.

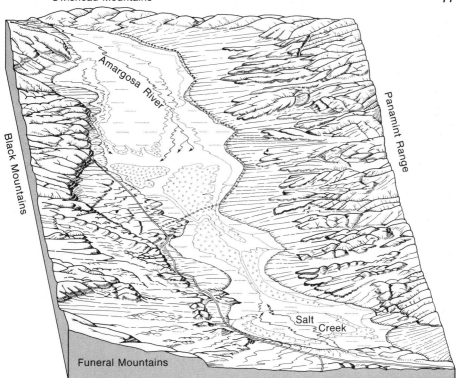

FIGURE 4.19
Block diagram of Death Valley, California, looking south. The diagram illustrates the different kinds of ground typical of the Basin and Range province. Gravel fans several miles long slope from the rocky ranges to the flat playa in the interior of the basins. Some of the playas are mud flats; others, like Death Valley, are salt pans. The triangle dot pattern indicates rough rock salt formed by evaporation of water in the ground; the horizontal dot pattern indicates rock salt smoothed by washing. The blank area is subject to flooding at the present time. The double broken line along the west edge of the Armagosa River flood plain indicates a dirt road. [From Hunt, 1966a.]

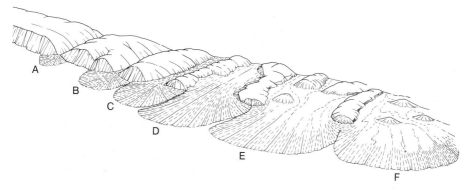

FIGURE 4.20
Some different forms of gravels fans in the Basin and Range Province. Newly formed fault scarps may be without fans or have small ones at the base (A). In the next stage the front is dented (B), and the fans may extend up the mountain valley (C). A block mountain being engulfed in its own debris may have a front that no longer is straight because of burial (D), and the fans may even wrap around spurs in the mountains (E). In the final stage (*pan-fan stage*) the gravel buries all except isolated hills (F). [After Davis, Proc. Nat. Acad. Sci., v. XI.]

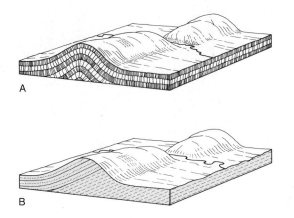

FIGURE 4.21
Water gaps on the Columbia Plateau (A) are formed where rivers cross anticlinally folded lavas. The landforms are topographically similar to those in the Valley and Ridge Province (B) in the Appalachian Mountains, but very different structurally. In the Applachians the ridges are formed at the outcrops of resistant formations turned up on the flanks of the folds, which are much broader than the ridges.

In the valley of the Great Salt Lake, impressive wave-cut terraces and deltas have formed up to a thousand feet above the present lake (Fig. 7.6). These are primarily constructional features dating from the Pleistocene Epoch, when Lake Bonneville—an ancient lake fed by glacial meltwaters—filled the valley to a depth of 1,000 feet. All the cities along the west foot of the Wasatch Mountains in Utah are on gravel deltas that were built into Lake Bonneville.

Columbia Plateau and Cascade Range

The Columbia Plateau consists mostly of elevated plains of nearly horizontal lavas mantled with wind-blown silt. Rivers, however, have cut canyons in the lavas. Unusual topographic features of the plateau include the volcanic cones almost 500 feet high that form the Menan Buttes along the Snake River Plain in southern Idaho, and the young volcanic cones and lava flows at Craters of the Moon National Monument at the north edge of the plain. The Snake River Plain is as flat as most of the central United States; Hell's Canyon, where the Snake River leaves the plain, is deeper than the Grand Canyon but lacks its spectacular landforms and colors.

An especially unusual feature of the Columbia Plateau is the Grand Coulee. This tremendous canyon, which is now dry except for the water turned into it artificially, has dry falls as high as those of Niagara and other features that are the result of work by a large river. Grand Coulee and Moses Coulee to the west were eroded during the Pleistocene when catastrophic floods discharged from Lake Missoula, a lake that once existed in Montana, along Clarks Fork. Lake Missoula was formed when the ice pushed southward in the Northern Rocky Mountains and dammed Clarks Fork. Fed by glacial meltwaters, this lake covered 3,300 square miles and was 2,000 feet deep at the ice front. Shorelines are numerous, closely spaced, and not conspicuously developed, as if the lake level fluctuated greatly. The lake repeatedly overflowed along the south edge of the ice barrier and discharged vast floods into the Spokane River and onto the Columbia Plateau. Evidently when the ice barrier was breached, the lake discharged catastrophically. The floods that swept westward produced the scablands (bare lava surfaces) on the Columbia Plateau and cut the tremendous gorges (coulees) in the lavas.

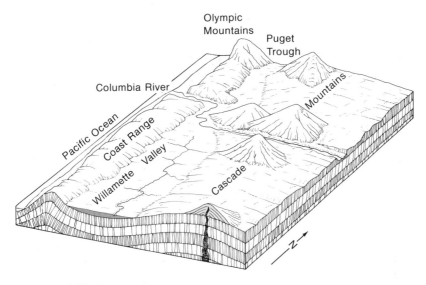

FIGURE 4.22
Diagram of part of the High Cascades, which are built of lavas arched upward. On top of the arch are numerous high volcanic cones (Mts. Rainier, St. Helens, Adams, Hood, Jefferson). West of the High Cascades is a structural depression that forms the Willamette Valley and Puget Trough. The Coast Ranges west of the trough are folded lavas and sedimentary formations; the Olympic Mountains are a domal uplift of ancient rocks. The Columbia River has cut canyons across the Cascades and the Coast Range.

In the Yakima Ridge area of the plateau, the rivers pass through upfolded ridges of lava in watergaps much like those of the Appalachians (Fig. 4.21).

West of the Columbia Plateau is the Cascade Range, an uplift of lavas surmounted by tremendous Pleistocene volcanoes that form the higher peaks, which have altitudes of 10,000 to 14,000 feet (Fig. 4.22). The ground is young and, for the most part, rocky. This barrier in Washington and Oregon is breached only by the Columbia River, which crosses it in a gorge about 100 feet above sea level. South of the gorge is the volcanic cone at Mt. Hood, altitude 11,250 feet. North of the gorge are two other cones: Mt. St. Helens, 9,700 feet, and Mt. Adams, 12,300 feet. The range extends northward into Canada and southward into California. Mt. Lassen and Mt. Shasta are the southernmost large volcanic mountains of this range.

At the west foot of the Cascades is a structural trough forming a broad flat plain—the Puget Trough in Washington and the Willamette Valley in Oregon. West of the plain, the high Olympic Mountains and the hills of the coast ranges end in sea cliffs at the Pacific Ocean.

Sierra Nevada and California Coast

In California, the lavas of the Cascade Range thin, and a tremendous mass of granite rises from under the lavas to form the Sierra Nevada—an uplifted block of granite. The uplift took place along a fault on the east side; the block is tilted toward the

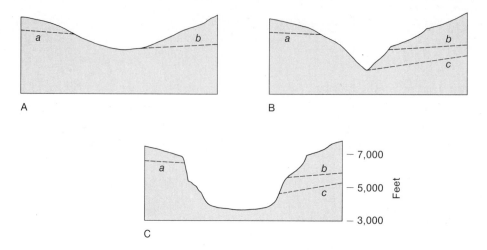

FIGURE 4.23

Profiles illustrating the development of Yosemite Valley. (A) In late Tertiary time the Merced River was in an open valley; *a* represents the profile of a tributary that was unable to cut downward as rapidly as the river. Downcutting of tributary *b* kept pace with that of the river. (B) Before the Pleistocene glaciations the Sierra Nevada was uplifted, and the river became deeply incised in a narrow V-shaped valley; tributary *b* was left hanging. Downcutting of tributary *c* kept pace with that of the river. (C) During the glaciations the valley was further deepened and widened, and tributary *c* was left hanging. [After Matthes, 1930.]

west. The structural history is similar to that of the Teton Range of Wyoming (Fig. 4.16, C). Summits range from 10,000 to 14,000 feet in altitude, increasing southward The western side slopes fairly smoothly to the Central Valley of California. In places, this slope is cut by deep glacial canyons, the most notable being Yosemite Valley.

The Sierra Nevada, like the Cascades, are young mountains, much of the uplift having occurred during the Quaternary Period. During the Pleistocene the summits were glaciated and the valleys deepened. Narrow U-shaped glacial valleys were cut into broad preglacial valleys, producing a discontinuity in the slopes known as a topographic unconformity (Fig. 4.23). At the western foot of the Sierra is a belt of older, deeply weathered ground that slopes beneath gravel fans and other deposits eroded from the mountains.

Drainage from the west slope of the Sierra and the Central Valley collects in the Sacramento and San Joaquin rivers and discharges into San Francisco Bay. The Central Valley, a vast alluvial plain, is bordered on the west by the Coast Ranges. This range, with summits less than 5,000 feet in altitude, and most under 3,000 feet, forms sea cliffs along much of the Pacific shoreline. The only plain along this coast is at Los Angeles at the western foot of the massive San Bernardino and San Gabriel Mountains.

The structural and topographic differences between the physiographic regions are among the important factors that affect the kind of ground in the different parts of the country. On a map showing kinds of ground in the United States, patterns of areal distribution would in large part coincide with the boundaries of the physiographic regions.

Steep slopes favor runoff and erosion, and on unstable bedrock steep slopes favor landsliding, as California homeowners have been slow to learn. As a consequence of such topographic and structural differences, the regions differ in frequency and extent of bare rock surfaces and the kinds and extent of surface deposits that cover the bedrock. In the dry western mountains, the mountain sides are being stripped by erosion and the sediments are being deposited in the flatter piedmont areas. In the Appalachians, either this is not happening, or it is happening very slowly; erosion on the floors of the broad valleys seems to be progressing at about the same rate as erosion on the mountain sides—as if some kind of equilibrium has been reached.

As far as weathering and the development of modern soils are concerned, we need be reminded only that steep slopes favor runoff whereas flat ground collects water. This has important consequences. On a plateau, for example, the flat upland may collect most of the rain water that falls on it. The hillsides bordering that upland receive practically the same rainfall but most of it runs off; hillsides average drier than uplands. The flat valley floor, on the other hand, not only collects most of the rain water that falls on it, but it also collects the runoff from the bordering hillsides.

Moist uplands, dry hillsides, and wet valley bottoms have very different ground climates, the subject of Chapter 5, and they support different kinds of vegetation, the subject of Chapter 6. As a consequence, they have different amounts of accumulated organic matter in their modern soils, different degrees of acidity or alkalinity, and different kinds and rates of weathering. Such differences within a region may be as great as the differences between regions. In the Appalachians these differences seem to be in equilibrium; in the Rocky Mountains they are not.

Mountains affect the climate in other ways. Where relief is great, high elevations are cooler than low ones. The tops have shorter growing seasons and are more subject to frost, excepting some mountain valleys that collect cold air that flows from higher ground. The topography affects the circulation of the atmosphere, causing most precipitation to fall on the windward sides of mountains, leaving their leeward sides nearly dry. Most of our deserts are situated leeward of high mountains.

BIBLIOGRAPHY

Atwood, W. W., 1940, The physiographic provinces of North America: Ginn, Boston, 536 pp. with map by Raisz showing landforms in the United States.
Bird, J. B., The physiography of arctic Canada: Johns Hopkins Univ. Press, 336 pp.
Bowman, Isaiah, 1909, Forest physiography: Wiley, New York.
Bretz, J. H., Smith, H. T. U., Neff, G. E., 1956, Channeled scabland of Washington: new data and interpretations: Geol. Soc. America Bull., v. 67, pp. 957–1049.

Bryan, K., 1936, The formation of pediments: Internat. Report 16th Geol. Cong., Washington, v. 2, pp. 765–775.

Calif. Div. Mines, 1951, San Francisco Bay counties guidebook: Calif. Div. Mines Bull. 154, 392 pp.

Calif. Div. Mines, 1954, Geology of southern California: Calif. Div. Mines Bull. 170.

Clark, T. H., and Stearn, C. W., 1960, The geological evolution of North America: Ronald Press, New York, 434 pp.

Cotton, C. A., 1949, Geomorphology: Wiley, New York.

Davis, W. M., 1909, Geographical essays: Reprinted 1954 by Dover Publications, 777 pp.

Denny, C. S., 1965, Alluvial fans in the Death Valley region, California and Nevada: U.S. Geol. Survey Porf. Paper 466.

Dury, G. H., 1959, The face of the earth: Penguin Books, Baltimore, 223 pp.

Embleton, C., and King, C. A. M., 1968, Glacial and periglacial geomorphology: St. Martins Press, New York, 608 pp.

Fenneman, N. M., 1931, Physiography of western United States: McGraw-Hill, New York, 534 pp.

Fenneman, N. M., 1938, Physiography of eastern United States: McGraw-Hill, New York, 689 pp.

Fryxell, F. M., 1927, The physiography of the region of Chicago: Univ. Chicago Press, 55 pp.

Hack, J. T., 196), Interpretation of erosional topography in humid temperate regions: Am. Jour. Sci. (Bradley Volume), v. 257-A, pp. 80–97.

Hansen, W. R., 1965, The Black Canyon of the Gunnison—Today and yesterday: U.S. Geol. Survey, Bull. 1191, 76 pp.

Hawley, J. W., and Gile, L. H., 1966, Landscape evolution and soil genesis in the Rio Grande region, southern New Mexico: Guidebook, 11th Ann. Field Conf., Rocky Mtn. Section, Friends of the Pleistocene.

Hunt, Chas. B., 1967, Physiography of the United States: W. H. Freeman and Company, San Francisco, 480 pp.

———, 1969, Geologic History of the Colorado River: U.S. Geol. Survey Prof. Paper 669.

Kroeber, A. L., 1947, Cultural and natural areas of native North America: Univ. Calif. Press, Berkeley, 242 pp.

Lemke, R. W., and others, 1965, Quaternary geology of the northern Great Plains, *in* The Quaternary of the United States: Princeton Univ. Press, pp. 15–27.

Leopold, L., Wolman, M. G., and Miller, J. P., 1964, Fluvial processes in geomorphology: W. H. Freeman and Company, San Francisco, 522 pp.

Lobeck, A. K., 1939, Geomorphology: McGraw-Hill, New York, 731 pp.

Love, J. D., and Reed, J. C., Jr., 1968, Creation of the Teton landscape: Grand Teton Nat. Hist. Assoc., 120 pp.

Mackin, J. H., and Cary, A. S., 1965, Origin of the Cascade landscapes: Wash. Div. of Mines and Geol. Inf. circular 41, 35 pp.

Matthes, F. E., 1930, Geologic history of Yosemite Valley: U.S. Geol. Survey Prof. Paper 160, 137 pp.

Powers, W. E., 1966, Physical geography: Appleton-Century-Crofts, New York, 566 pp.

Rich, J. L., 1935, Origin and evolution of rock fans and pediments: Geol. Soc. America Bull., v. 46, no. 6, pp. 999–1024.

Ruhe, R. V., 1956, Geomorphic surfaces and the nature of soils: Soil Sci., v. 82, pp. 441–455.

Salisbury, R. D., and Atwood, W. W., 1908, The interpretation of topographic maps: U.S. Geol. Survey Prof. Paper 60, 84 pp.

Shelton, J. S., 1966, Geology illustrated: W. H. Freeman and Company, San Francisco, 434 pp.

Thornbury, W. D., 1965, Regional geomorphology of the United States: Wiley, New York, 609 pp.

U.S. Geological Survey, 100 topographic maps illustrating physiographic features: Washington.

Upton, Wm. B., Jr., 1970, Landforms and topographic maps: Wiley, New York, 135 pp.

Williams, H., 1949, Ancient volcanoes of Oregon: Oregon State Syst. Higher Educ., Condon Lect. Publ., Eugene Ore.

Climate—both above the ground and within it—has a major influence on the distribution of plants and animals. Climate and the kind and number of organisms that live in and on the ground have strong effects on the processes of weathering and erosion, thereby contributing to differences in the kind of ground. The four distinctive stands of vegetation in this scene in the Colorado Rockies—shrubby slope, wet meadow, coniferous forest, and alpine summits—are the result of four different climates, four different kinds of weathering and erosion processes, and, consequently, four different kinds of ground.

5/ Climate on and in the Ground

Macro- and microclimates; some properties of air; radiation and conductivity; ground temperature; freeze and thaw; dew; precipitation, runoff, and ground moisture; climatic changes; bibliography.

MACRO- AND MICROCLIMATES

When out of doors we are acutely aware of the climate about our heads and shoulders. At our feet, however, the climate is very different from the one reported by the weather man. The ground may be hotter than the air at head height by 50° F. or more; in Death Valley, ground surface temperatures as great as 190° F. have been recorded. At an early age we learn not to walk barefoot on hot pavements. Conversely, the ground may be colder than the air about our heads. It may condense water vapor while the air at head height is dry, or it may be windless and shaded by vegetation while our heads and shoulders project into sunlight and wind. The climate of the air about us is referred to as *macroclimate;* climate on and in the ground is referred to as *microclimate.*

Climate is the major factor that controls the distribution of plants and animals. Climate and organisms (the subject of the next Chapter) greatly affect the processes of weathering and erosion, and so contribute to differences in the kind of ground.

The ground surface marks the interface between two very different physical systems, the one above being largely gaseous and the one below being largely solid. The unconsolidated materials of the ground constitute a transition layer in which gases derived from plant and animal metabolism and from decay of organic matter, are mixed with air entering the ground and occupying voids in it. Most soils contain more air than most surface deposits, and most surface deposits contain more air than most bedrock formations. A measure of this difference in air content is the difference in density of the materials. Rocks average about 2.7 times as heavy as an equivalent volume of water; surface deposits average little more than twice as heavy as an equal volume of water, and soils are only 1.5 times as heavy as water.

SOME PROPERTIES OF AIR

Air consists of nitrogen (78%) and oxygen (21%). Most of the remaining one percent is argon; only 0.03 percent is that extremely important constituent carbon dioxide (CO_2). These figures, however, are for dry air; water vapor is always present in the atmosphere. At 0°C (32°F) a cubic meter of air can hold 5 grams of water vapor; at 10°C (50°F) a cubic meter of air can hold 30 grams of water vapor.

When a given volume of air contains the maximum amount of water that it can hold at a particular temperature, it is saturated. The most commonly used measure of atmospheric moisture is *relative humidity*—the actual amount of water vapor in the air expressed as a percentage of the maximum amount the air could hold at the same temperature.

Air, of course, has weight; the upper layers of the atmosphere compress the lower ones, and half the mass of the atmosphere is compressed in a layer only $3\frac{1}{2}$ miles thick. Two-hundred miles above sea level, only traces of air exist. The average pressure of the atmosphere at sea level is about 14.7 pounds per square inch (roughly a ton per square foot). This is equal to the pressure of a column of water about 34 feet high or a column of mercury about 30 inches high. With each 900-foot increase in altitude, the pressure decreases by about $\frac{1}{30}$. On a mountain top 18,000 feet high, the atmospheric pressure is only a little more than 7 pounds per square inch, which means that there is only half as much air above 18,000 feet as there is between sea level and 18,000 feet. Thus altitude affects solar radiation and the temperature on and in the ground.

If not filled with water, pore spaces in the ground contain air. The soil atmosphere has a composition similar to the air above ground except that it has less oxygen (see p. 259) and more water vapor, carbon dioxide, and other of the gaseous products of the soil organisms. The carbon dioxide content of soil air, for example, may be 10 percent or more, and not the mere 0.03 percent that is in the air above ground. Soil air provides the oxygen needed by plant and animal life in the ground; for plant growth, ground must be well aerated, hence the agricultural practice of tillage. For engineering purposes, however, air in the ground lessens its supporting strength and must be excluded by compaction.

Air in the ground interchanges with that above ground as a result of changes in barometric pressure. Increases in barometric pressure cause air to be forced into the ground; decreases in pressure allow air to escape. The low pressure that commonly prevails before thunderstorms may cause earthy smells to rise from the ground.

Cold air is heavier than warm air and moves downhill to collect in low places, which are subject to early freeze and may have vegetation quite different from that on higher ground. Such differences affect soils.

RADIATION, CONDUCTIVITY

At low altitudes the compressed layer of air insulates the ground against solar radiation. At sea level something like 45 percent of the solar radiation reaches the ground surface. At high altitudes nearly 75 percent reaches the ground (Fig. 5.1), because less radiation is reflected by clouds and absorbed by water vapor (mountain

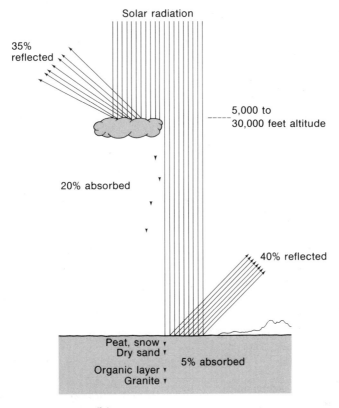

FIGURE 5.1
Approximate economy of solar radiation at sea level
at middle lattitudes, and relative depths of heat penetration in different kinds of ground. Above 30,000 feet
the air is rare and contains almost no impurities; it
cannot readily absorb radiant heat and consequently
is cold. Above the ground surface, heat is transfered
by *radiation*—that is, in the form of waves travelling
at the speed of light; absorption of the waves at a
surface causes it to warm. Below the ground surface,
heat is transfered by *conduction;* that is, warm molecules vibrate and pass some of that energy to adjacent
cooler and quieter molecules.

tops are good places to get a sunburn). Radiation that reaches the ground warms
the surface, and heat is conducted into the near-surface layers.

The amount of heating depends in part upon the kind of ground: different materials
have different reflectivities (p. 90), different heat capacities, and different heat con-
ductivities. The conductivity of granite is about 1 percent that of conductive metals
like copper and silver, the conductivity of the organic-rich layers of soil is less than
half that of granite, the conductivity of dry sandy ground is about a tenth that of
the organic layers, and the conductivity of peat and snow about half that of sandy
ground.

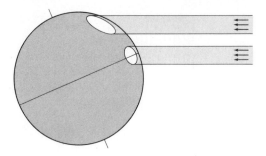

FIGURE 5.2
Near the earth's poles, the sun's rays must penetrate
more atmosphere and spread over more ground than
near the equator, hence polar ground is correspondingly colder.

Moisture greatly increases the conductivity of ground, and the water in ground
is the principal factor affecting the rate of warming. Two pieces of ground that are
alike except for moisture content will heat very differently. They will absorb the
same amount of heat, but the moist ground will seem cooler because the heat is
distributed throughout a greater mass. Clay ground retains moisture whereas sandy
ground may become dry, and consequently sandy ground reaches a higher temperature because the heat is concentrated in the shallow, less conductive, surface layer.

Although ground at high altitudes is greatly warmed by solar radiation; the air
next to that ground stays cool because it is rarefied (less dense) and contains few
impurities, and therefore does not readily absorb radiant heat. This cool air contributes to the maintenance of snowfields and glaciers on high mountains.

GROUND TEMPERATURE

Ground surfaces heat and cool more rapidly than do water surfaces because (1) water
surfaces reflect more radiation; (2) radiation penetrates water so that heat is distributed
downward; and (3) more heat is required to raise the temperature of water than of
other substances (which is one reason why the beans burn when the water boils off).
The difference in the rates at which land and water surfaces heat and cool creates
convection currents that account for the coastal land-sea breezes; the warmed air,
which is light, rises and becomes displaced at the lower levels by denser cool air.

At or near the ground, extremes of temperature are greater than they are a few
feet above the ground, which is where standard weather readings are recorded.
Ground temperatures are generally higher in daytime and lower at night than the
air temperatures a few feet above ground. Moreover, the average moisture content
is greater near the ground.

The ground in the southern part of the United States is warmer than the ground
in the northern part because of the angle of the sun's rays (Fig. 5.2). A major
consequence of this is the difference in length of growing season at different latitudes.
Moreover, for every $10°C$ ($18°F$) rise in temperature, organic chemical reaction rates
approximately double. This is to say that, other factors being equal, the rates of

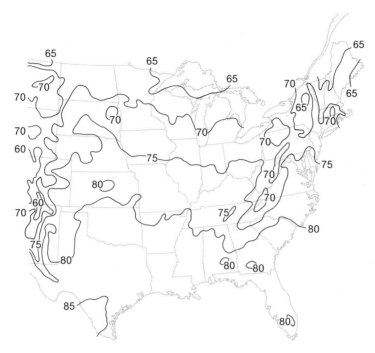

FIGURE 5.3
Average summer ground temperatures at depths of about 2 feet, as inferred from Weather Bureau temperatures for July (1899–1938) in the central and eastern United States. In most kinds of ground, temperatures a few feet below the surface are nearly constant and are approximately the same as the average seasonal temperature of the macroclimate of the locality. [After U.S.D.A.]

the reactions in the southern states, including the rates of weathering, are approximately double those in the northern states, and the rates in northern states are approximately double those in the arctic, where the ground is frozen 8 to 10 months each year. Just as refrigeration slows food spoilage, so does freezing slow soil development. Arctic soils are weakly developed, whereas soils in the southern states are strongly developed, and to considerable depths. Temperature is not the only influential physical variable, but it is probably a major one.

During the day, temperatures at the ground surface are high enough to be lethal to many seedlings. The effect of these high temperatures, however, does not penetrate very far into the ground. At a depth of 2 or 3 feet in most kinds of ground there is little daily variation in temperature. Below about 10 feet, there is little or no annual variation; the ground temperature there is essentially the same as the average annual temperature of the macroclimate of the locality. At the caverns in Virginia, at Mammoth Cave in Kentucky, and at Carlsbad Cavern in New Mexico, the average annual temperature of the macroclimate is about 55°F, and the temperatures in those caves are nearly constant at about 55°F. At most localities the temperature of the ground at a depth of about 2 feet is approximately the average *seasonal* temperature of the locality. Figure 5.3 illustrates the probable summer ground temperatures in various parts of the country as deduced from the average summer air temperatures.

90

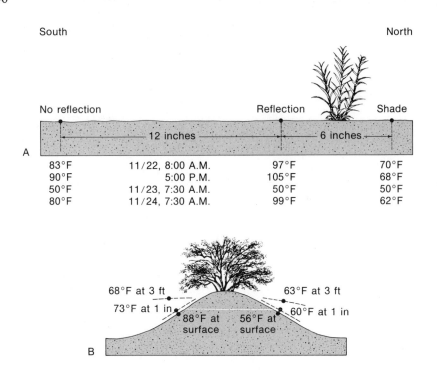

FIGURE 5.4
Examples of differences in ground temperature due to differences in exposure in Death Valley, Calif. [After Hunt, 1966a.]

Ground temperatures may vary greatly within short distances. In towns, snow melts from sidewalks before it melts from lawns. Figure 5.4 gives two examples of differences due to exposure. The reflectivity, or *albedo*, of the ground partly determines its degree of heating by solar radiation, and the heating determines the rate of drying. Dark ground heats more and dries faster than ground that is light in color. Differences of 30°F have been recorded for adjacent black and white sands on similar exposures. The high ground temperature recorded in Death Valley (p. 85) was measured on pale brown gypsum; some of the nearby surfaces blackened with desert varnish probably reach 212°F (100°C). Forests, grasslands, and deserts (other than those in which dark rocks prevail) differ little in albedo (Fig. 5.5). Ground in deserts heats more than ground in humid areas where there is forest or grass. In deserts there is little moisture to absorb heat and few clouds to reflect the sun's ray, hence solar radiation is greater.

A pebble or cobble lying on bare ground is subjected to three very different microclimates. The upper surface, exposed to the highest temperatures and the greatest changes in temperature, is in general being eroded. Around the pebble, at ground line, is a narrow band where temperatures are less extreme and where there is maximum wetting, by dew as well as by surface water. Even in deserts this narrow zone is densely populated with microorganisms—perhaps also with some megascopic

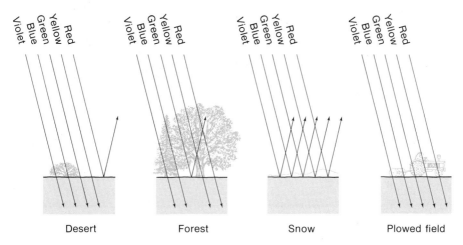

FIGURE 5.5
Every kind of ground is a color mirror that reflects some wavelengths and absorbs others. Red deserts and green eastern forests are warmed comparably, although they reflect different light rays. White light is reflected from snow but is absorbed by dark ground like that of a freshly plowed grassland soil. Evaporation is greater from dark ground than from light ground.

ones, algae. Temperatures on the underside of the pebble are moderate, and soil moisture condenses on it; this surface may have hollows dissolved in it.

Vegetation greatly affects the movement of air, which in turn affects moisture retention. The wind velocity above the canopy, whether of grass or forest, may be several times that at ground level. Summer temperatures near the ground commonly are more than 5°F cooler than the temperatures above the canopy; night temperatures tend to be higher and day temperatures lower under vegetation than above it, and relative humidity may be 10 percent higher under the vegetation. As a result, evaporation is greatly reduced in the shade of vegetation. Some examples of micro-climate variations at a grassland location are tabulated below.

	Temperature (°F)			
	Height above ground		Depth below surface	
	1 foot	1 inch	1 inch	1 foot
Summer				
Early afternoon	120°	100°	95°	72°
Early morning	62°	59°	69°	72°
Winter				
Late afternoon	16°	10°	10°	25°
Early morning	−32°	−15°	4°	25°

Source: Whitman and Wolters, 1967.

92

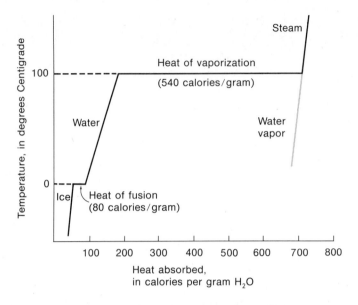

FIGURE 5.6
A gram of ice at 0°C absorbs 80 calories in changing to a gram
of water at the same temperature—an amount of heat that is only
a little less than the 100 calories needed to raise the temperature
of the water to boiling. A gram of boiling water absorbs 540
calories in changing to a gram of steam at the same temperature.
Evaporation of water at intermediate temperatures similarly absorbs heat and cools the air around it; the amount of cooling
depends on the pressure and temperature of the change.

Some temperature comparisons (°F) in the ground under grassland are given in
the following table.

	Mid-January	Early April	Mid-June	Mid-August
Ground surface	12°	40°	74°	79°
Ground 4 feet deep	36°	40°	58°	67°

Source: Whitman and Wolters, 1967.

In January the ground is colder at the surface than at depth, whereas in the summer
the ground is warmer at the surface than at depth.

Changes in the state of water on the surface or in the upper layers of the ground
materially affect the temperatures on and near the surface (Fig. 5.6). Evaporation
of dew or other surface water has a cooling effect because that change in the state
of water is accompanied by the absorption of heat—a phenomenon illustrated by
the desert water bag or by the cooling effect of woods, where water is changed to
vapor by the transpiration of plants. Temperatures in woods commonly are 5° to
10°F lower than those outside the woods; such differences in temperature are a

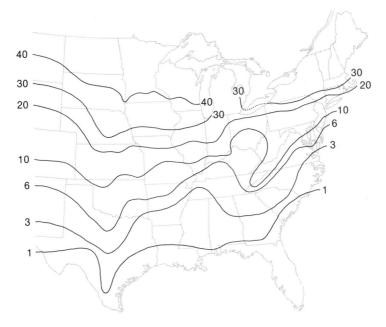

FIGURE 5.7
Depth of frost penetration in the eastern United States. In the northern states, all the soil layers freeze unless protected by snow cover; in the southern states only the uppermost layers freeze. [After U.S.D.A.]

measure of moisture availability. Conversely, the change in state of water from vapor to liquid is accompanied by a release of heat; thus the precipitation of dew or frost by cold ground raises the temperature of the layer of air immediately above it.

In brief, temperatures at the ground surface are more extreme than temperatures in the air above the ground, and the latter are more extreme than the temperatures within the ground. Excessive temperatures, high or low, can be lethal to plants and to animals in the ground. Temperature governs the capacity of air to hold vapor, whether the air is in the ground or above it, and it controls the rate of the biochemical processes involved in plant growth and weathering. Engineers must consider temperature too because it controls depth of freezing and frequency of freeze and thaw.

FREEZE AND THAW

The depth to which frost penetrates the ground (Fig. 5.7) and the effects caused by it vary greatly, depending upon whether the ground is heavily wooded or sparsely covered with vegetation and whether it is covered by snow. Where the ground is bare or only sparsely vegetated, a ground frost commonly consists of interconnected ice lenses that form a dense impermeable mass that extends a few feet into the ground. Under forest cover, ground may develop a porous and permeable, honeycomb-like frost, and the freezing is not as deep as under bare ground. Under snow cover, the ground commonly remains unfrozen.

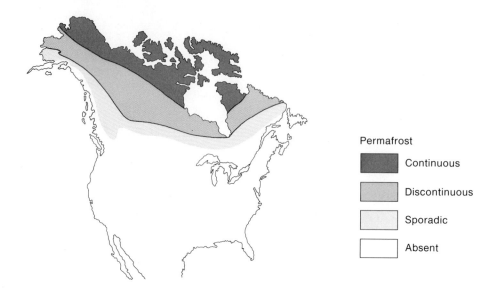

FIGURE 5.8
In northern Alaska and northern Canada the ground is below freezing temperature, and there is a continuous layer of permafrost. Southward, this layer is discontinuous, being interrupted by ground that averages above freezing. Farther southward is a belt with small patches of permafrost. The southern coast of Alaska, west coast of Canada, and southern Canada are without permafrost. In the areas of permafrost, the active layer thickens southward.

Water expands when it freezes, increasing its volume by almost 10 percent, which is why a deep frost causes heaving of the ground. The freezing of the upper few feet of the ground causes moisture to be drawn upward from the warmer unfrozen ground below. This moisture collects in lenses that approximately parallel the ground surface. The amount of heaving of the ground equals the cumulative thickness of these lenses. Because freezing draws water upward, the heaving can amount to as much as 6 inches in ground frozen to a depth of only 3 or 4 feet; this is many times greater than the heaving that could be caused by the original moisture, even assuming the ground was 25 percent saturated. The water content of the frozen layers may be two or three times greater than that of the ground before it was frozen (Trefethen, 1959, p. 387).

In arctic and subarctic latitudes much of the ground is permanently frozen, a condition referred to as *permafrost*. North of the Brooks Range in Alaska, the permafrost may extend as deep as 1,000 feet. To the south it thins, and in central Alaska becomes discontinuous. In southern Alaska permafrost is developed only in small local areas (Fig. 5.8).

The surface layer of ground with permafrost is subject to thawing during the summer. This surface layer that thaws is referred to as the active layer. When winter begins, the top of the active layer is first to freeze, and water becomes trapped between it and the underlying permanent ice. This water may be squeezed upward and dome the surface layer of the ground, forming what is known as a *pingo*. This phenomenon causes unusual engineering problems because the outbreak occurs where the capping layer of ice is thin, which can very well be under a building, highway, or airstrip. Buildings have become molds for the ice that filled them!

Frost boils, which disrupt highway pavements, are analogous to the pingos in the active layer of permafrost. As a pavement warms in spring, water frozen in the upper layers of the subgrade melt; because this water cannot drain downward into the still-frozen lower layers, the saturated layers of the subgrade become an unstable foundation for the pavement.

Frost damage, therefore, is of two kinds, the heaving that results from growth of the masses of ground ice, and the frost boils that develop when the upper layer melts.

Many structural features of ancient soils and surface deposits south of the border of the Wisconsinan drift have been attributed to frost heaving and permafrost conditions that prevailed at the time of the glaciation. Freezing and thawing produce various kinds of ground patterns (Fig. 10.7), boulder fields where resistant formations crop out on mountain sides, and contortions and involutions of the bedding in surface deposits and the layering of soil.

DEW

Probably the most important effect of ground temperature is control of the formation of dew, and, to a lesser extent, of frost. Dew has been explained by comparing the ground surface to the outside of a glass of ice water; air that comes into contact with the chilled surface of the glass is cooled, and water vapor condenses on the glass. But the ground-atmosphere relationship is not quite analogous to the chilled water glass, because dew can form on ground surfaces that are warmer than the air. Lawns are frequently covered with dew even though colder pavements are not.

Dew also has been interpreted as being soil vapor that rises from warm soil and condenses when it encounters a blanket of cold air at the surface. Dew on grass has been attributed in part to chilling of the air close to the ground, and partly to the condensation of transpired moisture. In support of the latter interpretation are the numerous records of plants that lost weight over night although wet with dew; in support of the former interpretation is the identification of airborne salts in dew.

The amount of dew collected on ground in arid and arctic regions has been estimated at about 0.5 inch annually, which constitutes a substantial fraction of the annual precipitation. In humid regions, the dew that collects on the ground may amount to 3 inches in a single year.

PRECIPITATION, RUNOFF, GROUND MOISTURE

Annual precipitation is another variable factor—probably the major factor—that affects ground climate and kind of ground. The effects of differences in precipitation, and its opposite, evaporation, are most noticeable between the eastern and western United States. Where the precipitation averages less than about 20 inches annually, the forests give way to grasses (Fig. 6.2). Where the precipitation averages less than

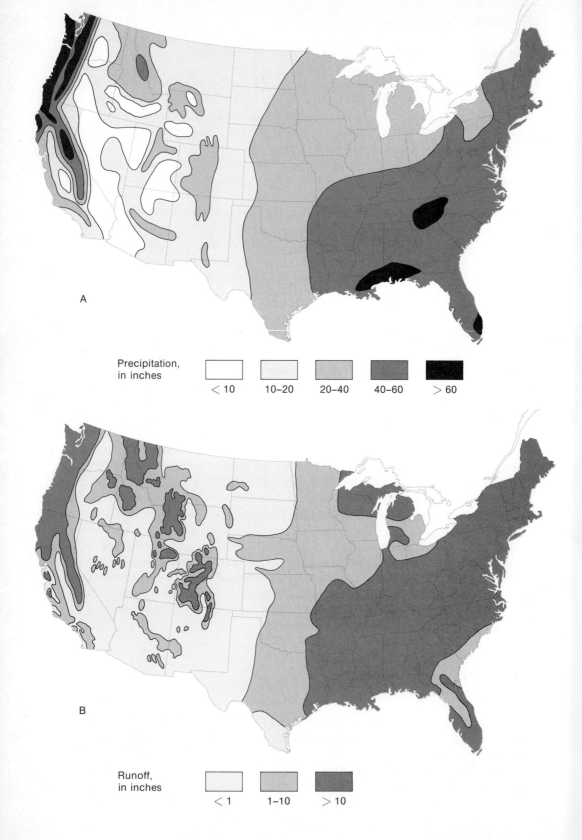

A

Precipitation,
in inches

< 10	10–20	20–40	40–60	> 60

B

Runoff,
in inches

< 1	1–10	> 10

about 15 inches, grasses give way to shrubs, and in the arid regions where the precipitation averages less than about 10 inches annually, the shrubs may be widely spaced, leaving much or most of the ground bare.

Where the annual precipitation averages less than 20 inches, the runoff—water running off the surface—averages less than about 1 inch annually (Fig. 5.9), and through-flowing, permanent streams do not develop. Were the Rocky Mountains nonexistent, the western part of the Central United States would have no through-flowing rivers, only dry washes that would discharge floods after rains. In contrast to the western part of the country, our Atlantic seaboard has many rivers whose annual discharge is as great as or greater than that of the Colorado River.

Where precipitation averages less than about 1 inch annually, there is not enough water on the ground to wet more than the surface layers. The water that falls on such ground seeps into it and is wholly contained in the pores of the surface layers. Where precipitation is greater than about 20 inches annually, the amount of water is sufficient to fill the pores of the ground and drain into the underlying bedrock. This has profound effects on the kind of ground, because large amounts of water seeping through the ground remove soluble constituents, causing the ground to become acid. Where the precipitation is adequate to wet only the upper layers of the ground, and is then soon evaporated, the soluble constituents are retained and the ground becomes alkaline. The 20-inch rainfall line (Fig. 5.9, A) closely approximates the boundary between acid and alkaline ground in the United States.

Besides amount of precipitation, the form and intensity of precipitation also affect the kind of ground. The precipitation may occur entirely as rain, as in southerly latitudes, or a considerable part may occur as snow, as in the north. Snow melt minimizes runoff and increases infiltration of moisture into the ground. In some regions, as in the eastern United States, the precipitation occurs fairly uniformly throughout the year. In other regions, such as the central United States, most precipitation occurs during the summer—the growing season. Our Pacific Coast receives most of its precipitation during the winter months.

Intense rainstorms produce runoff that can wash the ground and erode it, whereas the same amount of rainfall over a longer period of time would produce little runoff and allow the water to infiltrate the ground. The retention of moisture on or in the ground is affected by wind and temperature, and these in turn are affected by the vegetation. Near the floor of a forest, wind may be only half as great as at the tree tops, and a third as great as in the air above the tops. Such differences occur on grassland too, but at lesser heights above the ground. Shady ground, we all know, is cooler than sunny ground, and the moisture it holds is subject to slower evaporation.

FIGURE 5.9
Average annual precipitation (A) and average annual surface runoff (B) in the United States. The humid eastern part of the United States receives more than 20 inches average annual precipitation, and the average annual runoff is more than an inch. Lands with 10 to 20 inches average annual precipitation are classed as semiarid; if the precipitation is less than about 10 inches they are considered arid. Through-flowing streams cannot originate where the precipitation is less than about 20 inches; streams in the western United States originate high in mountain ranges; those that do not are, with few exceptions, ephemeral. Areas with high rainfall and high runoff are subject to severe and frequent flooding. [Part A after U.S.D.A.; part B after U.S.G.S.]

On the other hand, vegetation keeps some of the rainfall from reaching the ground. One can stay dry in a light sprinkle by taking shelter under a tree. When the rainfall becomes heavier, the dripping begins, but some of the water moves down stems and trunks. Much of this water evaporates and does not reach the ground. The importance of water on and in the ground is discussed more fully in Chapter 10.

CONTROLLING GROUND CLIMATE

Artificial heating may be used to protect high-value crops against frost. Candles of solid fuel burning with a low flame through a night can raise the temperature under fruit trees by 5°F or more. Smudge pots are more commonly used; the heat from the fires in the pots helps raise the temperature near the ground, but the main effect is produced by the smudge—a small cloud of smoke that protects the trees by reducing losses of long-wave radiation from the ground. Some small crops can be covered with cloth. The most common way of purposely modifying ground climate, however, is by irrigation or other watering of the ground.

In some farmed areas protection is needed against excessive wind. This is particularly so on the treeless Great Plains, where homes, barnyards, and farmed land are protected against the strong winds by trees grown to form windbreaks, or shelterbelts (Fig. 5.10). Shelterbelts provide a number of kinds of benefits. By slowing down wind speed at ground level they reduce the hazards of fire, wind breakage, blowdown, wind erosion, snow drifting, and excessive evaporation. Shelterbelts are an important part of land development and land improvement on the Great Plains, especially the northern Great Plains.

A shelterbelt consists of several rows of plants, commonly five to ten, each row being of a different kind and growing to different heights. Low, dense-growing shrubs are placed on the outer row exposed to the prevailing wind, and progressively taller-growing plants are placed in the succeeding rows. Each row therefore provides increasingly higher protection. In the central United States, the prevailing winds are from the northwest, and the rows generally are placed along the north and west sides of the property to be protected.

CLIMATIC CHANGES

Major changes have taken place in both the macroclimate and the climate of the ground, as attested by the Pleistocene glaciations and interglaciations. One of the effects of these climatic changes is the occurrence of deeply leached acid ground in parts of some deserts, where the present precipitation is not adequate to have caused the leaching. The effects of past climates are more conspicuous in those deserts that are the beds of now dry lakes, like the desert west of Great Salt Lake.

Even within the Holocene Epoch there have been climatic changes, as evidenced by advances and retreats of glaciers and by changes in peat bogs, in the thickness of tree rings, in lake levels, in alluviation, and in dune activity. In the western United States, just before A.D. 1, the climate was wetter than it is now. There were lakes at the now-dry playas; alluvium was deposited along streams, and there were small

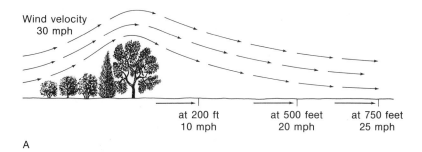

Wind velocity
30 mph

at 200 ft   at 500 feet   at 750 feet
10 mph   20 mph   25 mph

A

B

FIGURE 5.10
On the Great Plains. Shelterbelt around a ranch house and windbreaks around fields
protect against wind erosion when the fields are ploughed and against snow drifts
in winter.

glaciers at the heads of many valleys in the western mountains. About A.D. 1000,
the climate was drier and/or warmer; the glaciers melted away, the alluvium in the
valleys became eroded, and there was renewed dunal activity. This deterioration of
the climate was reversed again in the period 1600 to 1750 when conditions again
became somewhat wetter, and there were readvances of the glaciers.

In some of our western valleys, there is archeological evidence that streams dried
up in response to climatic change during the late Holocene. Archeological remains
found in prehistoric centers of occupation reveal that those who lived in these valleys
migrated upstream as the water supply dwindled (Fig. 5.11). The physical geology
supports this evidence. In Pleistocene time, for example, it is known that streams
draining the north side of the Henry Mountains in Utah were capable of transporting
coarse cobble gravel to the Dirty Devil River 20 miles from the mountains. In latest
Pleistocene time, only occasional floods were able to move gravel that far; today,
the cobble gravel is being deposited 10 miles short of the river.

Short-term wet and dry spells are recorded in the weather statistics of particular
localities. Even city dwellers are becoming aware of these as droughts force curtail-
ment of car washings and lawn sprinkling.

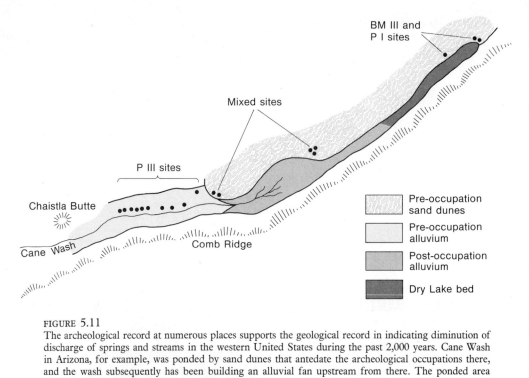

FIGURE 5.11
The archeological record at numerous places supports the geological record in indicating diminution of discharge of springs and streams in the western United States during the past 2,000 years. Cane Wash in Arizona, for example, was ponded by sand dunes that antedate the archeological occupations there, and the wash subsequently has been building an alluvial fan upstream from there. The ponded area now is a dry lake bed. Basketmaker III to Pueblo I sites, which date from about A.D. 700, are located on pre-occupation dunes at the edge of the dry lake. Upstream, on pre-occupation alluvium near Chaistla Butte, are Pueblo III sites that date from about A.D. 1200. Between these and the dry lake sites are mixed sites of intermediate age. Probably the population shifted up the valley as the flow in Cane Wash diminished. [After Hunt, 1955.]

The most evident effect of climate on ground conditions is reflected in the biota, especially the vegetation. As with people, some plants like warmth, others prefer the cold; some like dry environments, others prefer wet ones. Moreover, because of these preferences, or *adaptations* as biologists would say, the vegetation and the rest of the biota that lived in certain places in the geologic past changed along with the climate. The paleontological record, discussed in Chapter 2, holds the evidence of many such climatic changes. The next chapter discusses other aspects of the biota, especially the plant and animal populations and how they respond to and cause differences in the kind of ground.

BIBLIOGRAPHY

Antevs, E., 1938, Post-pluvial climatic variations in the southwest: Meteorol. Soc. Bull., vol. 19, pp. 190-193.

———, 1948, The Great Basin, with emphasis on glacial and postglacial times, vol. 3, Climatic changes and pre-white man: Utah Univ. Bull., v. 38, no. 20, pp. 168-191.

Blaney, H. F., 1955, Climate as an index of irrigation needs: *in* Water, U.S. Dept. Agriculture Yearbook, 1955, pp. 341-345.

Blumenstock, D. I., and Thornthwaite, C. W., 1941, Climate and the world pattern: *in* Climate and Man, U.S. Dept. Agriculture Yearbook, 1941, pp. 98-127.

Chang, Jen-Hu, 1958, Ground temperature: Blue Hill Meteorol. Observ., Harvard Univ., pt. 1, 300 pp., pt. 2, 196 pp.

Emiliana, C., 1958, Pleistocene temperatures: Jour. Geol., v. 63, pp. 538-578.

Fitton, E. M., and Brooks, C. F., 1931, Soil temperatures in the United States: Mon. Weather Rev., v. 59, pp. 6-16.

Geiger, R., 1957, The climate near the ground (2nd ed.): Harvard Univ. Press, Cambridge, Mass. 494 pp. Translated by M. N. Stewart and others.

Hunt, Chas. B., Recent Geology of Cane Wash, Monument Valley, Arizona: Science, v. 122, pp. 583-585.

Kendrew, W. G., 1942, The climates of the continents: Oxford Univ. Press, New York, 473 pp.

Richards, S. J., 1952, Soil temperature. Soil physical conditions and plant growth: Academic Press, New York, B. T. Shaw, Ed.

Smith, Guy D., and others, 1964, Soil-temperature regimes—their characteristics and predictability: U.S. Dept. Agriculture Soil Conservation Serv., TP-144, 14 pp.

Sutton, O. G., 1953, Micrometeorology, a study of physical processes in the lowest layers of the earth's atmosphere: McGraw-Hill, New York, 333 pp.

Trefethen, J. M., 1959, Geology for Engineers: Van Nostrand, New York, 632 pp.

The decay of organic matter on and in the ground is the result of biochemical rather than physico-chemical processes; that is, the decay is caused by various kinds of organisms—bacteria, actinomycetes, fungi, and algae—that feed on and derive energy from the dead organic matter. This biochemical process returns reusable nutrients to the soil. Some of the organisms, notably lichens and mushrooms, are big enough to be seen, like these mushrooms growing on the bark of a live tree. Tree bark, like fallen branches and leaf litter, is nonliving organic matter. As some trees grow, they shed their bark; but on most trees the bark cracks, and these irregular openings form convenient places for spores of fungi to lodge. The mushrooms in this picture have a growth habit similar to the so-called shelf mushrooms; unlike the more numerous ground dwellers, which have upright stems, these have eccentric stems. [Photograph by Ralph Weiss.]

6 / *Flora and Fauna*

The plant and animal population; microflora and microfauna; bacteria, actinomycetes, fungi, algae; macroflora and macrofauna, plant geography, some influences on soil; ground conditions reflected in plant growth, examples of effects of chemistry, physical conditions, sensitivity to ground conditions at limits of range of species; ground moisture and plant growth; water-loving plants, reflect both quantity and quality of water; drought-resistant plants; accumulator plants; floods and fires; bibliography.

The life that exists in and on the ground, principally in the organic-rich and uppermost of the soil layers, consists of (1) a microflora of bacteria, antinomycetes, fungi, and algae; (2) a macroflora composed chiefly of the roots of flowering plants; (3) a microfauna of protozoa; and (4) a macrofauna of earthworms, insects, arachnids, nematodes, and gastropods. In some ground, the macrofauna also includes a number of burrowing mammals, such as rodents, rabbits, and foxes. The kind of ground, plus other environmental factors such as topography, climate, and especially drainage, limit the kind and number of plant and animal populations; the organisms, in turn, have their marked affects on the soils, and can maintain them, deplete them, or improve them. The microorganisms move around within the pores of the ground; the macroorganisms push the soils aside.

MICROFLORA AND MICROFAUNA

Bacteria usually can be expected to outnumber all other plants, as well as all animals, that inhabit the ground. A gram of soil may contain a million or more. Bacteria have about the same size range as clay particles. Some are rod-shaped, as much as 10 microns long; others are cocci, rounded masses less than 0.5 micron in diameter (Fig. 6.1). Those that require an abundant supply of oxygen are called *aerobic* bacteria; those that can exist on a minimal supply are called *anaerobic*. Like the green plants, some bacteria are capable of manufacturing their own food from inorganic sources, and are thus called *autotrophs* (self-nourishers). They derive their energy from the oxidation of inorganic materials and obtain their carbon from the CO_2 in the air or moisture in the ground. Some autotrophic bacteria can live in environments that are extreme—with high salinity and great range of temperature, like the brine-saturated ground around Great Salt Lake.

Most bacteria do not obtain their food from inorganic sources, and for this reason are called *heterotrophs* (nourished by others). Some are *parasites,* feeding at the expense of other living organisms; others are *saprophytes,* feeding on dead organic materials and degrading them to inorganic ones that become available as nutrients. Some bacteria live symbiotically—that is, in intimate association with another living organism to the mutual benefit of both. Perhaps the better known of these are the nitrogen-fixing bacteria that live in the roots of legumes. The bacteria obtain carbohydrates and mineral food from the legumes, and the legumes use the nitrogen fixed by the bacteria to build proteins.

Actinomycetes, single-celled plants that develop coiled or branched hyphae (Fig. 6.1), are classified by some authors as bacteria, by others as fungi, and by still others as intermediates between the two. They average somewhat larger than the bacteria, and are almost as numerous in the ground as are bacteria. Actinomycetes are aerobic heterotrophs, and although a few are pathogenic (disease-causing), their chief function in the ground is in the decomposition of dead organic matter.

Fungi, which include the molds, may be as abundant in the ground as the bacteria. Some are parasites, though not necessarily pathogenic, and others are saprophytes. Like animals, fungi are incapable of photosynthesis. Most are microscopic in size, but a relative few are large—for example, the mushrooms. In general, fungi seem to be less tolerant of salinity than bacteria; in Death Valley they grow in briney ground containing as much as 12 percent dissolved salts (Hunt and Durrell, in Hunt, 1966, p. 61). Fungi are more tolerant of acid conditions than are the bacteria; some can utilize poorly aerated ground. Apparently all the soil fungi are heterotrophic.

Algae that live in the ground are mostly simple forms, and although they contain chlorophyll and are capable of photosynthesis, some live as deep as 4 inches in soils and evidently do not require sunlight. Desert algae are common on the undersides of transluscent stones, where they can receive diffuse light. Their moisture supply includes the dew that, even in the driest deserts, collects on such undersurfaces.

Lichens, which encrust bare rock surfaces, trees, and soil, are combinations of fungi and algae. The alga provides the foodstuffs needed by the fungus, and the alga derives the advantage of being shielded against excessive sunlight, desiccation, and mechanical injury. As a consequence, such symbiotic pairs can survive in some of the harshest environments on earth—hot, bare rock surfaces in the desert and near-freezing bare

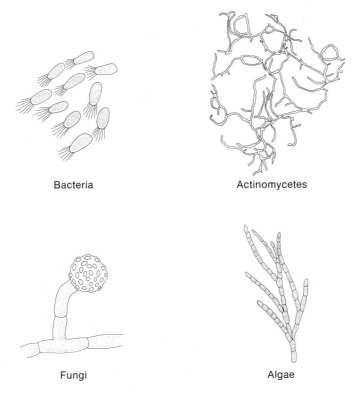

Bacteria   Actinomycetes

Fungi   Algae

FIGURE 6.1
Four kinds of microscopic plants living in the ground. *Bacteria* are unicellular and reproduce by fission. *Actinomycetes*, branching bacteria, look like a tangle of threads. *Fungi* are plants without seed, stem, leaf, or root and have no chlorophyll; reproduction is other than by fission. Some, like the mushroom, are megascopic. *Algae* lack seed, stem, leaf, or root, but do have chlorophyll.

rock surfaces at the edge of perennial snow. The role of lichens in the weathering of rocks and surface deposits is uncertain, but that lichens do contribute to weathering seems to be demonstrated by the etching of glass windows of old cathedrals where there are colonies of crustose species (Hale, 1961, p. 90).

The microfauna of protozoa, which includes the rhizopods (among them amoeba) and flagellates, feeds on the microflora, including the bacteria.

MACROFLORA AND MACROFAUNA

The macroflora faithfully reflects the climatic regions (Fig. 1.4). At the north is the arctic or tundra zone, which is treeless except for dwarfed willows. The vegetation is mostly mosses, lichens, and sedges. South of this is spruce-fir forest, the only forest that extends all the way across the continent. South of this, in the humid eastern United States, in succession is northern hardwood forest, central hardwood forest,

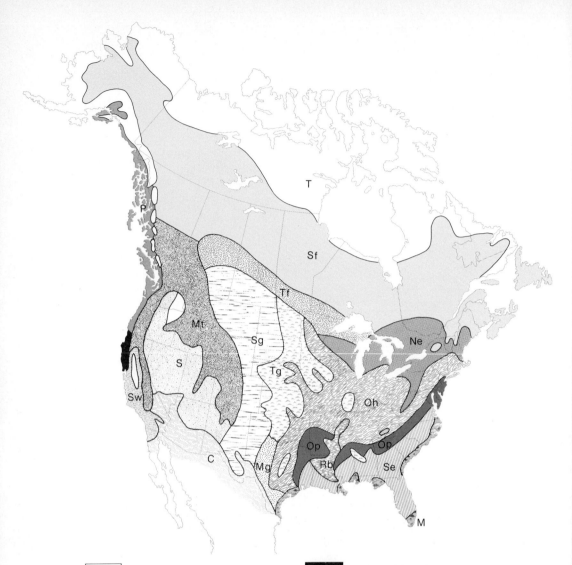

T	Tundra
Sf	Spruce-fir forest
Tf	Transition pine-aspen forest
Ne	Northeast hardwood forest
Oh	Oak-hickory forest
Op	Oak-pine forest
Se	Southeast pine forest
P	Pacific coast forest north: spruce-hemlock south: Douglas fir
Mt	Coast Range-Rocky Mountain conifer forest lower: pine-Douglas fir higher: spruce-fir summits: alpine meadow

	Redwood forest
Sw	Southwest broadleaf woodland
Sg	Short grassland
Tg	Tall grassland
S	Sagebrush shrubland
C	Creosote bush shrubland
Mg	Mesquite and desert grassland
Rb	Riverbottom cypress-tupelo-sweetgum
M	Mangrove swampland

FIGURE 6.2
Natural vegetation in North America.

and southeastern pine forest. Westward, near the 100th meridian and 20-inch rainfall line, the forests are replaced by grasses, and still farther west, where the rainfall averages less than about 15 inches annually, is desert shrub. Forests in the Rocky Mountains and along the Pacific coast are zoned conspicuously in two patterns, one conforming to the latitudinal climatic zones and the other to altitudinal zones. Both kinds of zoning are controlled chiefly by temperature—by extremes of temperature and by length of growing season. Forests in the Appalachian Mountains are similarly zoned, although less conspicuously. On top of the Great Smoky Mountains, near 6,000 feet altitude, is vegetation rather like that at sea level in northern New England.

Such different kinds of macroflora contribute different kinds and amounts of organic matter to the ground. Grasses have deep, short-lived roots, and organic matter is added to the deep layers of grassland soils as well as to the surface layers. Forest trees, on the other hand, have long-lived roots, and organic matter is added mostly to the surface in the form of fallen leaves and twigs. The quantity of organic matter in the ground ranges from almost nothing in deserts to 20 tons per acre in the southeastern United States and almost twice that in the Great Lakes region. These differences are accompanied by differences in the populations of organisms that are dependent on the organic matter on and in the ground.

Few soil animals, micro or macro, can feed on and digest the woody matter in soil—the cellulose, hemicellulose, and lignin. Even the pesty termites, of which there are many kinds, are but an apparent exception, for the work of digestion is actually done by protozoa that inhabit their stomachs. For the most part the organic compounds that make up woody tissue are reduced by the microflora, which in turn becomes food for the animal population. Among the macrofauna, some feed on growing plants or decaying plant matter; millipedes are an example. Others, such as the centipedes, are predaceous.

The macroflora contributes both to the physical condition of the ground and to its chemical properties. The physical effects include the development of microrelief (Fig. 6.3), the pushing aside of particles by roots, and the splitting of hard rocks where roots gain access to cracks. Another important physical effect is the blanket of litter that is so important in regulating ground temperature and retaining moisture. Still another is the hygroscopic effect exerted by particles of organic matter within the ground. Their greater surface area gives them a greater capacity for the absorption of moisture than clay particles, and they thus help keep the ground moist. This water, however, is held very tightly by the humus, and much of it is unavailable to plants.

The chemical effects of the macroflora include the extraction, by the roots, of chemical substances from the ground; their return to the ground in another form when plant parts decay; and the production of carbon dioxide, which is all important in weathering. Both the macro and micro plants that synthesize organic compounds liberate oxygen, and the ground consequently has a high oxidation potential. Under oxidizing conditions, organic matter is carbonized (in effect, burned) and converted to CO_2, but under reducing conditions, oxygen is deficient, little CO_2 can be liberated, and the carbon is released as volatile hydrocarbons (such as methane, CH_4).

Animal life contributes to the weathering and erosion of ground materials by aerating the ground and mixing the materials. Hoofed and other animals trample the surface layers, in places forming trails that discharge runoff and hasten erosion.

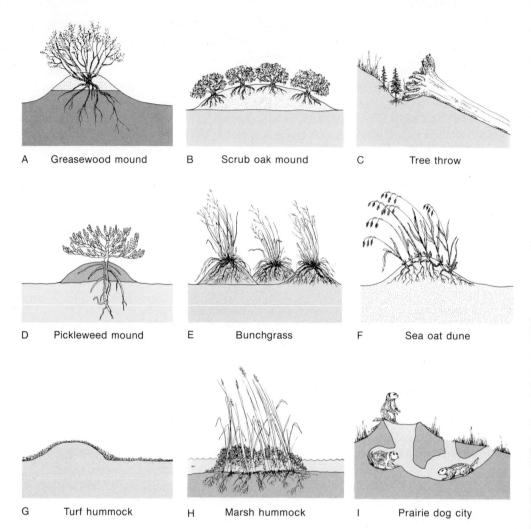

A Greasewood mound   B Scrub oak mound   C Tree throw

D Pickleweed mound   E Bunchgrass   F Sea oat dune

G Turf hummock   H Marsh hummock   I Prairie dog city

FIGURE 6.3

Some examples of microrelief produced by organisms. (A) Greasewood mound on alluvium; wind-blown sand and silt collected around the plant rests on a pedestal of alluvium where that surface has been lowered by erosion. Roots extend into the alluvium. (B) Sand mound under scrub oak on Colorado Plateau desert. Roots need not extend into the ground below the sand; moisture can be obtained from ground dew. Rain is insufficient to provide seepage through the sand. (C) Tree throws in northern Appalachian hardwood forest. The overturned ground provides well-aerated habitat for seedlings and contributes to downhill creep of the soil. (D) Pickleweed mound on saline playa. The plant grows on a mound of salt due partly to the accumulation of salt-rich plant stems and partly to the wick effect of the plant and its roots extending into the brine. (E) Bunchgrass on the Great Plains; the hummocks are largely organic matter. (F) Sea oat sand dune on a Florida barrier beach. (G) Turf hummock in the arctic; a moss capped mound of frost heaved mineral soil. (H) Marsh hummocks formed by fibrous peat collected around the plant. (I) Prairie dog city.

Burrowing animals—ants, worms, rodents—transport fresh mineral and rock particles to the surface and carry leaves, grasses, and other organic matter into the subsoil. Earthworms can so mix the layers in a soil as to change the type completely in a few years. Some ant nests in the tropics are said to have tunnels hundreds of yards long. The openings increase the circulation of air in the ground, increase the depth of water penetration, and increase the spread of carbon dioxide from which carbonic acid is generated.

GROUND CONDITIONS
REFLECTED IN PLANT GROWTH

Surface deposits and soils form the substrate in which plants root themselves, and this substrate is the principal source of plant food and water. Because the various species of plants that inhabit a particular climatic zone differ greatly in their requirements for water and nutrients, it follows that plant stands will reflect differences in ground conditions within such a zone. The differences in ground conditions from one place to another explain why we find grassy meadows in the midst of a Rocky Mountain conifer forest, swampland along stream bottoms crossing southeastern pine forests, sparse and dwarfed shrubs growing on shale next to lush growths of shrubs on sandy ground in western deserts, rich grasslands on loess in the midst of barren lavas (scablands) in the Columbia Plateau, dwarfed pine and oak on serpentine ground in the Maryland Piedmont, and jack pine on sandy outwash beside northern hardwoods on loamy moraines in the Great Lakes region.

Some differences in plant stands seem to be controlled chiefly by the chemistry of the ground. In Maryland, serpentine ground, which is magnesium rich but deficient in other elements, has dwarfed vegetation (Shreve and others, 1910). In the Gulf States, from Texas to Alabama, ground on limy formations is marked by belts of tall grass, whereas adjoining sandy formations support pine forest.

More commonly, the differences in plant stands seem to reflect physical differences in the ground. On the Piedmont Province, where deep ancient soils (saprolite) are developed on the metamorphic rocks, a number of strikingly different ground conditions are reflected in differences in the vegetation (Fig. 6.4). These ground conditions are

> 1. deep saprolite;
> 2. decayed rock under thin soil;
> 3. shallow weathered rock.

Each of these conditions varies depending on the kind of rock that has weathered, whether granite, schist, amphibolite, slate, or marble. In turn, any of these 15 different kinds of ground may underlie gravel. The effect that each of these kinds of ground has on the vegetation will be modified depending on topography—that is, whether the site is flat upland, gently sloping upland, or steep hillside. Still different are sites along valley bottoms. Vegetation along the sides of streams differs from that on the floodplains back from the stream—clearly a function of the availability of water—and both of these differ depending on the depth and kind of fill in the valley.

Alluvial floodplain		Gravelly valley side		Colluvium over deep saprolite		Shallow saprolite over gneissic bedrock	
Species	%	Species	%	Species	%	Species	%
Sycamore	30	Red maple	23	Sweet gum	32	Beech	26
Sweet gum	30	Black oak	20	Tulip tree	29	Black oak	16
Red maple	30	Hornbeam	18	Red maple	6	Red maple	13
Others	10	Sweet gum	14	Hickory	6	White oak	13
		Hickory	11	Black oak	6	Hickory	13
		Others	14	Others	21	Dogwood	10
						Others	9

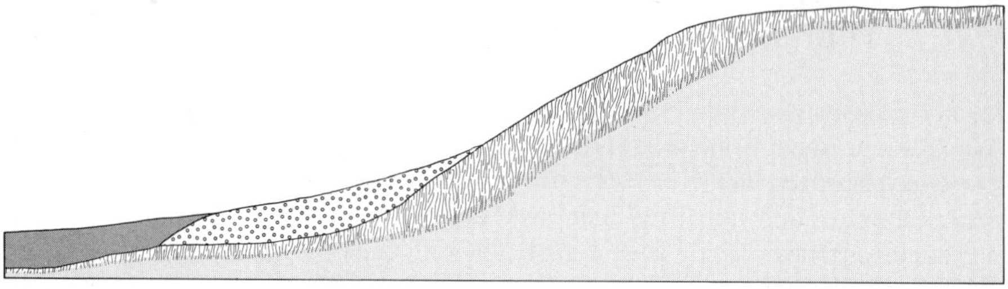

A

Thin alluvium		Deep saprolite		Thin gravel over deep saprolite	
Species	%	Species	%	Species	%
Hornbeam	32	Hickory	27	Tulip tree	50
Red maple	25	Dogwood	25	Black locust	22
Beech	21	Hornbeam	16	Sweet gum	11
Others	22	Black oak	14	Others	17
		Others	18		

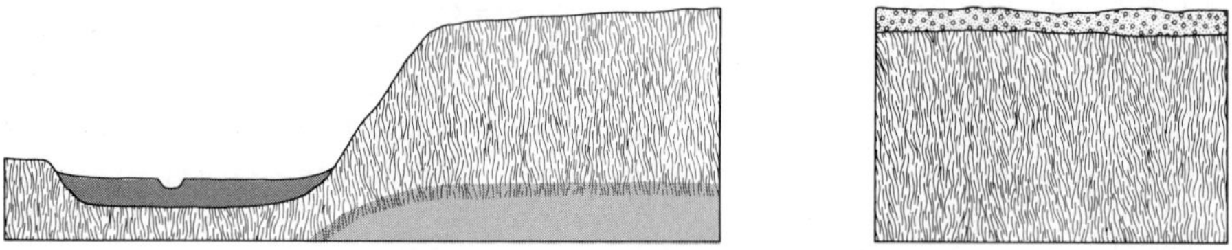

B   C

FIGURE 6.4
Differences in percentages of various plant species, evidently reflecting differences in ground conditions in the Piedmont Province, along Kennedy Expressway in Maryland. Geology and plant distribution as revealed during construction in 1962.

Northern hardwood forests on the Appalachian Plateaus in northern Pennsylvania (Goodlett, 1956, p. 56) are beech type forests growing mostly on nonrubbly ground. Rubbly areas within these forests are likely to have a red oak type forest. The most rubbly ground in the central hardwood, or oak, forests of that region is likely to have chestnut-oak type forest; ground still more rubbly than under the red oak type and thicker than under the chestnut-oak type supports stands of white oak (Fig. 6.5). Figure 6.6 is another example, from Connecticut, illustrating differences in plant stands on different kinds of glacial deposits.

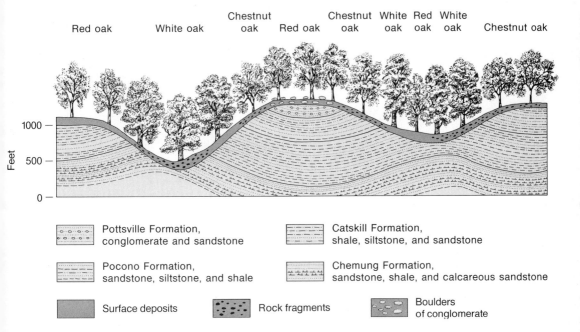

FIGURE 6.5
Idealized profile showing forest types, surface deposits, and bedrock in the central hardwood (oak) forest in northern Pennsylvania. [After Goodlett, 1956.]

In the Appalachian forests, toppling of trees builds mounds that are a vastly different habitat for the germination of trees than is the adjacent forest floor. The mounds have a high percentage of mineral matter in contrast to the leaf-littered, organic layer on the forest floor. The mounds favor germination of white pine, and it has been postulated that the white pines that were once abundant in northern Pennsylvania became established following windthrow, and that the absence of white pine in the present forests is partly the result of cessation of windthrow brought about by clear-cutting of the forests (Goodlett, 1956).

A site in the Hudson River Highlands (Fig. 6.7) is an example of the gradual filling of a pond and the advance of vegetation into it. Three kinds of forest are distinguished there, each seeking a preferred kind of ground. Mountain sides have red oak and chestnut oak; coves and valley bottoms have forests related to the northeast hardwoods; mountain tops that are rocky have pitch pine and scrub oak.

As climatic limits for a given species are approached, the species become increasingly sensitive to differences in kind of ground. This is not to imply, however, that species at the edges of their ranges are confined to the most favorable habitats; Griggs (1914) cites the chestnut as being limited to limestone in one part of the edge of its range and being limited to nonlimy ground in another. In the center of its range, a species may have the advantage, for example, of rapidity of growth, but at the edge of its range, where growth may be slower, it must find a particular kind of ground. Whatever the complex of causes, it seems to be true that at the edges of

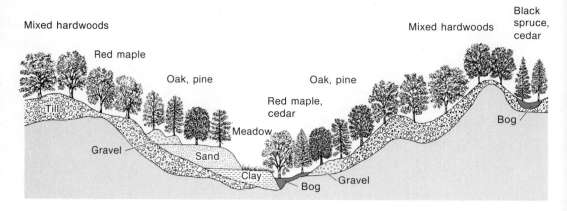

FIGURE 6.6
Relationship between vegetation and kind of ground in a glaciated Connecticut valley. Well-drained uplands have mixed hardwoods; excessively drained gravel and sand have oak and pine. Poorly drained ground, on clay, has meadow; upland bogs have black spruce and cedar; bogs in the alluvial valley have red maple and cedar. Rocky promontories have scarlet, chestnut, and black oak. [After Lunt, 1948.]

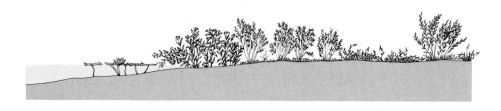

FIGURE 6.7
Vegetation on the Hudson River Highlands. (A) Pond-shore vegetation. As the edge of the pond is filled, the vegetation advances over it. [After Raup, 1938.]

their ranges, species are sensitive to differences in the kind of ground, and, in many localities, are "indicator plants."

As a result of the sensitivity of species to ground conditions at the limit of their range, the boundaries between the vegetation of the different climatic zones are surprisingly sharp at many places. One example is the sharp boundary along the Fall Line near Washington and Baltimore. Mixed oak and pine forest grows on the Coastal Plain and central hardwood forest on the Piedmont Province. Several species on the Piedmont Province, among them white pine (*Pinus strobus*), do not cross the boundary to grow on the Coastal Plain (Shreve et al., 1910), and several species on the Coastal Plain do not cross the boundary to grow on the Piedmont Province, among them loblolly pine (*Pinus taeda*), southern red oak (*Quercus digitata*), and holly (*Ilex opaca*) as a tree.

Another sharp boundary is between the woodland on the lower slopes of western mountains and the shrubland on the gravel fans around the foot of those mountains. As a consequence of the ground factor, boundaries between plant zones may be sharp even where boundaries between the climatic zones are gradational.

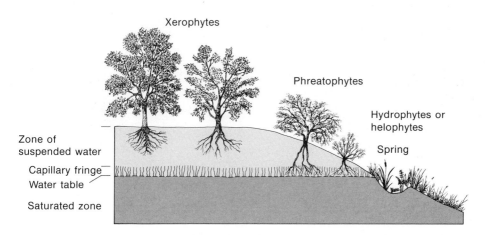

FIGURE 6.8
Plant stands differ greatly depending on the availability of water. The species of xerophytes on a particular piece of ground will depend on the frequency and severity of the droughts; the species of phreatophytes and of plants in the spring zone will depend on the salinity or alkalinity of the water. Helophytes have roots submerged in water but grow above it; hydrophytes grow wholly submerged.

GROUND MOISTURE AND PLANT GROWTH

Availability of moisture is probably the most apparent of ground conditions that are reflected by vegetation (Fig. 6.8). Plants that root above the water table (p. 235) are capable of surviving protracted periods of drought, and are called *xerophytes* (from Greek *xer*, "dry"). Ground may be dry at the surface, but the water table may be shallow enough to be reached by plant roots; plants that can reach the water table have a perennial supply. They are called *phreatophytes*. Wet ground in northern latitudes is marked by sedge, rush, and black spruce; in temperate latitudes by sycamore, birch, and willow; and in southerly latitudes by cypress. Where a water table intercepts the ground surface, forming a spring, still a third array of plants grows in the water—some with roots submerged and the foliage out of water (*helophytes*), and some wholly submerged (*hydrophytes*).

Water-Loving Plants

There are as many different kinds of stands of water-loving plants as there are modes of occurrence of water on and in the ground (Fig. 6.8). Perched water tables, like the examples in Figure 3.7, may be ephemeral. Plants growing in environments with ephemeral water tables differ from those growing where water tables are perennial, and differ from the surrounding xerophytes, which exist without the benefit of even an ephemeral water table. If in one area the recharge of an ephemeral water table is augmented by runoff from nearby ground, but in another area is dependent on infiltration of direct rainfall, the chances are that the floras of these two areas will be different. Recharge is minimized on ground that slopes enough to favor runoff of direct rainfall.

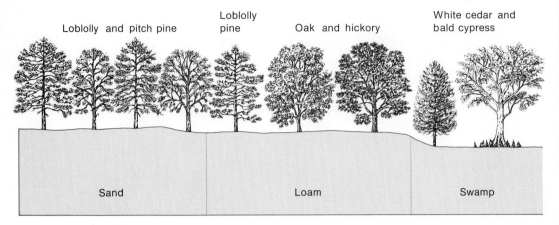

FIGURE 6.9
Diagrammatic transect showing relationship between some kinds of plant stands and different kinds of ground in the Coastal Plain of Maryland.

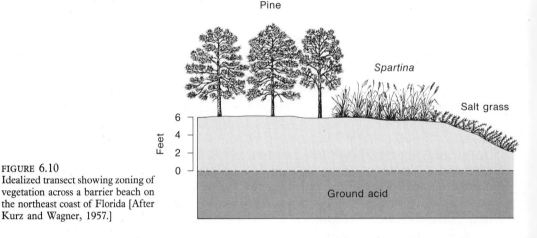

FIGURE 6.10
Idealized transect showing zoning of vegetation across a barrier beach on the northeast coast of Florida [After Kurz and Wagner, 1957.]

An impermeable layer perching ground water may not be horizontal but may dip so that the ground water moves deeply into the ground beyond the depth of plant roots. If the impermeable layer crops out and discharges its perched water into colluvium, its trace may be marked by a stand of plants that differ from those in the drier ground uphill and downhill.

Water-loving plants can be useful not only for indicating relative availability of ground water (both the depth and the dependability of the supply), but also can be useful for indicating quality of the water. Figure 6.9 illustrates changes in vegetation that reflect changes in availability and salinity of ground water on the Coastal Plain of Maryland. At the edge of the sea is beach sand that is weathered by alternate wetting and drying in salt water and abraded by frequent washing. Back of the beach is dune sand, in part stabilized by beach grasses. The dunes even support some scrub oaks. The dune sand is reddened and partly cemented by iron oxides, clay, and possibly salts. Back of the beach is an estuary where peat accumulates on a mud

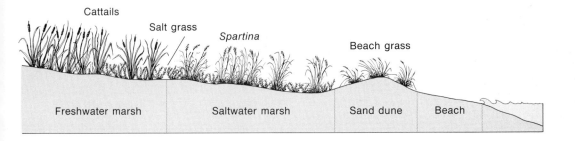

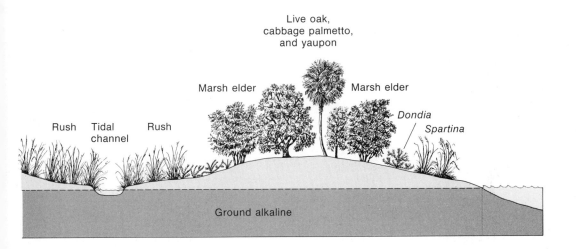

bottom; the kind of peat that forms is partly a function of the salinity of the water. On higher ground, where the water is not salty, there is cypress swamp. At these places the peat is woody, not fibrous like that formed where cattails and grasses grow. On the upland is beach sand deposited during a former high stand of sea level, and bordered by loam deposited in former estuaries. Pines grow on the sandy ground, mixed oak and pine on the loam.

In Florida, studies of beaches and related deposits and their vegetation have disclosed that the kind of vegetation depends partly on the age of the particular piece of ground and partly on the availability and salinity of the water in the ground (Fig. 6.10). A low barrier beach that is subject to washing by storm waves is practically without vegetation. Beaches higher than the reach of the waves become stabilized, or at least partly so, by a succession of plants. Seaward are perennials that can tolerate salt spray and that spread laterally. This spreading must be rapid enough to keep the plant above the drifting sand.

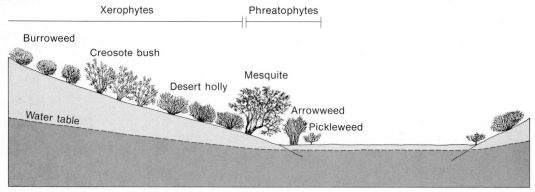

A

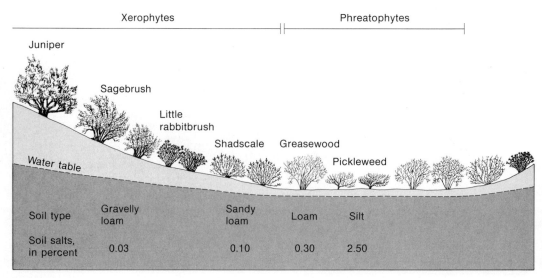

B

FIGURE 6.11

Transect across Death Valley (A) and across a playa in southwest Utah (B) showing the orderly zoning of the vegetation, which reflects availability and quality of water for plant growth. The center of Death Valley is too saline for flowering plants. In desert basins where ground water is deep, the xerophytes extend to the edge of the playa. [Part A after Hunt, 1966b; part B after Pieneisel, 1940.]

Inland from the barrier beach is a lagoon. The mainland beyond supports a still different kind of vegetation, grading to the southeast pine forests. The ground under the forest is older than on the barrier beach and is thoroughly leached. It is 98 percent insoluble matter and strongly acid. The ground at the barrier beach contains considerable shell matter, about 15 percent, and the limy ground is alkaline. The flora on the comparatively lime-rich, barrier beaches is related to the tropical floras; many or most of the species on the old ground on the mainland have northerly distributions.

In the Sea Islands Section of Georgia and South Carolina, the sea has advanced onto the land and, in places, has killed parts of the oak forest. It is presumed that this encroachment of salt water is due to downwarping, although shore erosion and eustatic rise of sea level may be contributing to it.

Even the deserts have water-loving plants—at the oases. These plants are zoned in an orderly way with respect to the quantity and kinds of salts in the desert water (Fig. 6.11).

In Death Valley honey mesquite (*Prosopis juliflora*) grows out to about the 0.5 percent brine line; beyond this is arrowweed (*Pluchea sericea*), which grows to about the 3 percent brine line—approximately the salt content of sea water. Beyond the arrowweed is pickleweed (*Allenrolfea occidentalis*), which can survive with its roots immersed in water containing about 6 percent salt—about twice the salinity of sea water. This is the maximum salt tolerance for flowering plants. With greater concentration even the algae, fungi, and bacteria are reduced in numbers. Zoning of plants according to differences in the salinity of ground water is common around the saline playas in the Great Basin—for example, around Great Salt Lake.

Agricultural crops also vary in salt tolerance. Some of the more salt- and alkali-tolerant crops are cotton, barley, tomatoes, and beets. Some of the less tolerant ones are peaches, oats, beans, peas, and sweet clover.

Drought-Resistant Plants

Even xerophytes indicate differences in the availability of moisture in desert ground. Given a certain amount of rainfall, the moisture available for plant growth on a particular piece of ground is determined partly by the runoff rate on the ground surface, by inhomogeneities in the subsurface that cause differences in permeability, and by position of the site relative to surrounding catchment areas that supply runoff to it (Figs. 6.12, 6.13). These relationships are most clearly revealed in arid regions, but they prevail also in humid areas wherever plant growth depends on suspended water.

In deserts, the availability of water to particular sites is largely controlled by runoff; in humid regions the different kinds of ground that are reflected in different kinds of plant stands may reflect soil fabric and soil structure, toughness of the ground for root penetration, and perhaps the distribution of nutritional elements. Whatever the causes, differences in kind of ground, whether in humid or dry regions, commonly are marked by differences in plant stands.

In the deserts in the Southwest, hillsides have an excessive rate of runoff and may be without vegetation; adjoining washes, which collect the runoff, may have considerable vegetation. Washes on the hillsides may have a sparse growth of vegetation that dies off during a succession of dry years when the washes do not collect sufficient runoff to maintain the plants.

In the Colorado Plateau the deserts on Cretaceous shale are marked by mat saltbush (*Atriplex corrugata*). Most of the ground is bare; the shrubs are concentrated along rills and other places where the ground moisture is increased because of runoff from neighboring areas. When there is a succession of wet years, curly grass (*Hilaria jamesii*) and some species of *Eriogonum* may grow with the mat saltbush, but these

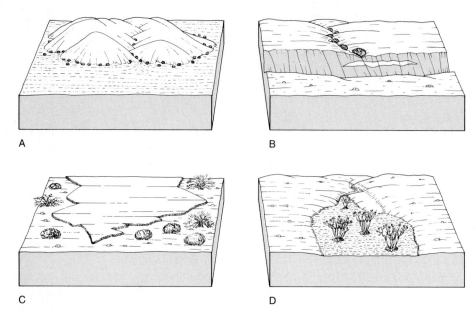

FIGURE 6.12

Four common kinds of inhomogeneous ground conditions affecting occurrence of water in the ground and plants on gravel fans in the southern Great Basin. (A) Hills surrounded by fan gravels are mantled by a shallow layer of loose detritus. Runoff is high, partly because of the slope of the hills; detritus collected at the break in slope forms permeable ground that receives and holds runoff from the hills. A ring of shrubs—desert holly in the example—grows in this detritus. (B) Old gravel fans have caliche layers firmly cemented with calcium carbonate or other salts (white lenticular layer). The caliche layers are impermeable and serve to perch an ephemeral water table. (C) Old gravel fans are mantled by smooth desert pavement, which has a high rate of runoff. The supply of water in such ground is low. The younger gravels lack the smooth pavement and have a low rate of runoff and a high rate of infiltration. In these gravels the water is recharged by runoff from the areas of desert pavement as well as by water originating on the young gravel surface. The permeable gravels support stands of perennial shrubs; the pavement areas are bare or have stands of a different and drier type of shrub than do the young gravels. (D) Some fans that are mantled by desert pavement have shallow depressed areas, a few feet wide and a few inches deep, that contain a high percentage of fine-grained sediments. These depressions, at the lower end of shallow washes, collect runoff from the pavement areas and, because of their permeability, hold the water better than the surrounding gravels. In the example sketched, the depression has a stand of deserttrumpet (*Eriogonum inflatum*). [After Hunt, 1966b.]

cannot survive an ensuing succession of dry years, and they die off. Their presence or absence with the mat saltbush is an indicator of the recent climatic trend.

Sandy deserts on the Colorado Plateau are marked by shadscale (*Atriplex confertifolia*), blackbrush (*Coleogyne ramossissima*), sand sagebrush (*Artemesia filifolia*), and winterfat (*Eurotia lanata*). In general, shadscale grows on sandy loam, blackbrush on loamy sand, and sand sagebrush and winterfat on ground sandy enough to develop dunes. Ephemeral, perched water tables in these areas are marked by stands of a scrub oak. These differences in stands of xerophytes are useful for distinguishing between areas with deep, modern dune sand and those where the modern dune sand thinly covers weathered early Holocene dune sand (Fig. 6.18).

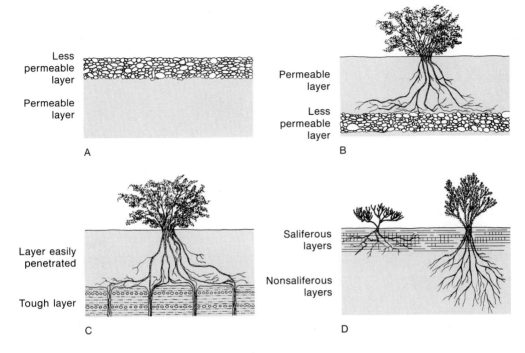

FIGURE 6.13

Inhomogeneities in the ground affect plant occurrence and depth, lateral spread, and general pattern of root systems. (A) An impermeable layer at the surface, such as well-formed desert pavement, causes runoff; such surfaces are bare. (B) A buried permeable layer collects water in a perched water table, where lateral spreading of roots is favored. (C) Buried tough layers, like caliche, may be broken by fissures that serve as channels for water seeping into the ground. Root systems in such ground have a rectangular habit; lateral root branches spread on top of the tough layer, and vertical ones extend downward along the fissures. (D) Saliferous layers alternating with nonsaliferous ones within the capillary fringe. Such ground commonly supports a mixed stand of phreatophytes with different salt tolerences. Roots of the salt-tolerant plants spread in the saliferous layers; roots of those that are not salt tolerant spread in the less saliferous layers. [After Hunt, 1966.]

In the western part of the Great Basin, Billings (1950) found that the vegetation growing on the chemically altered rocks is different from that growing on the same rocks where they are unaltered. In some western mountains, at the altitudes of the yellow pine forests (ca. 7,000 to 8,500 feet), highly xeric ground, such as thin, rocky colluvium on southern exposures, may be marked by stands of mountain mahogany (*Cercocarpus ledifolius*) or serviceberry (*Amelanchier alnifolia*) rather than pine. Damp ground with deep colluvium containing much organic matter commonly supports stands of aspen.

The examples cited have stressed differences in vegetation attributable to differences in availability of moisture, but the nutritional demands of different species also affect their distribution. This brings us to consider what are called accumulator plants.

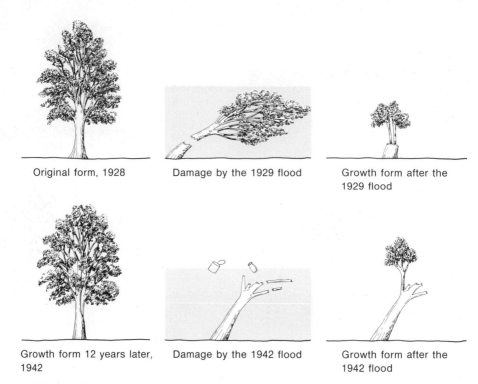

Original form, 1928

Damage by the 1929 flood

Growth form after the 1929 flood

Growth form 12 years later, 1942

Damage by the 1942 flood

Growth form after the 1942 flood

FIGURE 6.14
Four stages in the growth of a tree damaged by floods of the Potomac River. [After Sigafoos, 1964.]

ACCUMULATOR PLANTS

Some species of plants accumulate certain elements and can be useful for indicating the composition of the ground. In Death Valley, pickleweed accumulates sodium chloride, and arrowweed accumulates the sulfate of sodium where sodium is dominant and the sulfate of calcium where calcium is dominant. Plants that accumulate metals have been used for geochemical prospecting (Carlisle and Cleveland, 1958). During the uranium rush on the Colorado Plateau in the 1950's, botanical prospecting was successfully used to locate some buried uranium deposits (Cannon, 1960).

Some elements may be accumulated by plants in sufficient quantities to be toxic to range animals. The accumulation of selenium by locoweed is a notable example. Other elements may be toxic to certain plant species—for example, nickel and chromium in some serpentine ground. Some minerals favor plant growth, like the potassium-rich mineral glauconite, which occurs in certain formations along the Coastal Plain. Where those formations crop out, there are good farms and forests, but only scrubby oak and pine grow on neighboring formations that lack the glauconite.

FIGURE 6.15

Sketch of a flood-trained and buried cottonwood sapling that has given rise to a large clump of young trees. This example is representative of the process of flood-training as it occurs on the Little Missouri River. Downstream is to the right. [After Everitt, 1968.]

FLOODS AND FIRES

Trees along river bottoms that are subject to flooding become deformed by the flood waters, and their scars or deformed growth record both the date and height of the floods. Figure 6.14 illustrates successive stages in growth and deformation of an ash tree as a result of damage caused by four major floods on the Potomac River above Washington, D.C. (Sigafoos, 1964). Similar changes in tree growth due to flooding have been described along the Little Missouri River in western North Dakota (Fig. 6.15) (Everitt, 1968).

Fires have marked effects on the biota. On grassland or under forests *surface fires* burn only the litter and duff of the surface layers and the shrubs and small trees. These sometimes develop into *ground fires* and extend into the ground by burning roots. Ground fires may spread widely and be unnoticed. Some forest fires affect only the low-growing vegetation, but *crown fires* burn the tree tops. Destruction usually is not complete unless the three kinds of fire occur together.

Fires naturally tend to spread fastest down wind, more slowly laterally, and slowest toward the wind. In mountainous country, however, the heated air creates an uphill draft so that fires in forested mountains tend to move uphill even against the prevailing wind. At night, the litter on forest ground generally becomes damp, and this greatly slows the spread of surface fires; very early morning is the most favorable time of day to fight forest fires.

The hazards of grass and forest fires are greater in the semiarid western United States than in the humid East. Electric storms are much more frequent in the east than in the west, but the eastern storms generally are accompanied by rain, which protects the ground against many serious fires. In the western United States, storms are fewer but many are dry electric storms; when lightning strikes, a whole forest may become engulfed in fire. Seasonal distribution of precipitation is a factor, too.

In California, for example, most precipitation comes during winter; the dry summers are seasons of grave fire hazard.

Rains following a burn may cause severe erosion, and at least greatly muddy the streams draining the burned area. Smoky the Bear reminds us that we can contribute to minimizing grass and forest fires by rather simple precautions—putting out one's cigarette, match, or camp fire. But fires caused by lightning have always been a factor affecting the biota, and always will. The fact that the number of species in western forests is small compared to the number in eastern forests may be due in part to the role played by fire in restricting the plant species to those that can seed quickly. If the forest could have evolved unmolested by fire perhaps the species there would be much more numerous.

Next we look at the ground under the plant cover.

BIBLIOGRAPHY

Billings, W. C., 1950, Vegetation and plant growth as affected by chemically altered rocks in the western Great Basin: Ecology, v. 31, no. 1, pp. 62–74.

Braun, E. L., 1955, The phytogeography of unglaciated eastern United States and its interpretation: Bot. Rev., v. 21, pp. 297–375.

Cannon, H. L., 1960, The development of botanical methods of prospecting for uranium on the Colorado Plateau: U.S. Geol. Survey Bull. 1085-A, pp. 1–50.

Carlisle, D., and Cleveland, G. B., 1958, Plants as a guide to mineralization: Calif. Div. of Mines Spec. Rept. 50, 31 pp.

Chaney, R. W., 1956, The ancient forests of Oregon: Condon Lect. Publ., Univ. Oregon Press, 56 pp.

Cloudsley-Thompson, J. L., 1967, Microecology: St. Martin's Press, New York, 64 pp.

Deevey, E. S., Jr., 1949, Biogeography of the Pleistocene: Geol. Soc. America Bull., v. 60, p. 1315–1416.

Everitt, B. L., 1968, Use of the cottonwood in an investigation of the recent history of a flood plain: Am. Jour. Sci., v. 266, pp. 417–439.

Flowers, S., 1934, Vegetation of the Great Salt Lake region: Bot. Gaz., v. 95, pp. 353–418.

Gilman, J. C., 1945, A manual of soil fungi: The Collegiate Press, Ames, Iowa.

Gleason, H. A., and Cronquist, A., 1964, The natural geography of plants: Columbia Univ. Press, 420 pp.

Goodlett, J. C., 1956, *in* Denny, C. S., 1956, Surficial geology and geomorphology of Potter County, Pennsylvania: U.S. Geological Survey Prof. Paper 288, 72 pp.

Griggs, R. F., 1914, Observations on the behavior of some species on the edges of their ranges: Bull. Torrey Botanical Club. v. 41, pp. 25–49.

Hale, M. E., Jr., 1961, Lichen handbook: Smithsonian Institute, Washington, D.C., 178 pp.

Hansen, H. P., 1947, Postglacial forest succession, climate and chronology in the Pacific Northwest: Am. Philos. Soc. Trans., v. 37, pt. 1, 130 pp.

Hunt, Chas. B., 1966, Plant ecology of Death Valley, California: U.S. Geol. Survey Prof. Paper 509, 66 pp.

Jackson, R. M., and Ray, F., 1966, Life in the soil: Studies in Biology no. 2, Edw. Arnold, London.

Kevan, D. K. McE., 1962, Soil animals: Philosophical Library, New York, 237 pp.

Kurz, H., and Wagner, K., 1957, Tidal marshes of the Gulf and Atlantic Coasts of northern Florida and Charleston, South Carolina: Florida State Univ. Studies 24, 168 pp.

Lovering, T. S., 1958, Accumulator plants and rock weathering: Science, v. 128, pp. 416–417.

Lunt, H. A., 1948, Forest soils of Connecticut: Conn. Agricultural Exper. Sta. Bull. 523.

Lyford, W. H., 1963, Importance of ants to Brown Podzolic Soil genesis in New England: Harvard Forest Paper No. 7, 18 pp.

Martin, P. S., 1958, Pleistocene ecology and biogeography of North America: Zoogeography, Am. Assoc. Adv. Sci., Washington, D.C., pp. 375–420.

Meinzer, O. E., 1927, Plants as indicators of ground water: U.S. Geol. Survey Water Supply Paper 577, 95 pp.

Oosting, H. J., 1956, The study of plant communities (2nd ed.): W. H. Freeman and Company, San Francisco, 440 pp.

Raup, H. M., 1938, Botanical studies in the Black Rock forest: The Black Rock Forest Bulletin, No. 7, Cornwall-on-the-Hudson, 161 pp.

Robinson, T. W., 1958, Phreatophytes: U.S. Geol. Survey Water Supply Paper 1423, 84 pp.

Shantz, H. L., and Piemeisel, R. L., 1940, Types of vegetation in Escalante Valley, Utah, as indicators of soil conditions: U.S. Dept. Agriculture Tech. Bull. 713, 46 pp.

Shreve, F., 1951, Vegetation of the Sonoran desert: Carnegie Inst. Wash, Publ. 591, 187 pp.

Shreve, F., and others, 1910. The plant life of Maryland: Maryland Weather Service, v. 3, Johns Hopkins Press, Baltimore, Md., 533 pp.

Sigafoos, R. S., 1964, Botanical evidence of floods and flood-plain deposition: U.S. Geol. Survey Prof. Paper 485-A, 35 pp.

Thorp, J., 1949, Effects of certain animals that live in soils: Sci. Monthly, v. 68, pp. 180–191.

Waksman, S. A., 1952, Soil Microbiology: Wiley, New York, 356 pp.

Warming, E., 1909, Oecology of plants: Oxford, Clarendon Press, 422 pp.

Surface deposits, the parent material for modern soils, are of many kinds, depending on whether they originated in cold-wet, cold-dry, warm-wet, or warm-dry environments. Some are products of the Pleistocene glaciations; others formed in glaciated regions. The Upper Mississippi Valley, for example, is blanketed with glacial deposits; where the drainage was dammed by the ice there were once vast lakes that collected quiet water deposits. Farther south, along the Ohio, Missouri, and Mississippi Rivers, the meltwaters overflowing from the lakes left alluvial floodplain deposits. As each flood ebbed, these became bare mud flats, and dust blown eastward from them covered the uplands with loess. These surface deposits of loess, alluvium, lake beds, and glacial drift differ considerably from one another but are obviously closely related. Southwestern deserts have similarly related, but still different, suites of surface deposits. One kind, illustrated here, is desert pavement, an amazingly smooth pavement of closely packed stones, a common feature on gravel fans in that region. The stones originally were rounded stream gravels, but they have lain there so long that they are disintegrating. [Photograph by John Stacy, U.S.G.S.]

7 / Surface Deposits
PARENT MATERIALS OF MODERN SOILS

Distribution, kinds, ages; transported deposits, glacial deposits, boulder trains and erratics, drift, till, stratigraphy; lake deposits, kinds, Lake Bonneville, Death Valley; eolian deposits, loess and dunes and their stratigraphy; alluvial deposits, floodplains and terraces, stratigraphy; colluvium and related gravity deposits, creep, talus, landslides, avalanches, mudflows; shore deposits; volcanic deposits, importance of volcanic ash; residual deposits; saprolite, ancient red soils on sedimentary formations, bauxite deposits; caliche, evaporites; cave deposits; desert varnish; peat; clinker; bibliography.

DISTRIBUTION

Despite the fact that surface deposits constitute the most valuable layer of the earth, the United States has no satisfactory map of them. Figure 7.1, which first appeared in *Physiography of the United States,* is an attempt at such a map of the whole country. The United States Geological Survey has published a magnificent, highly useful, and informative map of the bedrock geology, but has neglected the ground around us. This ground needs more attention, especially with the increased crowding on it. It may be hoped that publication of Figure 7.1, with its evident shortcomings, will provoke the Geological Survey into assembling and publishing a map of the surface deposits that is as fine as their map of the bedrock geology.

Surface deposits are of many kinds—many more kinds than are indicated on the map, as this chapter will attempt to show. Most of the deposits are transported—by water, wind, or ice; some are sedentary. The deposits also are of many different ages.

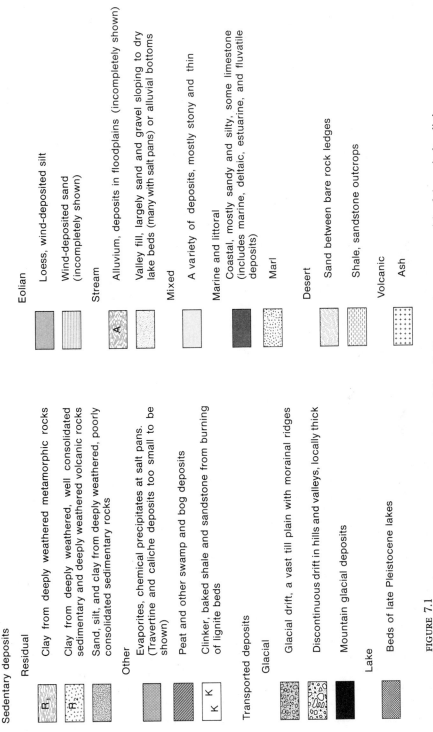

Sedentary deposits

Residual

Clay from deeply weathered metamorphic rocks

Clay from deeply weathered, well consolidated sedimentary and deeply weathered volcanic rocks

Sand, silt, and clay from deeply weathered, poorly consolidated sedimentary rocks

Other

Evaporites, chemical precipitates at salt pans. (Travertine and caliche deposits too small to be shown)

Peat and other swamp and bog deposits

Clinker, baked shale and sandstone from burning of lignite beds

Transported deposits

Glacial

Glacial drift, a vast till plain with morainal ridges

Discontinuous drift in hills and valleys, locally thick

Mountain glacial deposits

Lake

Beds of late Pleistocene lakes

Eolian

Loess, wind-deposited silt

Wind-deposited sand (incompletely shown)

Stream

Alluvium, deposits in floodplains (incompletely shown)

Valley fill, largely sand and gravel sloping to dry lake beds (many with salt pans) or alluvial bottoms

Mixed

A variety of deposits, mostly stony and thin

Marine and littoral

Coastal, mostly sandy and silty, some limestone (includes marine, deltaic, estuarine, and fluvatile deposits)

Marl

Desert

Sand between bare rock ledges

Shale, sandstone outcrops

Volcanic

Ash

FIGURE 7.1
Surface deposits in the United States, except Alaska and Hawaii. These are the parent materials of the agriculturalist's soils; most are late Pleistocene or Holocene, and ages overlap. [After *Physiography of the United States* by Chas. B. Hunt. W. H. Freeman and Company. Copyright © 1967.]

Holocene deposits, which are rather limited in extent, include alluvium deposited along the floodplains of streams, beach and dune sand, and hillside deposits (*colluvium*) in hilly and mountainous terrain. In addition there are minor (in extent) Holocene deposits at marshes, lakes, and springs, and there are some lavas and volcanic ash. The most extensive surface deposits are the products of the glacial and pluvial climates of the Pleistocene Epoch. The oldest deposits are sedentary, the deep, clayey residual soils (R_1 and R_2 in Fig. 7.1). The principal residual deposits are pre-Wisconsinan; many are pre-Quaternary, and some pre-Cenozoic.

We look first at the transported deposits, keeping in mind that each kind may have had any or all of several origins. A particular deposit may consist of

1. Material eroded from unweathered bedrock and freshly deposited as such.

2. Material eroded from slightly weathered bedrock.

3. Material eroded from strongly weathered and decayed bedrock.

4. Material eroded from an older surface deposit or soil.

These different ultimate origins of the materials commonly are recorded by the distinctive textures and mineralogy of the particular deposit. Most surface deposits consist of materials eroded from older deposits (4 above); unweathered bedrock (1) has provided the least. In North America the surface deposits having the greatest extent are the glacial deposits, but in the world as a whole, the residual deposits produced by deep weathering, notably in the tropics, may be more extensive.

TRANSPORTED DEPOSITS

Glacial Deposits

Glacial deposits cover most of the continent north of Long Island and north of the Ohio and Missouri Rivers (Fig. 7.1). They are easily divisible into older and younger deposits that differ greatly in their degree of weathering and erosion. The younger deposits, those of the Wisconsinan glaciation, are almost as little weathered as when they were deposited by the ice and its meltwaters. Moreover, they are only slightly eroded and still preserve the original constructional landforms. The older deposits, those laid down during the Pre-Wisconsinan glaciations, are deeply weathered. They are much dissected by erosion, and the landforms consequently are of erosional rather than depositional origin.

Deposits of the Wisconsinan Glaciation vary greatly in composition from one part of the glaciated country to another depending on the kind of bedrock from which they were derived and on the rates of ice advance and retreat. New England hills and mountains, for example, are of hard crystalline rocks, and these rocks were the source of the large and small boulders, or *glacial erratics,* that are widely strewn over parts of New England. The boulders were used to build the stone walls so characteristic of that section of the country. The granite boulders on the north shores of Long Island were carried there by glacial ice advancing southward across New England. Similarly, in central New York, boulders of Precambrian rocks from the Adirondacks have been transported southward and deposited on Paleozoic formations.

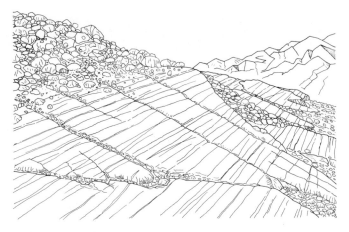

FIGURE 7.2
Glacial striae on a bare rock surface; at the left they are overlain by glacial
till. On some kinds of rock, such as quartzite, striated surfaces may even
preserve a glaze or other polish.

Plymouth Rock is one of our better known glacial erratics. Another well-known
erratic, a granite boulder near Leyden, Massachusetts, is about 15 miles south of
its probable source. It measures 10 feet in length, 6 feet in width, and $5\frac{1}{2}$ feet in
height. Even though it weighs 20 tons, it is so well balanced that it can be swayed
by hand.

In the Central United States, the glacial drift contains few boulders; there were
no mountains in the source areas to provide them. This was not true, however, of
the Superior Upland: Precambrian rocks were transported southward from there and
deposited with the glacial debris that rests on Paleozoic formations. In Montana,
southward advance of the ice across the Missouri Plateau is recorded by stones that
were plucked from the intrusive laccolithic mountains (Sweetgrass Hills, Bearpaw
Mountains, Little Rocky Mountains), transported southward, and deposited in trains
extending southward from the mountains. Such trains show the direction of ice
advance; so too do the glacial striae (Fig. 7.2).

Not only is the glacial drift in New England characteristically stony, so also are
the modern soils developed on the drift. The composition of the drift and of the
soils varies greatly depending on the bedrock overridden by the ice. Along the
Connecticut River valley, for example, the drift is reddened with debris from the
Triassic formations, whereas the drift on and composed of the crystalline rocks east
and west of the valley has gray tones.

Another example is in northern Pennsylvania and western New York, where the
ice advanced across many miles of shale, sandstone, and limestone. The glacial drift
in that area is far less stony or bouldery than in New England. Moreover, south
of Lake Ontario the variations in the amount of calcareous matter in the till reflects
the pattern of the bedrock formations across which the ice advanced (Fig. 7.3).

Along the coast of New England, the glacial deposits are interlayered with marine
deposits that in places are now about 100 feet above sea level (Sammel, 1963). Until
very recently, the rise of the coast there has progressed more rapidly than the eustatic
rise of sea level, and both the marine and glacial deposits have been uplifted. In

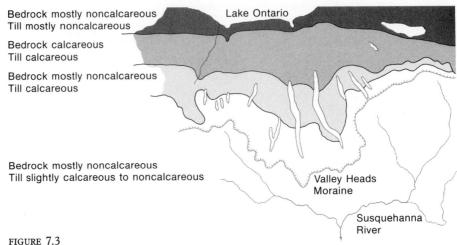

Bedrock mostly noncalcareous
Till mostly noncalcareous

Bedrock calcareous
Till calcareous

Bedrock mostly noncalcareous
Till calcareous

Bedrock mostly noncalcareous
Till slightly calcareous to noncalcareous

Lake Ontario

Valley Heads
Moraine

Susquehanna
River

FIGURE 7.3
Distribution of calcium carbonate in Wisconsinan till and in bedrock in south-central New York. In the belt where bedrock and till are calcareous, half the till fragments are carbonate rocks. South of this is a belt where the bedrock is mostly noncalcareous, but the till, derived from the north, contains up to 30 percent calcareous matter in the matrix. Farther south the till generally contains less than 5 percent calcareous matter. [After Denny and Lyford.]

the past few thousand years, however, sea level has risen faster than the land. The evidence for this is found in submerged archeological sites, such as the ancient fishweir that now lies beneath 16 feet of water at Boston, and a lithic site in the Taunton River, Massachusetts.

In places the ice retreated gradually by a process called stagnation. That is, ice that had filled a valley might melt at the sides and become surrounded by areas from which ice had melted. The size, shapes, and spacing of these open areas largely controlled the kind of deposits collected there. Some of them contained lakes; others were drained by streams of glacial meltwaters. The hilly New England country favored ice stagnation, whereas the more open country of the central United States favored retreat of the fronts of the lobes of ice. The central United States, therefore, is characterized by end moraines (Fig. 4.13).

The mantle of glacial deposits, collectively known as *glacial drift*, includes some deposited by the ice itself (unstratified—that is, without bedding) and some deposited by glacial streams (stratified). Unstratified drift (till) is a chaotic, unsorted mixture of boulders, cobbles, and smaller stones in sand and clay. Ground moraines (Fig. 4.9), which are irregular blanket deposits, usually have compositions closely reflecting that of the bedrock formations, indicating that most of the material has not moved very far. Till also occurs as terminal or end moraines along former ice fronts, where sand, gravel, and boulders were pushed into irregular heaps by the advancing ice and where stratified drift, deposited in fans fronting the melting ice, collapsed when the ice front melted.

The sequence of principal deposits laid down by the continental ice sheets in different parts of the United States and their probable correlations are given in Table 7.1. The indicated correlations between the regions are uncertain, and there are additional uncertainties within the eastern states. For example, the Harbor Hill

TABLE 7.1
Deposits of continental glaciers in the United States

	Western United States		Central United States	Eastern United States	
	Coast of Washington	Columbia Plateau		West Pennsylvania and New York	New England, Long Island, New Jersey
Wisconsin — Upper Wisconsin	Vashon drift	Okanogan till	Valders drift Mankato drift Cary drift	Hamburg and Marilla drifts Valley Heads drift	Lexington till; Buzzards Bay and Sandwich moraines in New England; Harbor Hill, Ronkonkoma moraines on Long Island and Nantucket
Wisconsin — Lower Wisconsin	Admiralty drift	Spokane till	Tazewell drift Iowan drift Farmdale drift	Binghampton drift Olean drift	
			Illinoian drift		
			Kansan drift	Pre-Olean drift	
			Nebraskan drift		Jameco formation and underlying Mannetto gravel on Long Island; Jersey drift in New Jersey ?

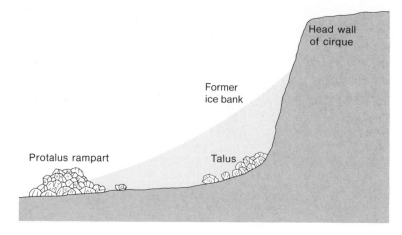

FIGURE 7.4
Topographic profile of the head wall and floor of a glacial cirque, illustrating formation of a protalus rampart. The rampart of boulders and cobbles was built at a time when perrenial snow or ice was contained in the cirque and blocks of talus from the head wall rolled down the bank.

moraine on Long Island (Fig. 4.6, Table 7.1) has been correlated with the Olean and classed as Early Wisconsinan; it has also been correlated with the Valley Heads moraine and classed as late Wisconsinan.

In the western United States a bewildering array of local names has been applied to the deposits of the mountain glaciers. Almost every mountain range has its own set of names. Some mountain valleys are reported to show many stages of glaciation, whereas other valleys are reported to show only a few; at first, this appears to be the result of badly confused stratigraphy. The confusion, however, arises because of lack of detailed knowledge of the substages and not of the major stages, for the general relationships between the major stages are reasonably clear. Generally the four major units can be distinguished, and there is a minimum of disagreement about their correlation. The stages and the most used names are given in Table 7.2. Being glacial deposits, they contain almost no fossil remains, and the correlations are based on the physical geology, especially on the chronology of the outwash as recorded by overlap of successive deposits. Intermountain correlations are based in part on differences in the degree of weathering of the deposits.

Deposits of pre-Wisconsinan age generally are weathered quite differently from those that are younger because the near surface layers of pre-Wisconsinan deposits characteristically contain a high percentage of decomposed, rotten stones. Moreover, these old deposits are much eroded and retain little of their original morainic or other form. A dozen local names have been used for these deposits, but they all have been assigned to the pre-Wisconsinan and are so referred to in this book.

There is much disagreement about which stages of the early and late Wisconsinan glaciation are represented in the deposits of the mountain glaciers, but this need not cloud the fact that in most places two major sets of deposits can generally be distinguished. The older deposits, correlated with the early Wisconsinan, commonly contain a high percentage of firm stone having a weathered rind; the younger deposits,

TABLE 7.2
Names most generally used for designating glacial deposits in the mountains of the western United States, their correlation with the deposits in the Middle West, and the nomenclature used in this book

		Middle west	Rocky Mountains	Sierra Nevada	Nomenclature used in this book
Holocene		Cochrane	Historic Temple Lake	17th–19th centuries Little Ice Age	Historic Middle Holocene
Pleistocene	Wisconsinan	Valders Mankato Cary	Pinedale	Tioga	Late Wisconsinan
		Tazewell Iowan Farmdale	Bull Lake	Tahoe	Early Wisconsinan
		Illinoian Kansan Nebraskan	Buffalo	El Portal (Sherwin) Aeolian Buttes	Pre-Wisconsinan

correlated with the late Wisconsinan, commonly contain fresh rock with little more than a surface stain of iron and manganese oxide (desert varnish; see p. 158). Rather than attempting to use the more than 30 local names that have been applied to these deposits, they are grouped in this book as either early or late Wisconsinan (Table 7.2).

Two stages of Holocene glacial deposits can generally be distinguished in and close to the glacial cirques. The younger, mostly bouldery deposits located within the cirques (Fig. 7.4) are evidently no more than a few hundred years old, for the boulders are mostly free of any surface stain.

In some places, the older Holocene deposits are morainic—that is, they were formed by ice shove, and may be located outside the cirque. Their surfaces are darkly stained with desert varnish, but the boulders are fresh rock, like the late Wisconsinan deposits. Physiographic position also helps separate the earlier Holocene deposits from the late Wisconsinan ones, which are situated beyond the Holocene deposits.

Lake deposits

Lakes and ponds and the sediments laid down in them are second only to glacial drift as characteristic features of glaciated regions. The northeastern and north-central parts of the United States abound with lakes and ponds, including the Great Lakes, and these are only the southernmost parts of a vast expanse of lake country that extends northward from there along both sides of Hudson Bay. During Pleistocene time the lakes were greater in size and in number than they are today, and they were of many kinds, depending on whether they formed

1) in front of the continental ice sheets as a result of rivers being dammed by ice;

2) in basins and river valleys overdeepened as a result of scour by the ice;

3) on ground moraine;

4) in a structural basin, like the more than 100 closed basins in the Basin and Range Province;

5) in sinkholes where formations are readily soluble (for example, limestone or gypsiferous ground), as in Florida;

6) along rivers at cut-off meanders or depressions in deltas;

7) back of landslides;

8) back of lava flows or in volcanic craters; or

9) in depressions that are in part due to wind erosion, as on the Great Plains.

The now-dry beds of the extensive lakes that formed in front of the continental ice sheets, especially in the Great Lakes region, are composed mostly of fine-textured sediments, the lake bottom muds. In addition, the ground is flat, and consequently the drainage is very poor. Swamps are numerous, and many former swamps are now represented by deposits of peat. Such ground can pose difficult foundation and other engineering and sanitary problems.

In the hillier New England region, lakes were numerous around the stagnating blocks of valley ice, but these lakes were small. Around the sides were deposited coarse sediments, deltaic sand and gravel forming kames, kame terraces, and other outwash (Fig. 4.9). The lake bottoms collected muds. This ground is poorly drained and may contain small lakes, ponds, or swamps.

Curiously, lake deposits are as characteristic of the deserts in the Great Basin as they are of the glaciated parts of the country. Death Valley (Fig. 4.19) and the hundred or more other desert basins in that arid region were produced by folding and faulting. Lakes and ponds are numerous there at the present time, but they were vastly more numerous and much larger during the Pleistocene (Fig. 7.5). Many of the Pleistocene lake basins are now dry, and their beds are playas, which are usually dry but subject to flooding in wet periods.

Largest of the Pleistocene lakes in the Great Basin was Lake Bonneville (Figs. 7.5, 7.6) in northwestern Utah (Gilbert, 1890). Great Salt Lake, Utah Lake, and Sevier Lake are remnants of it. Lake Bonneville was 1,000 feet deep and covered 20,000 square miles. It was fed by meltwaters from the Wisconsinan glaciers and snowfields in the Middle Rocky Mountains; the low rate of evaporation that prevailed under the cool climate of the Pleistocene Epoch probably contributed to maintenance of the lake. Sediments more than 100 feet thick were deposited in Lake Bonneville by the streams draining into it from the Rocky Mountains. The deposits include

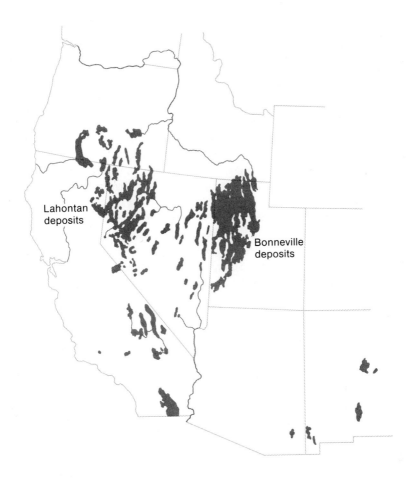

Lahontan deposits

Bonneville deposits

FIGURE 7.5
Map of the Basin and Range Province, showing Pleistocene lake beds (for a list of these ancient lakes and a brief discussion of them see Geol. Soc. America Bull., vol. 33, pp. 541-552, 1922). [After Meinzer, 1923a.]

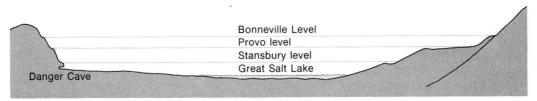

Bonneville Level
Provo level
Stansbury level
Great Salt Lake
Danger Cave

FIGURE 7.6
Cross section of Lake Bonneville deposits at Great Salt Lake. Danger Cave lies only 50 feet above lake level.

deltas and embankments deposits at the shorelines, gravel bars, sand bars, spits, and fine-grained lake bottom sediments. That the lake was contemporaneous with the Wisconsinan glaciation in the bordering mountains is shown by the interbedding of the shoreline deposits with glacial outwash.

Growth of the lake ended when it overflowed its northern rim and discharged into the Snake River in southern Idaho. At Red Rock Pass, the point of overflow, a gorge was quickly eroded in valley fill to a firm ledge of bedrock 300 feet below the lake's former high level, and the lake stood at this new level for a long time. During this time, the Provo stage of the lake, extensive deltas formed at the mouths of each of the big valleys draining from the ice and snowfields. Each of the deltas covers many square miles. Urban development in Utah has largely been determined by these surface deposits, the state's largest cities—Provo, Salt Lake City, and Ogden—and many of its smaller towns are built on these deltas.

As the Pleistocene waned and the glaciers retreated, the lake level gradually fell. Remains of Pleistocene animals—elephants and camels—have been found in the nearshore deposits of the Provo Stage, but no articulated skeletons of Pleistocene animals have been found in deposits much younger. By the time the water level had fallen to within 50 feet of the present level of Great Salt Lake (about 470 feet lower than that of the Provo Stage), the Pleistocene animals had become extinct and both the fauna and flora consisted only of modern species. Deposits in Danger Cave and other caves on the western edge of the Great Salt Lake Desert (Fig. 7.6), contain the fossil remains of Holocene animals and of early men, dated by the radiocarbon method as being 11,500 years old. The cave deposits consist of gravel at the base overlain by sand that contains campfire sites and chipped flint. Above this are alternating beds of eolian dust (loess) and plant matter and some beds containing angular blocks of limestone scaled from the roof of the cave. The absence of lake deposits above the basal beds of the cave shows that the cave has been above the level of the lake since it was first occupied. The salt pan on the Great Salt Lake Desert began forming when the Bonneville flats were bared by the drying of Lake Bonneville. As the lake dried, the lessening of the load on that part of the earth's crust allowed the central part of the basin to rise and the shorelines on the former islands there are about 150 feet higher than the eastern and western shores, providing an example of crustal rebound due to unloading.

Another large ancient lake, called Lake Lahontan, developed in western Nevada; Pyramid Lake, Lake Winnemucca, and Walker Lake are remnants of it. Even Death Valley contained a lake several hundred feet deep. The salt pan there illustrates the general features of the numerous natural salt pans that formed in the Great Basin as the lakes dried.

The salt pan in Death Valley ranges from a few inches to a few feet thick. At the center, the salts in the crust and in the brines are mostly chlorides. These are surrounded by a narrow zone in which the salts are mostly sulfates, and these in turn are surrounded by a zone containing small amounts of carbonate salts. This zoning, which reflects the solubility of the salts, can be illustrated by evaporating brine in a dish. When this is done (Fig. 7.7, A), the first salts to precipitate are carbonates (c), which form at the edge of the dish and across the bottom. As the water level drops and the salinity of the brine increases, sulfates (s) are deposited.

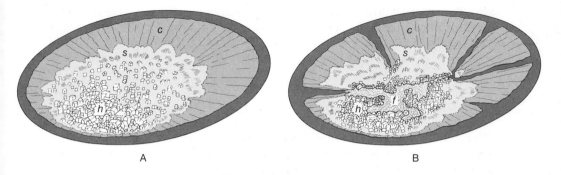

FIGURE 7.7
Diagram illustrating the chemistry and history of the salt pan in Death Valley, California. (A) The salt pan is analogous to the residue that forms in an evaporating dish. As a brine evaporates, the least soluble salts, carbonates (*c*), are the first to crystallize. As the brine becomes more concentrated, the next more soluble salts, sulfates (*s*) crystallize, and finally, when maximum concentration is reached, the chlorides (*h*) crystallize. In Death Valley and in many of the other basins of the Great Basin, notably Salt Lake, the valley floors were faulted and tilted as the salts were deposited, hence they are crowded against one edge of the pan. (B) Occasional floods of fresh water reaching the valleys today remove the salts from the flooded areas and redeposit them in secondary concentric zones as the floodwaters evaporate. [After Hunt, 1966.]

Finally, when maximum salinity is reached, chlorides (h), are deposited, principally sodium chloride (halite).

The Death Valley salt pan and the others in the Great Basin including the one in the Great Salt Lake Desert consist of such concentric zones. But these are areas of active earth movement, and the valley floors were tilted while the salts were being deposited, so the rings are crowded against one side of the basin (Fig. 7.7, A). Since the time of tilting, influxes of fresh water have reworked the salts by washing them from areas subject to flooding and redepositing them in irregular growths around those areas (Fig. 7.7, B). The tilting of the Great Salt Lake Desert and crowding of the rings westward is presumably due to the isostatic rise of the Bonneville Basin.

Variations in composition of salts reflect such factors as differences in the source and composition of waters entering a salt pan, position in the pan, whether the evaporation is from surface water or ground water, whether it is standing or flowing, and seasonal and longer-range changes in temperature, precipitation, and evaporation.

Lakes are numerous in some parts of the country that have never been glaciated. Sinks are common in limestone country; if there is much rainfall, the water table will be high and many of the sinks will become lakes, as in Florida. Along the floodplains of meandering rivers, meanders are occasionally cut off, forming *ox-bow lakes*. On big deltas, like that of the Mississippi, lakes are numerous partly because of the irregular distribution of sediment and partly because of differential compaction and settling of the deltaic muds. In mountainous country, landslides may create lakes; an example is Earthquake Lake, west of Yellowstone National Park. A landslide, triggered by the 1959 earthquake, dammed the Madison River where it cuts through the Madison Range and formed a deep lake in the canyon. Lava flows have also dammed streams and formed lakes. Harney and Malheur Lakes in southeastern Oregon formerly drained into the Snake River by way of the Malheur River, but a lava flow dammed their outlets. Lakes form in volcanic craters and calderas (faulted craters); Crater Lake in Oregon is a superb example. On the southern Great Plains wind erosion scours out depressions, locally known as buffalo wallows, and ephemeral ponds form in these depressions.

Eolian Deposits, Loess and Dunes

Much of the central United States, including most of the glaciated areas, is blanketed with loess, a deposit of silt that originated as wind-blown dust. At times of thaw, when the glaciers melted, the rivers discharged vast floods; as these ebbed, the valley bottoms were left as broad mud flats. Prevailing winds from the west moved across the barren flats, whipping up clouds of dust and depositing them on the uplands. An example is the belt of loess along the east side of the Mississippi River in Kentucky, Tennessee, and Mississippi (Fig. 7.1). In places the loess is tens of feet thick; in general, it becomes thinner and finer-grained eastward from the valleys that were the sources (Fig. 7.8). Some loess was deposited west of the rivers, but the deposits there are neither as extensive nor as thick as those to the east.

Interbedded with the loess are water-laid layers of gravels or sand that evidently record rainstorms and washings that interrupted the accumulation of dust. In the loess are abundant rodent burrows and bones of rodents, camels, and elephants. Loessial soils are among our most fertile, for they contain a rich variety of minerals. Most loess is highly calcareous, or was so before being weathered. A difficulty with loessial soils is that they are highly susceptible to erosion.

Loessial deposits in the Mississippi and Missouri river valleys average finer grained than the similar deposits of comparable age on the High Plains. In Colorado and in western Nebraska and western Kansas, the silty loess grades into dune sand. This change in grain size has been attributed to a regional difference in weathering—that is, to the greater development of clay on the moist Central Lowland than on the semiarid plains (Jenny, 1941, p. 132). But the stream deposits in the valleys, which were the source for the loessial deposits, become coarser westward toward their source in the Rocky Mountains, and this factor together with the prevalence of strong winds on the treeless uplands of the Great Plains probably accounts for most of the decrease in sand content eastward. The stratigraphy of these loess deposits and their relationship to the till sheets is illustrated in Table 7.3 (p. 140).

Loess and other eolian deposits are extensive in parts of the plateaus and basins west of the Rocky Mountains. These deposits include a high percentage of volcanic ash. A deposit of pre-Wisconsinan loess in the eastern part of the Colorado Plateau forms an arcuate belt around the west side of the San Juan Mountains (Fig. 7.1). The source of this deposit seems to have been the deserts to the southwest.

Dunes are a common feature of many deserts, and their kinds and shapes are governed by the strength, direction, and persistence of the winds, by the ability of plants to grow on them or to survive partial burial, and by the maintenance of the supply of sand. Where the winds blow predominantely from one direction, as on the Navajo Reservation and on the Great Plains in eastern Colorado, the dunes are

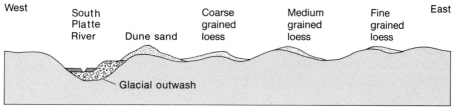

West   South   Coarse   Medium   Fine   East

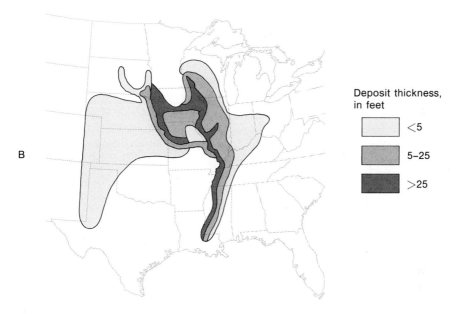

A

B

Deposit thickness, in feet

<5

5–25

>25

FIGURE 7.8

(A) Diagrammatic section illustrating how loessial and other eolian deposits thin and become finer grained and discontinuous leeward from their source. Dune sand was deposited on the bluff east of the source, the South Platte River. Farther east, silt (loess) was deposited, particularly on the eastern, lee sides of the hills. (B) Loess and other eolian deposits in the central United States. The loess thins leeward from its sources along the major river valleys that discharged outwash in Pleistocene time. Westward, toward the Rocky Mountains, the eolian deposits on the Great Plains are sandy.

crescentic in plan and convex toward the windward side (barchan dunes). They may be connected by sand trails aligned in the direction of the wind. Active dunes on the deserts may migrate off the formation that supplies the sand, and when this happens the dunes become smaller and smaller and literally blow away.

Dunes are widespread on the Great Plains in Nebraska, in northeastern Colorado, on the Colorado Plateau northeast of the Henry Mountains and on the Navajo Reservation, and in many of the basins in the Basin and Range Province. The dunes at White Sands National Monument are unusual in that they are composed of grains of gypsum.

TABLE 7.3

Relationship of Pleistocene till sheets to loess and other deposits
southward from the drift borders in the western plains states

South Dakota	Nebraska	Kansas

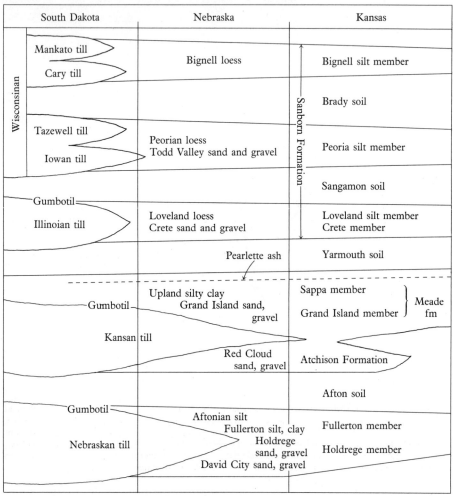

Wisconsinan

Mankato till
Cary till

Bignell loess

Bignell silt member

Sanborn Formation

Brady soil

Tazewell till
Iowan till

Peorian loess
Todd Valley sand and gravel

Peoria silt member

Sangamon soil

Gumbotil

Illinoian till

Loveland loess
Crete sand and gravel

Loveland silt member
Crete member

Pearlette ash

Yarmouth soil

Gumbotil

Upland silty clay
Grand Island sand, gravel

Sappa member

Grand Island member

Meade fm

Kansan till

Red Cloud sand, gravel

Atchison Formation

Afton soil

Gumbotil

Aftonian silt
Fullerton silt, clay
Holdrege sand, gravel
David City sand, gravel

Fullerton member

Holdrege member

Nebraskan till

Source: From Lugn (1935); Condra, Reed, and Gordon (1950); Schultz, Reed and Lugn (1951); Frye and Leonard (1952); Flint, (1971); and Wright and Frey (1965).

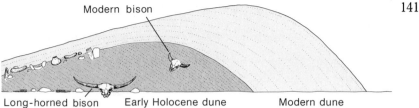

Modern bison

Long-horned bison   Early Holocene dune   Modern dune

FIGURE 7.9
Diagrammatic cross section of sand dunes on the Colorado Plateau illustrating their stratigraphy. Artifacts and sites of the late Holocene pottery-making Indians overlie the early Holocene dune, but are buried under the loose sand of the modern dune. The early Holocene dune is weathered and stabilized; in and under it are the sites and remains of pre-pottery Indians.

Very commonly in our western deserts, dunes of two ages can be distinguished. The older dunes, dating back to earliest Holocene or late Pleistocene time, are stabilized, stained reddish, and buried by the younger dunes. Many of the archeological sites that have yielded traces of early man in this country are associated with these ancient dunes. In general, the sand that makes up the modern dunes is derived from the reworking of older ones. The stratigraphic relationships are illustrated in Figure 7.9.

Loess deposits and sand dunes are widespread on the Columbia Plateau too, where the principal loess is known as Palouse Soil. The scablands of bare lava were formed where this loess was eroded from the plateau by floodwaters from Lake Missoula. A more extensive but thinner deposit of loess covers the lavas of the Snake River Plain in southern Idaho. Scattered around the Wasatch Mountains, on the lee side of the Lake Bonneville Basin, are remnants of deeply weathered loess that are overlapped by (and therefore antedate) the Lake Bonneville deposits. The loess must have been blown from the Lake Bonneville basin during a dry period in pre-Wisconsinan time.

In the northeastern United States, for example in New England, loessial deposits are common, but they are thin and not conspicuously different from the underlying glacial drift because frost-heaved cobbles and pebbles are admixed with the loess. The thinness of the deposits compared to those in the Mississippi River valley probably reflects differences in the regimen of the glacial meltwaters. In the Great Lakes region, the glacier and the lakes in front of it discharged into a few large river channels from an ice front nearly a thousand miles long; in the northeastern United States, the meltwaters discharged from stagnating ice in many small channels along a short front.

Alluvial Deposits

Alluvium is sediment transported and laid down by streams. According to the Soil Survey Staff (1951), a third of the world's food supply comes from crops grown on alluvial soils. Alluvium forms *deltas* where streams empty into lakes or seas, and forms *alluvial fans* where streams debouch from mountains onto flat country. Flood-plain alluvium is deposited when a stream overflows its banks. In general, the coarse

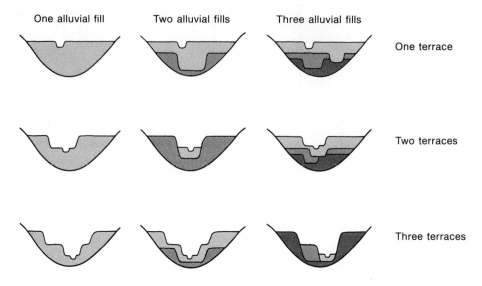

One alluvial fill   Two alluvial fills   Three alluvial fills

One terrace

Two terraces

Three terraces

FIGURE 7.10

Examples of valley cross sections, showing some common relationships between different alluvial fills and the alluvial terraces. [After Leopold and Miller, 1954.]

debris is deposited where the current is most swift, near the banks, and the finer sediments are deposited on the floodplain, away from the banks. In this way a stream may build itself a natural levee.

Because climates have changed greatly during the Pleistocene and Holocene, stream discharges have changed. During wet periods streams were bigger. Streams that discharged from Pleistocene glaciers or that were fed by snowfields, were very much larger and capable of transporting more coarse debris than are the streams of today, which are fed only by runoff from rain. As a consequence of the fluctuations in discharge, episodes of alluviation alternated with episodes when the alluvial deposits were eroded.

Figure 7.10 gives some examples of the kinds of relationships between alluvial fills and the erosion that removed parts of them. In the left column is a valley in which a single fill of alluvium was deposited, and subsequently cut into by the stream to form three erosional terraces in a single alluvial deposit. The center column illustrates possible combinations where two alluvial fills were dissected by episodes of downcutting. The right column illustrates three fills interrupted by episodes of erosion. The form of the terraces may be the same in all three examples, but the stratigraphy of the deposits and the valley histories differ greatly.

The modern agricultural soils that have developed on the terraces are also likely to be quite different. If there is only one terrace, it is likely to be composed of alluvium so young that it has not weathered noticeably since it was deposited. Where there are several terraces, the older ones are likely to have well developed soils. In order to understand the Holocene history of such a series of alluvial deposits and the history of the soils on them, the stratigraphy must be determined. A common and useful method of correlating alluvial terraces is on the basis of height above

TABLE 7.4
Generalized alluvial and dunal chronology in the western United States and their
paleontologic, archeologic, and geologic age

Geologic Age		Alluvial, dunal chronology	Paleontology	Archeology
Holocene	Late	Erosion, active dunes	Modern	Historic remains
		Alluvium (Naha Alluvium)		
		Erosion, active dunes		Prehistoric pottery
	— 1 A.D. —			
	Middle	Alluvium (Tsegi, Piney Creek, Calamity Alluviums)	fauna	Prepottery
	— 3,000 B.C. —			
	Early	Erosion, active dunes		
	— 11,000 B.C. —			
Late Wisconsinan		Alluvium (Jeditto, Neville Alluviums)	Mammoth, camel and other Pleistocene forms.	

streams. But the method has uncertainties because, depending on the valley history,
a younger terrace deposit may be either lower or higher than an older one. Along
some drainage courses, terraces may even cross one another. For example, along the
South Platte River, at Denver, Colorado, the Holocene alluvium is 30 to 40 feet
below the terrace of Late Wisconsinan gravel, but up the tributaries to the east the
difference decreases, and in a few miles the Holocene alluvium is on top of and across
the late Wisconsinan alluvium.

Another common difficulty in alluvial stratigraphy is the occurrence of young layers
on top of older ones. Such layers may be washed onto an old fill from the bordering
hillsides or be deposited by exceptionally high floods that sweep over the ancient
fill. The mapping of alluvial deposits requires identifying the unconformity between
the ancient fill and deposits overlying it.

Alluvial fills and terraces can be seen in most valleys in the semiarid parts of
western United States. Many of the deposits have been named; their chronology is
given in Table 7.4. This alluvial stratigraphy, which can be correlated with dune
activity, is based on superposition of the deposits and their paleontology and archeology.

Where there are fills of three or more ages, the oldest is generally late Pleistocene
and is likely to contain bones of such animals as mammoth and musk ox; bones
of camel and longhorn buffalo also may be found, but they are more abundant in
the loessial deposits on the uplands. The younger fills are Holocene, and can be

distinguished stratigraphically by their archeological content. In the western United States, the older of the Holocene fills predates the invention of pottery and the bow and arrow (Fig. 2.6).

The older of the Holocene alluvial deposits were laid down sometime in the period 3000 B.C. to A.D. 1, when the climate in the western United States was wetter than it is today. Lakes formed in many of the desert basins. The deposits filled arroyos that had been cut into the late Pleistocene alluvium during early Holocene time, when the climate was drier and perhaps warmer than today's—a time that has been variously named, as the altithermal by some and as the hypsithermal by others. In this book, however, it will be designated merely by age, early Holocene. The comparatively wet period during which the alluvium was deposited in the arroyos has been called the "Little Ice Age," but in this book it too will be designated by age, middle Holocene. (The past 2,000 years are latest Holocene.)

In the western United States there was another period of arroyo cutting around the twelfth and thirteenth centuries, and a final period of alluviation after the Spaniards had entered the Southwest. The present cycle of arroyo cutting started during the latter part of the nineteenth century.

In both the humid east and the arid west, but particularly in the Basin and Range Province, streams issuing from the mountains deposit gravel in great fans apexing at the canyon mouths and sloping valleyward (Fig. 4.19). In the Basin and Range Province some fans are 6 to 8 miles long and 1,500 to 2,000 feet high. Some are built outward from the bases of mountains, as illustrated in Figure 4.20. Others extend back into the mountain and bury the lower stretches of the valley. Still others fill the valley and partly bury the foothills, as at the right in Figure 4.20; such fans are commonly the result of structural tilting. The stratigraphy of gravel fans is less well known than that of alluvial deposits. Almost certainly the rate of fan development is partly a function of climate; in wet periods fan deposition is probably accelerated; in dry periods it is probably slowed. Evidence in much of the Great Basin suggests that fan deposition there has slowed greatly during the Holocene.

Colluvium and Related Gravity Deposits

Colluvium is the name applied to the mantle of loose debris that works its way down hillsides by *creep*, a slow process also called *mass-wasting*. Some creep is caused by the heaving and thawing action of frost; when ground freezes, it heaves at right angles to the sloping surface, and when it thaws the heaved part settles vertically and thus moves downslope. Creep is also caused by wetting and drying of clays, by solution and crystallization of salts, by the growth of roots, by burrowing animals, and by flowage of saturated ground (*solifluction*). An impact of any kind on particles that rest on a slope will cause them to be displaced downhill. The impact may be caused by the falling of a tree, by a rolling stone that has become dislodged, by rill wash, or by rain or hail. Large boulders may be moved downslope by the washing away of fine grained materials supporting them. Colluvial deposits generally are a chaotic mixture of coarse and fine sediments, but in places there may be some sorting, with an accumulation of stones at the surface.

Mass-wasting, both by frost heaving and by solifluction, is especially active in

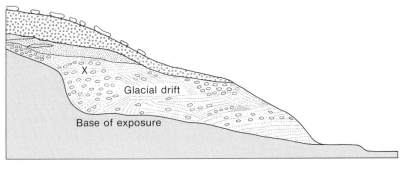

Colluvium   Transitional beds

FIGURE 7.11
Colluvium overlying stratified glacial drift, gravel, and sand in northern Pennsylvania.
At on locality, at the position marked X, the articulated remains of a Pleistocene pig
were recovered. The colluvial slopes are stable at the present time and are thought
to have formed soon after the glacial deposits were laid down, while the ice was
retreating. [After Denny, 1963.]

arctic and high-altitude environments. In arctic regions, measured rates of displacement range from about 1 cm to more than $3\frac{1}{2}$ cm per year—rates that would produce displacements of 10 to more than 35 meters in a thousand years. In humid-temperate latitudes, the process is much less active, and in the arid regions of today almost inactive, although in the past, when the climates there were less arid, it was sufficiently active to form widespread deposits that are now being eroded.

In the eastern United States, especially in the northern Appalachian Highlands, much of the colluvium is thought to have formed during deglaciation. The colluvium is younger than the glacial deposits and overlies them (Fig. 7.11). Apparently the colluvium was deposited shortly after the glacial deposits were formed, because the latter show no signs of being weathered before they were buried, and downhill movement of the colluvium has greatly slowed.

In the arid southwestern part of the United States most hillsides are mantled with colluvium, and in many places colluvial deposits of as many as three ages may be distinguished. These differ not only in their relative geomorphic and stratigraphic positions but also in their degree of weathering. The oldest date from the late Pleistocene, a comparatively wet period, and the colluvium correlates with the late Pleistocene alluvium and lakes. The next younger colluvium correlates with the middle Holocene alluvium (Table 7.4) and lakes. That these deposits antedate the occupation of the area by Indians that made pottery and used the bow and arrow is shown by the existence of their occupation sites on those deposits. Younger colluvial deposits, including some that are historic, are much less extensive than the older ones, which provides further evidence that the process has greatly slowed under the present arid climate. The stratigraphy of these deposits is important in studies of cliff retreat and other changes that affect the development of the landscape and its soils.

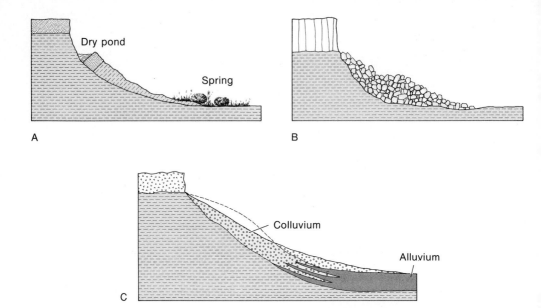

FIGURE 7.12

Three kinds of deposits related to landsliding, and some of the ways they may be dated. True landslides (A) move as more or less intact blocks on one or more shear planes which are concave upward, and the slid block is rotated backward. Commonly there is a poorly drained depression at the upper end and a spring at the toe of the slid block. Talus heaps (B) form where blocks from a resistant ledge or cliff can accumulate at the foot of the cliff or steep slope. Fine sediments may have been mixed with the boulders and washed out, leaving a jumble of the boulders. Most of the larger accumulations of talus are middle Holocene or older and are darkly stained with desert varnish in dry parts of the country. Colluvium (C) is formed by slow downhill creep of fine and course hillside debris. Old deposits antedating the present cliffs intertongue with middle Holocene or older alluvium. Younger colluvium graded to the present cliff fills gullies in the older deposit and overlaps the alluvium. Archeological sites with pottery are on both colluvium and alluvium; pre-pottery sites are in the alluvium and colluvium.

A special kind of downhill creep of colluvial and soil materials is caused by the falling of trees, especially on mountain slopes in the eastern hardwood forests. In parts of the Appalachian Highlands, hillside mounds formed by fallen trees are a common feature of the landscape and could accelerate the downhill creep of the colluvium and soil. This effect has been described by Goodlett (1956, p. 64), who writes that "The toppling of trees breaks the root mat, lays bare the underlying surficial mantle, and piles it into mounds. The act of toppling moves the material downslope for distances of several feet. After decay of the binding roots and erosion of the finer material (by rain drops) slope wash, frost, and creep carry some material back into the pit, but most of it is carried farther downslope. If such slopes [in northern Pennsylvania] have been forest-covered for at least 10,000 years, the continued toppling of trees downslope must have caused downslope movement of the mantle at a rate faster than that produced by all other processes combined."

Still other gravity deposits are formed by landslides and avalanches (Fig. 7.12), in which whole masses are moved along shear planes. Landslides and avalanches occur wherever slopes are oversteepened, whether artificially or by erosion. They develop in hardrock as well as in surface deposits and in arid as well as in humid climates.

FIGURE 7.13
Mudflows may overwhelm homes or highways. In California, unstable steep slopes soaked by winter rains are particularly susceptible to such earth movements.

They are not limited by altitude or latitude; some landslides form even under the sea. Most landslide shear planes are concave upward; the sliding block rotates as it moves down the shear plane, leaving a depression above the slide (Fig. 7.12, A). Landslides may become avalanches where the slid mass breaks into a chaotic assemblage of broken blocks because of catastrophically sudden movement. Avalanches are thought to slide on a cushion of air. Campers beyond the limit reached by the Madison Canyon slide were subjected to a strong blast of wind that overturned trailers; excavations at avalanches that descended onto glaciers reveal delicate ice features characteristic of the surface of glaciers preserved under the debris of the avalanche.

Another form of deposit in which gravity is a main factor is the accumulation of *talus* at the foot of cliffs or other steep slopes (Fig. 7.12, B). The wedging action of frost or of roots detaches blocks and slabs from cliffs, and they fall or roll freely to the foot of the cliff or slope and accumulate there in a heap.

Where there is much shale or other fine grained material, rains may soak the ground sufficiently to cause the flow of masses of mud, even on gentle slopes. *Mud flows* (Fig. 7.13) are neither liquid nor solid, but a mixture of both; because of their high density, they are capable of transporting large boulders. The deposit that results from a mud flow is an unsorted mass of boulders and cobbles in a fine-grained matrix. Where mud flows move onto open ground, they are in plan shaped like a tear drop, tapering toward their source. When constricted in a narrow valley, mud flows may form a steep sided narrow ridge in the middle or along one side of a valley. The mechanics of displacement involve movement between particles, rather than along one or a few shear planes.

Landslides and avalanches may be triggered by earthquakes, but most are due to wetting and drying, especially if there is much freeze and thaw, gradually increasing the strain to the point of rupture. As a consequence they are common in the periglacial belt—the belt peripheral to the Pleistocene and later glaciers. Boulder fields are common; some may be the result of catastrophic boulder slides. Others seem to have moved slowly, by creep, their boulders having became bared as erosion removed the matrix of fine-grained sediments that originally lubricated them. Numerous tests indicate that the boulders in most of these fields are now stable; only in isolated places beyond the limit of the Pleistocene glaciers has it been possible to detect recent movement.

Shore Deposits

Surface deposits along shores represent a wide range of environments. Part of the ground is dry, loose sand that along some shores forms dunes; part is tidal zone, alternately wet and dry; part is sandy and submerged offshore; part is muddy and submerged under estuaries; and part is wet or swampy back of or around the edges of the estuaries. Each of these environments is itself zoned where it grades into its neighbor. The tidal zone, for example, includes low ground that is submerged by all except the lowest tides, and high ground that is dry except at the highest tides.

The organisms that inhabit the shore zone are as varied as the environments, vividly described by Rachel Carson in her book *The Edge of the Sea*. On the rocky New England coasts, above the high-tide line, is a layer blackened with algae. At the high-tide line, this algal mat is home for periwinkles. Below this are barnacles, and below them grows a seaweed, rockweed, that at high tide forms an upright jungle and at low tide lies prostrate. Near the low-tide line are the seaweeds, Irish moss and kelp. Each of these parts of the tidal zone has its own rich and distinctive flora and fauna.

Shores that are not rocky are similarly zoned. The parts exposed to surf are sandy, and the sand is piled in ridges paralleling the shore. The ridge forming the shore becomes a barrier beach, and back of it is likely to be an estuary. Estuaries are subject to the fluctuations of the tides but are protected from surf, and the ground in them is generally muddy. Back from the estuaries, water tables are high and the lowest parts are swampy.

In estuaries fresh water is mixed with salt water, and the salinity of the ground depends upon the mixture. Most estuaries are brackish and grade to fresh water at the mouths of the streams. There are exceptions to this, however. For example, along the coast of southern Texas, streams are small and the evaporation rates are high; Laguna Madre, back of Padre Island, is more saline than the ocean. The floras and faunas of estuaries are zoned with respect to the salinities, for some species are salt tolerant whereas others are not.

Inland from the shore along the coastal plains, one is likely to find sandy ridges representing old beaches that were built when sea level was higher than it is now. Back of them may be low places with poorly drained ground representing the ancient estuaries. Many of these are marshy or swampy.

Most beaches are built of the familiar quartz sand. Some, as in Florida, are composed largely of calcium carbonate in the form of fragmented remains of shells and corals. Still others, in volcanic regions, consist of sand-size fragments of volcanic materials.

Volcanic Deposits

Quaternary volcanic rocks are widespread in the western United States (Fig. 7.14). Chief among the eruptives are the volcanics of the Cascades, those of the Columbia Plateau and Snake River Plains eastward to and including Yellowstone Park, and those of the Basin and Range Province and Colorado Plateau. The lavas are not

FIGURE 7.14
Map of the western part of the United States showing areas in which Tertiary or Quarternary volcanic rocks are at or near the surface. The extensive lava beds, especially in Idaho, Washington, and Oregon, consist chiefly of basalt. No Tertiary or Quaternary volcanic rocks are exposed in the eastern part of the United States. [After Meinzer, 1923a.]

usually included with surface deposits, because they are consolidated rocks, but, interbedded gravels and volcanic ash are included.

That many of the eruptions occurred in Pleistocene time is indicated by the interbedding of the volcanic materials with glacial or other deposits containing the fossil remains of such Pleistocene animals as mammoth, mastodon, or camel. At Mount Mazama, the ancestral volcano in which Crater Lake formed, glacial deposits that were interbedded with the lavas became exposed by the Holocene eruption that created the lake. To the east, ash from this Holocene eruption has buried Indian artifacts that predate the use of pottery and the bow and arrow.

One of the better sections of interbedded Quaternary volcanics and glacial deposits is near the Mono Craters, California. The earliest pre-Wisconsinan till there (Aeolian Buttes Till, see Table 7.2) overlies volcanic rocks that may be early Pleistocene, the till is overlain by a bed of volcanic ash known as the Bishop Tuff. The ash bed is in turn overlain by a younger pre-Wisconsinan till (Sherwin Till), which is overlain by early Wisconsinan till (Tahoe). Basaltic lava rests on this till and is covered by the late Wisconsinan (Tioga) till and rhyolitic eruptives that represent the continuation of volcanic activity into Holocene time.

One of the most complete stratigraphic sections of Quaternary deposits in this country is the interbedded sequence of lavas and sediments on the Snake River Plains; the stratigraphy is summarized in Table 7.5. At Lake Bonneville, volcanic eruptions occurred during the Provo stage of the lake, and the eruptives are interbedded with those lake deposits. Near the bottom of Grand Canyon is a Pleistocene lava flow that is of more than passing interest because of its age. The flow has been dated radiometrically as 1.8 million years old, which indicates that the canyon was within 50 feet of its present depth when the lava was erupted. One of the most recent eruptions in the western United States took place at Sunset Crater in the San Francisco volcanic field, near Flagstaff, Arizona, and buried a prehistoric Indian pueblo (ca. A.D. 1100) under 6 feet of cinders; this is America's Pompeii.

The surface of young lava flows is for the most part quite rough, hence the use in the Southwest of the Spanish term *malpais* ("bad ground"). The upper parts of the flows contain closely spaced vesicles formed by bubbles of gas that were in the once-frothy lavas; the rock is like porous cinder. The denser rock below is broken by fissures, or cracks, that formed as the lava cooled. The vesicles and fissures make the rocky ground highly porous and permeable. The roughened surfaces gradually smooth out as they weather and as they receive and collect wind blown dust and organic matter from the decay of plants growing in the crevices.

Volcanic ash is an important constituent of the ground in much of western United States. The ash was produced in great quantity by many eruptions at many volcanic centers and was spread all over the west by winds. Because the ash is very fine grained and composed of unstable glassy material, it weathers readily, and its mineral matter becomes available for plant nutrition. Volcanic ash is an important contributor to the fertility of soil in many of the world's tropical areas, where the ground is subject to intensive leaching (for example, Java, Central America, and parts of Africa). The repeated eruption of volcanic ash restores the fertility of the soil.

SEDENTARY DEPOSITS

Residual Deposits

Three widely separated regions of the United States are characterized by extensive residual deposits called *saprolite,* or *residuum.* One region consists of the Piedmont Province and adjoining areas in the southeastern United States (R_1 in Fig. 7.1). The second is that part of the central United States that lies south of the Ohio and Missouri rivers, and the third is the Pacific Northwest (R_2 in Fig. 7.1). These deep and ancient deposits are the products of the chemical decomposition of hard rocks.

TABLE 7.5
Section of Quaternary rocks on the Snake River plain

Pleistocene	Holocene		Stream alluvium, talus, landslide debris.
			Basaltic lava flows.
			Older alluvium, talus, landslide debris.
			——— Canyon reexcavated to present gradient and depth ———
	Upper Pleistocene		Melon gravel; bouldery gravel in bars more than 200 feet thick and terrace deposits 300 feet above present canyon floor; related to overflow of Lake Bonneville into the Snake River. Considerable fauna of large vertebrates; small molluscan fauna.
			——— Canyon reexcavated to present gradient and depth ———
			Canyon filled by Sand Springs and Bancroft Springs basalt.
			——— Canyon and tributary valleys deepened to present grade ———
			Crowsnest gravel; terrace deposits within the present canyon.
			Thousand Springs basalt with thin lenses of gravel containing small molluscan fauna.
			Sugar Bowl gravel in terrace deposits 25 feet thick 400 feet above Snake River.
			Madson basalt.
	Middle Pleistocene		——— Entrenchment of modern Snake River canyon and tributary canyon ——— begins; development of thick caliche ends.
			Black Mesa gravel and sand; deposited on pediments cut to a grade more than 500 feet above present Snake River.
			——— Broad Valley entrenchment ———
			Bruneau Formation; basaltic lava flows and lake deposits with interbedded gravel; contains 25 species of freshwater mollusks, only one of which is extinct. Maximum thickness 800 feet.
	Lower Pleistocene		——— Local development of caliche; canyon cutting begins; main period of block faulting ends
			Tuana gravel; river channel and flood plain deposits.
Pliocene and Pleistocene			Glenns Ferry formation; lake, stream channel, and flood plain deposits with volcanic ash and basaltic lava flows. More than 2,000 feet exposed. Contains 87 species of lacustrine mollusks, 94 percent of which are extinct; flood plain facies contains vertebrate fauna of Blancan types.
Pliocene			Chalk Hills formation; vertebrate fauna.

Source: After H. E. Malde and H. A. Powers.

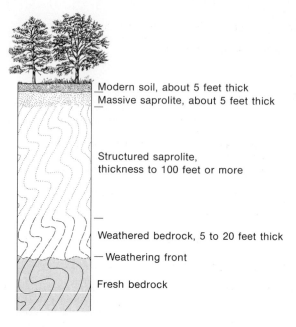

Modern soil, about 5 feet thick
Massive saprolite, about 5 feet thick

Structured saprolite,
thickness to 100 feet or more

Weathered bedrock, 5 to 20 feet thick
Weathering front

Fresh bedrock

FIGURE 7.15
Typical "ghost" layering in saprolite in the Piedmont Province
includes weathered bedrock, structured saprolite, massive saprolite, and the modern soil.

Many date from Tertiary time, and some from the Cretaceous. Saprolites grade downward from iron-rich aluminous clay at the surface to unaltered bedrock at depth (Fig. 7.15). In some places they are more than a hundred feet deep. Similar deposits are preserved in isolated places in other parts of the country.

The layering that is typical of residual deposits is perhaps best developed in those parts of the Piedmont Province where the parent materials are granite, gneiss, and schist. Above the fresh bedrock is a layer of weathered bedrock. This layer, variable in thickness, is tough and must be broken with a hammer, but the rock is sufficiently weathered that, when struck by a hammer, it doesn't ring like fresh rock but gives a dull thud. It is discolored brown or yellow with hydrated iron oxides, especially along partings. Clayey alteration of minerals in the rock can be seen with a micro-scope, but the minerals are still firm. Density of the weathered rock is 5 to 10 percent less than that of the fresh rock. In dense rocks this weathered layer is thin; in porous types it may be many feet thick.

Above the weathered bedrock is a layer called *structured saprolite*, because it perfectly preserves the structure of the parent rock, but the mass is altered largely to clay stained with iron oxide, and its density is only half that of the original rock. The alkalis (sodium and potassium) and alkaline earths (calcium and magnesium), which form highly soluble salts, have been almost entirely removed by leaching. More than 90 per cent of this layer is alumina, silica, iron, and combined water.

These deposits illustrate the confusion over the term "soil." The structured saprolite and the underlying layer of weathered bedrock are not soil to agriculturalists because they are far too deep for plant growth. The layer of structured saprolite is soil to an engineer because it can be excavated without blasting, but the underlying weathered bedrock must be blasted and so is not soil in engineering parlance. To geologists, both are orderly layers of an ancient weathering profile, and are therefore considered soil.

In many places the structured saprolite is capped by a layer of *massive saprolite* a few feet thick. This material has about the same general appearance and feel as the structured saprolite except that it is massive—that is, without structure. The boundary with the underlying structured saprolite is a gradational zone that is usually no more than $\frac{1}{2}$ to 1 inch thick. Any quartz veins present in the structured saprolite would end upward at this boundary, although scattered pieces of vein quartz would be widely distributed in the massive layer above the veins. On hillsides this massive layer probably is produced mainly by mass wasting, or creep; on flat uplands it is probably produced by frost heaving or by the mixing action of roots and burrowing animals. Massive saprolite consists 95 per cent or more of alumina, silica, iron, and combined water; almost all the alkalis and alkaline earths have been removed. Some parts are as white as snow, but most are stained with iron oxide and are brightly colored red, purple, brown, or yellow. In places the iron oxides are concentrated in nodules, fissure veins, or blanket veins.

One would expect that the massive saprolite would be generally present. Its absence in places suggests that the structured saprolite has been truncated by erosion.

On the Coastal Plain, on parts of the Valley and Ridge Province, and on the Ozark Plateau, ancient deep red soils are extensively developed on sedimentary formations. Those developed on limestone formations are brilliant red and similar to the well-known European *terra rossa*. All these soils are pre-Wisconsinan and many are pre-Pleistocene. In the southern part of the Appalachians, on the Cumberland Plateau, remnants of the old soils are moderately extensive, but they become relatively smaller and scarce northward, probably because during the glacial stages the old soils in the north were subject to severe frost heaving and were eroded. Some remnants, though, have been found under glacial drift and these record the considerable antiquity of the weathering; in places the weathering can be shown to be as old as Tertiary.

The ancient red soils that developed on limestone formations differ from saprolite in that the weathered materials seem to be unconformable on the limestone and do not grade downward to the rock. One has the impression that these soils have undergone much shifting by washing and slumping into sinks in the limestone. Moreover, old red soils on nearly pure limestone could not have developed from such parent material; there would be no source for the clay or iron oxides. The soils may have developed from impure limestone that has been eroded; they may be due in part to weathering of loessial materials.

Of special interest economically as well as scientifically are the bauxite deposits in the southern states. Bauxite, a hydrous aluminum oxide, is valuable as an ore of aluminum. These deposits formed as a result of intensive weathering of the sort that produced the saprolite on the Piedmont Province. The rock formations on which these residual deposits developed are older than the Wilcox Group, which is of Eocene

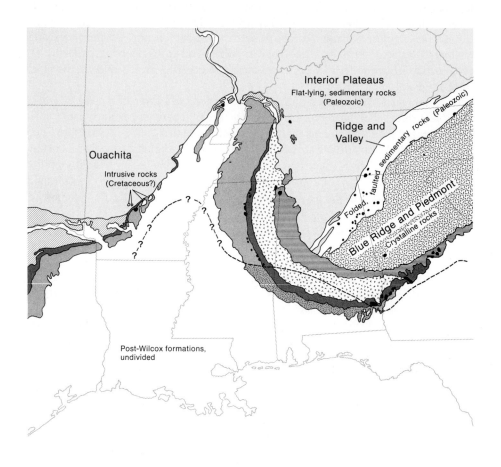

Interior Plateaus
Flat-lying, sedimentary rocks
(Paleozoic)

Ridge and Valley

Folded, faulted sedimentary rocks (Paleozoic)

Blue Ridge and Piedmont

Crystalline rocks

Ouachita

Intrusive rocks
(Cretaceous?)

Post-Wilcox formations,
undivided

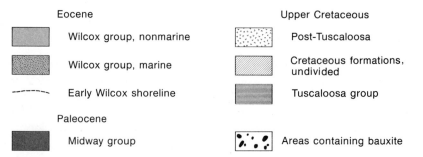

Eocene		Upper Cretaceous	
	Wilcox group, nonmarine		Post-Tuscaloosa
	Wilcox group, marine		Cretaceous formations, undivided
	Early Wilcox shoreline		Tuscaloosa group
Paleocene			
	Midway group		Areas containing bauxite

FIGURE 7.16

Bauxite-bearing areas of the southeastern United States, showing their relation to geologic formations and to major physiographic divisions. The principal bauxite deposit is in Arkansas, in the vicinity of the Cretaceous intrusives. Areas seaward of the Early Wilcox shoreline are post-Eocene. [After Gordon, Tracey, and Ellis, 1958.]

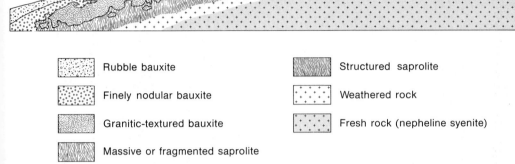

Rubble bauxite

Finely nodular bauxite

Granitic-textured bauxite

Massive or fragmented saprolite

Structured saprolite

Weathered rock

Fresh rock (nepheline syenite)

FIGURE 7.17
Sections of bauxitic saprolite in Arkansas. [After Gordon, Tracey, and Ellis, 1958.]

age (Fig. 7.16); bauxite and related clayey sediments derived from the bauxitic saprolite are contained in the Eocene formations. This stratigraphy clearly shows that the weathering occurred after the sediments of the Midway Group (Paleocene) were deposited and before deposition of the sediments of the Wilcox Group.

The largest bauxite deposits are in Arkansas. Some were formed as saprolite on buried hills of nepheline syenite, an igneous rock rich in aluminum. At the base of the bauxitic saprolite is weathered igneous rock, partly altered to an aluminous silicate clay mineral, kaolin. Above this is a layer of structured saprolite that consists mostly of kaolin near the bottom and bauxite near the top (Fig. 7.17). At the surface is rubble that is moved down the hillside by creep. Other deposits formed along streams and were derived by erosion of the first two kinds of deposit; these stream deposits are found in beds of the Wilcox Group.

The bauxite deposits in the Eastern Gulf Coastal Plain are smaller than those in Arkansas. They form pods a few feet to several hundred feet long, separated from one another by sand or sandy clay. Others are found in limestone sinks in the Valley and Ridge Province; many of these are cone-shaped. Some are elongated parallel to the joints in the limestone. Fossil plant materials associated with some of these indicate an age correlative with the deposits on the Coastal Plain. In the tropics there are extensive surface deposits of what is called laterite, genetically related to our bauxite deposits (see p. 191).

Along the Atlantic coast, saprolite that developed on the crystalline rocks is overlapped by Cretaceous formations—by the Tuscaloosa Formation in North Carolina and by the Patuxent Formation near Baltimore. Just north of Baltimore the Patuxent was derived in part by erosion of the saprolite, and the boundary between the reworked sediment and the parent saprolite is difficult to identify.

Some saprolite deposits are clearly Cretaceous or older, but most of them are younger. Similar residual deposits have formed by weathering of Tertiary formations and even of pre-Wisconsinan Quaternary formations. We look further at residual deposits in Chapter 9.

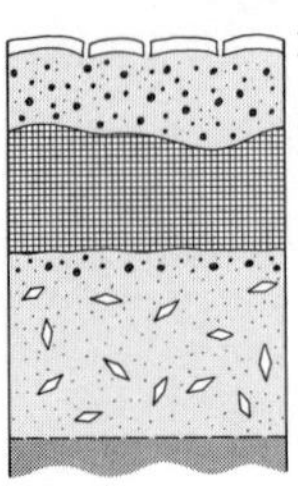

Layer	Salt	Thickness, in inches
Crust	Calcium carbonate	0.1
Silt	Sodium sulfate	1–6
Rock salt	Sodium chloride	12
Silt	Sodium sulfate	1–2
Silt	Calcium carbonate	> 12
— Water table		

FIGURE 7.18

Diagrammatic section showing caliche layers. As saline ground water evaporates, it rises upward through the soil, depositing the least soluble salts first and the most soluble (sodium chloride) last. Upward from the sodium chloride layer, the reverse sequence of salt layers is the result of downward-filtering rain waters.

Caliche

"Caliche" is a Spanish-American term for certain kinds of crusty deposits formed by the accumulation of salts. Most of them contain calcium carbonate, but in some places they contain calcium sulfate or other salts. They are characteristic of the deserts and semiarid lands of the West, where they formed at various times during the Quaternary Epoch. At least four quite different kinds can be distinguished, although the criteria for distinguishing them are only partly satisfactory.

One kind is the result of surf action at the shorelines of Pleistocene lakes. Such deposits are common as a cement in gravels or other shoreline deposits, but some are nearly pure masses of calcium carbonate. The composition and structure of the precipitates differ at different levels of shoreline in particular lake beds because of changes in lake conditions, and can be used for correlating remnants of the deposits.

A second kind of deposit formed in places where water rose by capillarity above the water table (Fig. 7.18). Examples are the thick caliche layers exposed at or near the surface in much of the west and the firmly cemented layers of gravel on old gravel fans. Like the shoreline deposits, these deposits are relics of times when the water table was higher than it is today. As such, they are subject to erosion and, when so modified, are not easily distinguished from the other kinds of caliche.

A third kind of deposit is represented by tufa or thick layers of travertine at springs (Fig. 7.19), especially at warm springs like those in Yellowstone Park. Deposits of this kind are forming at the present time, but during the wet periods of the Pleistocene the spring discharge was greater than it is now and the deposits accumulated at a faster rate. In the Great Basin, stone tools used by Indians that lived during the early Holocene or late Pleistocene have been found on top of such deposits, suggesting that most of the deposits there are Pleistocene rather than Holocene.

The lime-enriched layers in soils in the western states are a fourth kind of caliche deposit. In arid and semiarid regions, none of the small amount of rain that penetrates the ground ever reaches the water table. The water that does seep into the ground is held by capillarity until it evaporates. As a result, the soluble salts in the upper

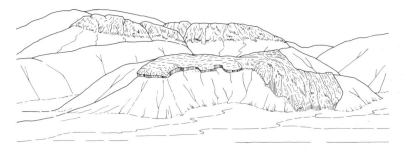

FIGURE 7.19
Some travertine mounds are demonstrably old. This one in the Great Basin was
deposited along a fault in the hills but now is dry. The travertine was deposited from
the hill to the valley floor in the foreground, but has been dissected by the gully
in the middle distance. Paleo-Indian artifacts (early Holocene) on the mound identify
it as a late Pleistocene deposit.

layers of the soil, especially the carbonates and sulfates, are dissolved and redeposited
below, forming a lime-enriched layer. Under some pre-Wisconsinan soils these
carbonate layers may be as much as 20 feet thick.

In their original forms the different kinds of caliche are readily distinguished, but
when eroded or otherwise modified they are not. Two clues help distinguish deposits
formed by rising water (capillary water) from those formed by water that enters the
ground by gravity. One is that veins of the salts deposited by capillary water tend
to thin and branch upward, whereas veins deposited by water moving downward
tend to thin and branch in that direction. The other clue is the distribution of the
kinds of salts; the more soluble salts tend to be above the less soluble ones in deposits
formed by capillary action, whereas the reverse is true in deposits formed by down-
ward-moving water.

One of the more complete studies of caliche has been made in New Mexico, where
Gile and others (1966) found orderly, age-correlative sequences in gravelly and
nongravelly ground. The sequence reported in gravelly ground is given as follows:

> Middle and Late Holocene caliche: occurs as filaments and flakes on pebbles, forming
> thin discontinuous coatings 6 to 18 inches below the surface.
>
> Early Holocene caliche: continuous between pebbles but not hardened; coatings on
> pebbles thicker than in the later Holocene deposits.
>
> Late Pleistocene caliche: continuous and partly cemented; permeability noticeably
> reduced.
>
> Older, but still Pleistocene, caliche: develops laminar structure and has firmly cemented
> layers.

In nongravelly materials the middle and late Holocene caliche forms faint carbonate
coatings on sand grains and thin filaments in veins. Early Holocene caliche develops
nodules. Late Pleistocene caliche thoroughly impregnates the whole layer, which is
indurated. The older Pleistocene caliche develops a laminar structure of nearly pure
calcium carbonate, and some of the layers are cemented.

Evaporites

Evaporite deposits are accumulations of salts that form where saline or alkaline brines are subject to considerable evaporation. Caliche deposits that developed in the capillary fringe above a water table are one kind. The term "evaporite" embraces the whole range of salts. Most evaporite deposits are residues from the drying of ephemeral lakes or ponds and as such are described with other lake deposits (pp. 136–137). Like certain of the caliche deposits, though, sedentary accumulations of salts, particularly sodium chloride and sodium sulfate, may form in the part of the ground that is moistened by the capillary fringe of a saliferous water table. The most common evaporites in desert regions, notably in the Basin and Range Province, are natural salt pans. Artificial salt pans along coasts are used for producing table salt.

Cave Deposits

The most conspicuous and best known of the cave deposits are the calcium carbonate "icicles," technically known as *stalactites*, that extend downward from the roofs of limestone caverns. These deposits of dripstone develop because water carrying the calcium carbonate in solutions descends through cracks, and the calcium carbonate is deposited as droplets of water evaporate from the tip. On the floor of the caves, though, are other deposits that, for our purpose, are more important.

The stratigraphy of deposits on cave floors is surprisingly similar at different latitudes and at different altitudes and contributes to the stratigraphy by which soils and surface deposits are dated. Commonly, at the base is gravel, clearly recording a stage of substantial water flow and probably the flow that opened the cave. Overlying this is clay or ochre deposited when the flow diminished. This in turn is capped by *stalagmites*, mounds of calcium carbonate deposited when flow had diminished to a mere drip from the roof. Probably most of the stalactites hanging from the roof formed at this stage. Overlying the stalagmites is dust. Many cave floor deposits contain a high percentage of organic matter, especially bird and bat guano.

Remains of Pleistocene animals are common in the stalagmite and lower layers; the layers above the stalagmite have yielded remains of a modern fauna. Early human artifacts, when present, underlie the stalagmite. The Holocene layers overlying the stalagmite contain artifacts of the Holocene occupations. Among the caves that have yielded such a sequence of deposits and of fossils and archeological remains are: Sandia Cave, New Mexico; a cave near Carlsbad, New Mexico; Ventana Cave, Arizona (not limestone); and Gypsum Cave, Nevada. A summary description of these and references to original descriptions are given in U.S. G.S. Bulletin 996-A.

Desert Varnish

Desert varnish is a black or dark brown stain of iron and manganese oxides on rock surfaces. As the name implies, the varnish is most conspicuous in deserts, but the staining is by no means limited to them, for it covers rock surfaces in humid regions too. In arid and semiarid regions the stain is very useful in Holocene stratigraphy

FIGURE 7.20
Desert varnish. The stain of iron and manganese oxide, known as desert varnish, is being deposited today wherever seeps discharge onto rock walls. Along canyon walls the varnish also coats dry rock surfaces, but these are old deposits that are being eroded; blocks of the varnished cliff that have fallen leave scars of unvarnished rock. The stain was deposited sometime before A.D. 1; it therefore antedates prehistoric Indian pueblo dwellings built against desert-varnished walls.

because surface deposits older than about 2,000 years are widely stained whereas younger deposits are stained only where there is much moisture.

The stain is usually no thicker than a coating of varnish, but in places is as much as a millimeter or two thick. The proportion of iron to manganese ranges from about 1/1 to about 10/1. Although it is least common on carbonate rocks the stain is found on every type of rock—on the top and sides of individual stones, on vertical or overhanging cliffs, and on sunlit or shaded surfaces.

The stain is being deposited today wherever water seeps onto rock surfaces. In the humid eastern United States, it is forming along railroad cuts and in tunnels; in the drier western states, it is forming where there are seeps along canyon walls. That moisture is needed to transport the iron and manganese onto the rock surfaces is evident; a major question is how the stain developed on such extensive, now-dry surfaces in the deserts.

There is abundant and good archeological evidence throughout the western United States and in parts of the old world indicating that the varnish on these dry surfaces was deposited more than about 2,000 years ago (Fig. 7.20). The same record appears in arid regions in other parts of the world. Study of the pyramids and other monuments in Egypt indicate that there has been practically no deposition of desert varnish in 2,000 years, slight deposition in 5,000 years, and considerable deposition of the dark stain on older stonework. Also, desert varnish is lacking on the youngest historic and protohistoric glacial deposits in cirques, but is present on older ones.

Trace elements occurring with the iron and manganese stain indicate that the varnish on stones seated in soil or colluvium is derived largely from that material, whereas the varnish on large bedrock exposures comes from weathered parts of the rock. Airborne materials are probably an insignificant source.

The occurrence of desert varnish in railroad cuts in the eastern United States and along tunnel walls shows that it can be deposited quickly if conditions are optimal. One of the necessary requirements is water; the stain that covers extensive dry surfaces in our deserts dates back at least to the pluvial period in the middle Holocene. The varnish may have formed rather quickly; biochemical tests have indicated that the selective deposition of the iron and manganese at live seeps is hastened by the activity of microorganisms in the slime that forms in the seeps. Perhaps in the middle Holocene there was sufficient dew on the widespread desert pavements that covered gravel fans and flat surfaces of cliffs to maintain colonies of similar microorganisms.

Peat

Peat is the partly carbonized organic residuum produced by greatly slowed decomposition of roots, tree trunks, twigs, seeds, shrubs, stems, mosses, and other vegetation, the decomposition being slowed because the ground is saturated with water and oxygen excluded. The plant structures generally are still visible. Pure peats contain less than 4 percent inorganic matter (moisture free basis), and they burn freely when dry.

Peat develops where conditions favor profuse accumulation of plant debris and protect it against bacterial and chemical alteration, as happens in poorly drained depressions, which become peat bogs (Fig. 7.21). These conditions exist in the glaciated parts of the country and in the southeastern United States in limestone sinks and other depressions where, because of the abundant moisture and warm climate, plant growth is profuse. Common peat-forming plants are trees, heath shrubs, sedges, grasses, mosses, pondweeds, waterlilies, reeds, cattails, algae, and ferns. Eight common types of peat ground are (1) pondweed basin, (2) grass-sedge marshes, (3) sphagnum-heath bogs, (4) cedar swamps, (5) spruce swamps, (6) tamarack swamps, (7) gum swamps, (8) cypress swamps. Where there has been subsidence of the coast, or rise of sea level, salt-water peat may accumulate over fresh-water peat.

In the north, peat may form in bogs that are virtually treeless but for patches of tamarack or black spruce; the growth is principally sphagnum moss and heath shrubs or grasses and sedges. In contrast, marshes have more open water, whether they are salt-water marshes near the coast or fresh-water marshes along rivers or lakes. Swamps are similar to bogs but are overgrown with trees.

There are three main topographic types of peat deposits—basins in which peat accumulates in marshes or ponds; built-up deposits including climbing bogs, in which peat forms on flat or gently sloping moist ground not covered by water; and composite areas consisting of peat built on filled basin deposits. In marshes or ponds there first is a growth of submerged and floating plants, and then a mat of sedges and rushes extends into the open water. Mosses growing with the sedges and rush cause the

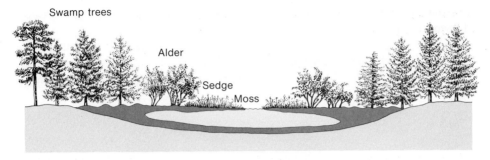

FIGURE 7.21
Cross section illustrating development of a peat bog. Over the surface of a pond of water is a mat of floating and submerged plants on which sedges may be able to take root, and the sedges in turn provide the still thicker substrate for larger plants like alder. As the mat of vegetation thickens, swamp trees may move onto the ground.

mat to thicken, and various water-loving shrubs can root in it even though the mat is floating on water. The lush growth, by accelerated transpiration, can dry what remains of the pond, and swamp trees may move into it. Built-up peat is composed of herbs, sphagnum, and heath shrubs and may form on top of deposits of aquatic plants and raise the surface of the deposit above the general level of the surrounding country.

Some peat deposits that have been forming since the close of the Wisconsinan glaciation are 18 feet thick; their rate of formation, therefore, must be about 2 inches per century.

Pure peat generally consists of about 10 percent solid matter and 90 percent water. The specific gravity generally is less than 1, and this depends on the texture, which in turn depends on the kind of plants that formed the peat; these may be:

1. Turfy peat, composed of slightly decomposed mosses; yellowish brown; soft, spongy and elastic. Specific gravity under 0.25; a cubic foot weighs less than 15 pounds.

2. Fibrous peat, brown or black, less elastic than the turfy kind, fibers either of moss, grass, roots, leaves or wood distinguishable by eye; specific gravity 0.25 to 0.65; a cubic foot weighs 15 to 42 pounds.

3. Earthy peat lacks fibrous structure and dries into earthlike masses; specific gravity 0.40 to 0.90; a cubic foot weighs 25 to 56 lbs.

4. Pitchy peat is dense when dry and breaks with smooth lustrous fracture; specific gravity 0.60 to 1.03; a cubic foot weighs 38 to 65 lbs.

Colors of peat range from yellow through shades of brown to black. Reds, white, or grays in peat generally are due to inorganic matter. A ton of peat contains only 200 pounds of solid matter, the rest is of water. Attempts to remove the water by compression have not been successful; peat is dried by evaporation.

The heat value of moisture-free peat ranges from 7,000 to 10,000 B.T.U., which is to say that the heat value of a ton of dry peat equals that of 1.3 tons of wood or a half ton of bituminous coal. In the United States, peat is used as a soil conditioner; it holds moisture and helps loosen the soil. In other parts of the world, notably

Scotland and Ireland, peat is used as fuel. Other potential uses include production of chemicals, paper, and packing materials.

Some peat bogs make important agricultural ground when drained, especially those that have formed where the ground is not excessively acid and those in which silt and clay are mixed with the peat. Such ground can produce high-value truck crops.

Clinker

Where coal beds that contain highly volatile hydrocarbons crop out at the surface, they may be ignited by spontaneous combustion, by lightning, or by ground fires. The coal burns not only along the outcrop but as much as hundreds of feet back into the ground, depending on the availability of oxygen. The temperature of combustion is high enough to bake overlying shales to a hard rock stained red by iron oxide. The baked shale and the ash from the burned coal bed are together called "clinker." This tough rock covers hundreds of square miles in the Tertiary lignite fields of the Missouri Plateau, where it is much used for railroad ballast and highway subgrades.

BIBLIOGRAPHY

Alden, W. C., 1953, Physiography and glacial geology of western Montana and adjacent areas: U.S. Geol. Survey Prof. Paper 231, 200 pp.

Atwood, W. W., and Mather, K. P., 1932, Physiography and Quaternary geology of the San Juan Mountains, Colorado: U.S. Geol. Survey Prof. Paper 166, 176 pp.

Bagnold, R. A., 1941, The physics of blown sand and desert dunes: William Morrow, New York.

Bergstrom, R. E. (editor), 1968, The Quaternary of Illinois: Univ. Illinois, College of Agriculture, Spec. Pub. 14., 179 pp.

Blackwelder, E., 1931, Pleistocene glaciation in the Sierra Nevada and Basin Ranges: Geol. Soc. America Bull., v. 42, pp. 865-922.

Bryan, Kirk, 1954, The geology of Chaco Canyon, N.M.: Smithsonian Misc. Coll., v. 122, no. 7, 65 pp.

Condra, G. E., Reed, E. C., and Gordon, E. D., 1950, Correlation of the Pleistocene deposits of Nebraska (revised): Nebraska Geol. Survey.

Denny, C. S., 1956, Wisconsin drifts in the Elmira region, New York, and their possible equivalents in New England: Am. Jour. Sci., v. 254, pp. 82-95.

————, 1965, Alluvial fans in the Death Valley region, California and Nevada: U.S. Geol. Survey Prof. Paper 466, 62 pp.

Elias, M. K. (editor), 1944, Symposium on loess: Am. Jour. Sci., v. 242, pp. 225-303.

Flint, R. F., 1955, Pleistocene geology of eastern South Dakota: U.S. Geol. Survey Prof Paper 262, 173 pp.

————, 1971, Glacial and Quaternary geology: Wiley, New York, 822 pp.

Frye, J. S., and Leonard, A. B., 1952, Pleistocene geology of Kansas: Kansas State Geol. Survey Bull. 99, 230 pp.

Gilbert, G. K., 1890, Lake Bonneville: U.S. Geol. Survey Mon. 1, 438 pp.

Gile, L. H., Peterson, F. F., and Grossman, R. B., 1965, The K horizon: a master soils horizon of carbonate accumulation: Soil Science, v. 99, pp. 74-82.

————, 1966, Morphological and genetic sequences of carbonate accumulation in desert soils: Soil Science, v. 101, pp. 347-360.

Gordon, M., Tracey, J. I., and Ellis, M. W., 1958, Geology of the Arkansas bauxite region: U.S. Geol. Survey Prof. Paper 299, 268 pp.

Hack, J. T., 1941, Dunes of the western Navajo country: Geog. Rev., v. 31, no. 2, pp. 240-263.

Horberg, C. L., and Anderson, R. C., 1956, Bedrock topography and Pleistocene glacial lobes in central United States: Jour. Geology, v. 64, pp. 101-116.

Horberg, C. L., 1954, Rocky Mountain and continental Pleistocene deposits in the Waterton region, Alberta, Canada: Geol. Soc. America Bull., v. 65, pp. 1093-1150.

Hunt, Chas. B., 1966a, General geology of Death Valley, Calif; U.S. Geol. Survey Prof. Paper 494-A, 162 pp., 494-B, 138 pp.

Jahns, R. H., and Willard, M. E., 1942, Late Pleistocene and Recent deposits in the Connecticut Valley, Mass: Am. Jour. Sci., v. 240, pp. 161-191.

Kaye, C. A., 1967, Kaolinization of bedrock of the Boston, Massachusetts, area: U.S. Geol. Survey Prof. Paper 575-C, pp. 165-172.

Krinitzsky, E. L., and Turnbull, P., 1967, Loess deposits of Mississippi: Geol. Soc. America Spec. Paper 94, 64 pp.

Leopold, L. B., and Miller, J. P., 1954, A postglacial chronology for some alluvial valleys in Wyoming: U.S. Geol. Survey Water Supply Paper 1261, 90 pp.

Leverett, F., and Taylor, F. B., 1915, The Pleistocene of Indiana and Michigan and the history of the Great Lakes: U.S. Geol. Survey Mon. 53, 529 pp.

Lugn, A. L., 1935, The Pleistocene geology of Nebraska: Nebraska Geol. Survey Bull. 10, 223 pp.

Moss, J. H., 1951, Late glacial advances in the southern Wind River Mountains, Wyo.: Am. Jour. Sci., v. 249, pp. 865-883.

Perkins, E. H., 1935, Glacial geology of Maine: Maine Tech. Exper. Sta. Bull. 30.

Ruhe, R. V., 1969, Quaternary landscapes in Iowa: Iowa State Univ. Press, 255 pp.

Scott, G. R., 1963, Quaternary geology and geomorphic history of the Kassler quadrangle, Colorado: U.S. Geol. Survey Prof. Paper 421-A, 70 pp.

Sharpe, C. F. S., 1938, Landslides and related phenomena: New York.

Snowden, J. O., Jr., and Priddy, R. R., 1968, Geology of Mississippi loess: Miss. Geol. Ec. and Topo. Survey, Bull. Ill.

Soper, E. K., and Osbon, C. C., 1922, The occurrence and uses of peat in the United States: U.S. Geol. Survey Bull. 728, 207 pp.

Varnes, D. J., 1958, Landslide types and processes: Chap. 3 *in* Eckel, E. B. (editor), Landslides and engineering practice: Natl. Research Council, Highway Research Board Spec. Rept. 29, pp. 20-47.

Wright, H. E., Jr., and Frey, D. G. (editors), 1965, The Quaternary of the United States: Princeton Univ. Press, 922 pp.

Virtually all modern soils are developed on surface deposits—rather than on the upper layers of bedrock that have weathered in place, which is the usual explanation. Among the important things that must be taken into account in determining the fitness of land for a particular use are, not just the soil, but also the depth to the water table (here marked by the standing water), the stoniness of the ground, and the slope. Some ground is level, some smoothly sloping, some irregularly hilly. Shown here is Brown Soil on floodplain gravel along the Cache la Poudre River on the Great Plains in Colorado.

8 / Modern Soils

Agricultural bias in soil science, importance of understanding non-agricultural soils; soil profiles, the layers, contrast between acid and alkaline soils, saturated ground, importance of organic matter, boundaries between soil layers, shorthand system for designating the layers; soil classification, concept of zonal soils, intrazonal and azonal soils, summary descriptions of modern soils, soil catena; The Seventh Approximation; soil distribution in North America, control by climate, contrasts in arctic, forest, grassland, and desert soils; light colored acid soils, some effects of the parent materials; arctic soils, dark forest soils, grassland soils, soils on the western mountains, reddish soils in the Southwest, Red and Yellow soils, lateritic soils, Ground Water Podzol; bibliography.

The agriculturalists' concept of "soil" is the subject of what has come to be called *soil science*—a discipline that has given rise to a highly developed technology concerned with the classification and use of soils for agricultural purposes. This highly successful technology has brought about a great increase in the yield per acre of ground. In the past half century, the productivity of some kinds of ground has been doubled, principally perhaps by improving varieties of crop plants, but also by the application of fertilizers, the practice of terracing and other methods of land leveling, by improved techniques of drainage and irrigation, and by planting of windbreaks. Soil scientists also have been concerned about the loss of soil as a result of overgrazing and deforestation and about the accumulation of alkali in irrigated land.

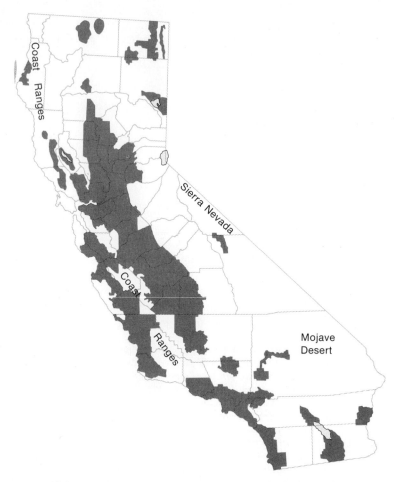

FIGURE 8.1
Soil survey in California, 1960. Soil surveys generally have been concerned almost
exclusively with agricultural lands; nonagricultural lands have received little study.
[After U.S.D.A.]

The emphasis that soil scientists have given to land suitable for agriculture is
exemplified by Figure 8.1, a map of the areas in California that have been covered
by soil surveys. Among the excluded nonagricultural lands are the redwood forests
of the northwest coast, the Sierra Nevada, and the deserts in the eastern part of the
state. Yet parts of the areas where the soils have not been studied are seriously subject
to erosion, notably in the high-rainfall redwoods area, but also in the deserts where
our exploding population is fast breaking the crust that holds the desert soils in place.

This emphasis on agricultural soils parallels geologists' emphasis on mineralized
areas. For years geologists mapped, remapped, and mapped again the areas in the
Basin and Range Province where mining was or had been a major activity. Until
about 1950, approximately 90 percent of the work of the U.S. Geological Survey
in that area was focused upon the few square miles of the mining districts; recently,

however, as increasing attention has been given to intervening areas where there is little or no mining, many new geological discoveries are being made that contribute to mining and prospecting as well as to general advancement of the science. A broader approach is also needed for the adequate study of surface deposits and soils. A technology dealing with the earth can probably afford to limit itself to a small part of the earth's surface, but a science cannot.

SOIL PROFILES

Most modern soils have developed on surface deposits, whether transported or residual. In some areas, modern soils have developed on poorly consolidated bedrock, but almost none have developed on hard-rock parent materials. In areas where the waters are charged with strong mineral acids, such as sulfuric acid, hard rocks have, in places, weathered to modern soils, but such exceptions are insignificant. In this chapter we will consider only those soils that have developed on poorly consolidated parent materials, mainly on surface deposits during Late Pleistocene and Holocene time.

The uppermost layer of a modern soil consists of a mat of plant débris grading downward into a mixture of decomposed organic matter and weathered parent material. As water filters through the decomposed organic matter, various organic acids are formed, including carbonic acid. Although this acid is a weak one, it is continuously being replaced by decomposition in the organic layers and being carried downward into the mineral matter of the layer below them, where it reacts with and alters the minerals. Some of the products of alteration are soluble and are moved downward in solution. Colloidal material may be moved downward in suspension.

The layer from which mineral matter is removed is referred to as the *leached, eluviated,* or *A Horizon* (Fig. 8.2). As the water continues moving downward, the composition and concentration of the dissolved matter change, and part of the dissolved matter is redeposited in a still lower layer called the *illuviated,* or *B Horizon,* referred to here as the layer of deposition. The organic layer, leached layer, and layer of deposition together constitute the *solum* (plural, *sola*). Other changes occur below the layer of deposition, but they are minimal, and the profile grades downward into weathered or fresh parent material, sometimes referred to as the *C* and *D Horizons,* respectively. The full descriptive names are used in this book. Soil horizons will be referred to as soil layers.

In humid regions, soils are acid. Such soils are sometimes called *pedalfers*—a name chosen to emphasize the removal of aluminum (*al*) and iron (*fer*) from the leached layer. The leached iron (in the form of oxides) and aluminum (in the form of silicates; e.g., clays) are precipitated at depths of one or two feet in the layer of deposition, probably because the soil water decreases in acidity as it filters downward. The process is called *podzolization,* and the resulting soils are called *podzols.* A consequence of the process is that the layer of deposition is darker and denser than the leached layer.

The alkalis and alkaline earths are leached to even greater depths, and are carried away from the soil by drainage through the parent material. Such an acid profile is an *open chemical system,* because the soluble constituents are removed from it as water filters downward or laterally from the soil. The significant feature of pedalfers

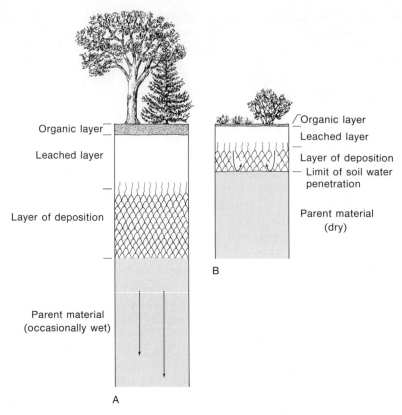

FIGURE 8.2
Profiles of two major kinds of late Pleistocene and Holocene soils. (A) Acid soils, characteristic of humid regions, develop on open systems from which soluble constituents are
removed by water draining from the bottom of the profile. (B) Alkaline soils, characteristic
of arid and semiarid regions, develop on closed systems, from which water is insufficient
to drain all the way through the ground. Three locally important variants of these two
major kinds of soils are: (1) mixed layer soils, in which the layers are mixed by frost
action (especially in arctic regions) or by burrowing animals (especially in grassland); (2)
bog soils in which organic matter may accumulate to considerable thicknesses, forming
peat; and (3) those saline or alkaline soils caused by ground water rising into the profile
and depositing salts.

is that moisture is sufficient to wet the soil to its capacity (*field capacity*) and to
allow excess water to drain downward to the water table and so remove the soluble
constituents.

To the degree that the water table is shallow and the ground water is acid or
alkaline, there will be alteration along the stratum (*aquifer*) in which the ground water
is being discharged. Such *ground-water weathering* is manifest in many places by
iron oxide stain along geologic contacts.

In semiarid and arid regions moisture is sufficient only to wet the upper layers
of the ground, and the water does not seep through to the water table. Vegetation
is of course sparse compared to that of humid regions. There is less organic matter

169

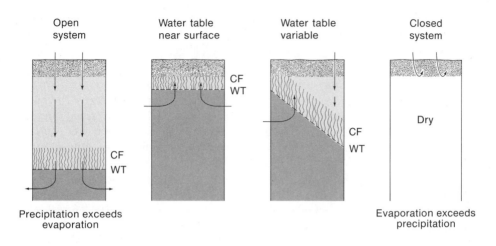

FIGURE 8.3
Water regimes, which affect soil development, are controlled partly by the precipitation-evaporation ratio and partly by the height of the water table (WT) and capillary fringe (CF). Direction of water movement is shown by arrows. At the two extremes are the open and closed systems that form, respectively, acid and alkaline soils (see Fig. 8.2). Conditions are mixed where water tables are sufficiently shallow for the capillary fringe to moisten the soil or where the height of the water table fluctuates seasonally. [After Yaalon, 1963.]

in the soils, the leaching is correspondingly less intense, and the CO_2 in the soil atmosphere is less. All this contributes to precipitation of carbonates. Such constituents as are leached are transported downward only as far as the water penetrates, where *all* the dissolved matter is precipitated as the water evaporates and the ground dries. Carbonates, whether contributed by the organic matter or parent material, form slightly soluble salts with the alkalis and alkaline earths.

The deposition of these forms a layer in which carbonates, especially calcium carbonate, have accumulated; soils that have such carbonate layers have been called *pedocals.* This process is known as *calcification* or *carbonation.* Such limy soils are alkaline and have lost no constituents; indeed, part of the carbonate derived from the organic matter has been added. Chemically these soils are *closed chemical systems.* Indeed, to the degree that they retain the carbonate from organic matter and the airborne salts and dust that are continuously being added, they are cumulative systems.

Other important soils that are less common have developed where the processes of the open and closed systems are mixed, chiefly as a result of movement of ground water (Fig. 8.3). Where water tables are high (Fig. 8.3,B), the lower soil layers or even the entire soil may be saturated, and the saturated layers have a reducing rather than an oxidizing environment. In humid regions such layers, when composed largely of mineral matter, are known as *gley.* The colors of the gleyed layers, which commonly are mottled, are brown, olive gray, or grayish blue, result from mixtures of amorphous organic matter and of iron that is largely in the ferrous state. In arid and semiarid regions salts accumulate.

There are all degrees of water saturation. The water may be free-standing and form a shallow pond, or the top layers of the soil may be dampened only by moisture in the capillary fringe above ground water that saturates the deeper layers. The saturated layer may be due to ground water perched on an impermeable layer, and the saturation may occur seasonally (Fig. 8.3,C). Or again, the ground may be only slightly permeable and kept damp or wet by infiltration of water running off adjoining areas. In this latter situation, sediment may be added to the top of the deposit, which, in humid regions, is then known as *accretion gley.* Depending on the degree and kind of saturation, there are differences in the organic layer, leached layer, and gleyed layer. Perennially wet ground might be gray; ground that is seasonally wet is more likely to be mottled. In arid regions where the water table fluctuates, salts become deposited through the zone of fluctuation.

The importance of the organic matter in soils and the importance of differences in the kind of organic layers on soils can hardly be overemphasized. It is the organic matter that does the work and that makes soil a unique layer of the earth. The organic matter gets into the soil and becomes part of it in several ways: by the death of roots, especially in grasslands; by infiltration as colloidal matter in acid soils; by the churning and mixing activities of burrowing animals, especially earthworms; and by churning and mixing caused by frost action, the crystallization of salts, and the swelling of clays.

The boundaries between the layers of a soil are described by their distinctness and width. The boundary between one layer and another is considered abrupt if the transition takes place within an inch, clear if within 1 to $2\frac{1}{2}$ inches, gradual if between $2\frac{1}{2}$ and 5 inches; and diffuse if more than 5 inches in width. The boundary may be smooth and essentially a plane, wavy with pockets wider than they are deep, irregular with pockets deeper than they are wide, or discontinuous.

Soils are enormously variable. They differ in composition and arrangement from point to point on both small and large scales—from a mineral grain to a coating, from the inside to the outside of a structural unit, and from one part of a field to another. Some of the changes are sharp; others are gradational. In general, the complexity of arrangement increases with age. For these reasons, soils are as difficult to sample as rock formations. Moreover, ground-up bulk samples yield products that bear little or no relation to what a root or bug encounters in its habitat.

Designating the Layers

A shorthand lettering system has been developed for designating the soil layers by letters—O, A, B, C, etc.—and has been broadened by the use of numerical subscripts to designate gradational layers:

O	organic layer	B_1	B layer gradational with the A
A_1	organic rich A layer	B_2	layer of maximum deposition
A_2	layer of maximum leaching	B_3	B layer gradational with the C
A_3	A layer gradational with the B	C	Weathered parent material

The shorthand system has been broadened further by the use of lower case letters to designate special properties of a layer. An accumulation of calcium carbonate in a layer of deposition would be designated B_{ca}, the subscript "ca" indicating the calcium carbonate. If the calcium carbonate accumulation is in the lower part of the layer, the numerical subscript is followed by the letter subscript—that is, B_{3ca}. An accumulation of calcium sulfate in the C layer would be indicated by C_{cs}. Some of the subscripts used by the U.S. Department of Agriculture are given below; the system differs in detail from the usage in some other countries, but efforts are being made through the International Soil Science Congress to develop a unified system.

b   Indicates a soil layer buried by a surface deposit. A leached layer buried under a sand dune would be indicated A_b.

ca   An accumulation of calcium carbonate (see above).

cn   An accumulation of concretions, usually of iron, manganese and iron, or phosphate and iron.

cs   An accumulation of calcium sulfate (gypsum).

f   Frozen ground; applicable in areas of permafrost in Tundra soils.

g   A waterlogged (gleyed) layer.

h   An unusual accumulation of organic matter.

ir   An accumulation of iron.

m   An indurated layer, or hardpan, due to silication or calcification.

p   Layer disturbed by plowing; a plowed leached layer would be indicated A_p.

sa   An accumulation of soluble salts.

t   An accumulation of clay.

With such shorthand symbols it becomes unnecessary to describe the layers of a soil. A particular grassland, alkaline soil, for example, need only be described as having an A_p, B_2, B_{3ca}, C_{ca}, C_{sa}, C profile.

Such shorthand is very useful in field notes and for communicating with fellow specialists, but is hardly good practice (although a common one) in publications that are expected to be read and understood by nonspecialists. In today's world there is a good deal of unnecessarily unintelligible writing under the guise of specialization, and the tendency is not confined to the literature on soils; we all are guilty. Well-understood words can usually be substituted for symbols and jargon in technical writing. Sometimes telegraphese is needed to conform to limitations of space and to avoid excessive repetition, but when needed, the terminology should be handily explained.

Generally acid

	Tundra
	Podzol with gley
	Podzol
	Mountain
	Gray Brown Podzolic
	Sol Brun Acide
	Red and Yellow Podzol
	Groundwater Podzol
	Alluvium
	Old lake beds with gley

Neutral or transitional

	Brunizem or Prairie
	Rendzina

Generally alkaline

	Chernozem
	Chestnut
	Brown
	Gray Desert
	Red Desert
	Red Prairie, Red Chestnut, Red Brown
	Salt Marsh
	Sand dunes

FIGURE 8.4
Soil map of North America. [After U.S.D.A.]

SOIL CLASSIFICATION

Concept of Zonal Soils

The late Pleistocene and Holocene soils being emphasized in this chapter are usually called *Zonal Soils* because the development of their soil profiles closely corresponds to the climatic and vegetation zones in which they are found (Figs. 8.4, 8.5). In terms of origin, however, the principal characteristics of some of the soils, especially the red and yellow soils south of the glacial drift and late Pleistocene loess, are largely inherited and are not comparable to the young soils developed north of the drift border. Even these, though, show some orderly variations with latitude.

In addition to the zoned, or zonal, soils, there are some whose profiles are poorly developed or even absent. Agriculturalists refer to these as Azonal Soils; in this book, they will be referred to as particular kinds of surface deposits. In most parts of the United States, surface deposits that have been laid down or exposed by erosion during the past 2,000 years are characterized by Azonal Soils. Common examples are late

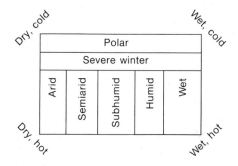

FIGURE 8.5
Schematic relationship of climate, vegetation, and soils. (Top) Distribution of climatic types. (Middle) Distribution of vegetation types on the climatic base. (Bottom) Distribution of soil groups on a climatic base. [After Blumenstock and Thornthwaite, 1941.]

Holocene alluvial deposits and dunes or severely washed areas on older deposits. Few deposits in the United States that are middle Holocene or older are without distinctive soil layering.

Still other soils have characteristics determined by a local condition, such as poor drainage or composition of the parent materials. These are classed as *Intrazonal Soils.* The three principal local conditions that cause the development of Intrazonal Soils are:

1. Excess water due to poor drainage, surface flooding, or a high water table (*hydromorphic soils*—for example, Gley, Bog, Half Bog Soils).

2. Excess salt or alkali caused by the presence of brines, sea water, or excessive evaporation (*halomorphic soils*—for example, salt or alkali crusts or efflorescence).

3. Excess limy material, as in marls or caliche (*calcimorphic soils*—for example, Rendzina and Brown Forest Soils).

Hydromorphic soils are extensive in the northern glaciated regions and along the Atlantic and Gulf coasts. In arid and semiarid regions, hydromorphic soils are found mostly in small areas at springs; and these soils are usually halomorphic too. Hydromorphic soils are extensive around some playas in the deserts of the Basin and Range Province, and, like those around springs, are also halomorphic. Calcimorphic soils develop in humid regions where the parent material is easily weathered marl or shaly limestone, but they also develop in arid and semiarid regions where the parent material is caliche.

The classification of soils as zonal, azonal, and intrazonal, is an elaboration of a similar classification developed early this century by a Russian soil scientist, V. V. Dokuchaiev, and was until recently the classification used by the U.S. Department of Agriculture and the cooperating state agencies. The classification as of 1949, which is still widely used (see M. Baldwin, C. E. Kellogg, and J. Thorp, 1938; and J. Thorp, and G. D. Smith, 1949), is shown in Table 8.1.

This is the classification that will be followed here; it has earned a good deal of international usage. Some of the less important Great Soil Groups, however, are lumped with the more important ones. The principal soils are summarily described in Table 8.2.

The distinctions between zonal, azonal, and intrazonal soils are necessarily arbitrary and not always consistent, like the geologists' distinctions between sedimentary, igneous, and metamorphic rocks. Whether a granite is an igneous or metamorphic rock depends partly on which granite is being considered and partly on which geologist is being consulted. Red and Yellow Podzolic Soils are classed as Zonal Soils, but since their principal *mineral* characteristics are inherited from the ancient residuum on which they formed (p. 150), they could as well be considered Intrazonal Soils.

Some of the gradational conditions are illustrated by the concept known as the *soil catena.* Soils on a hillside receive moisture from sites up the hill, and even though they have developed on the same parent material, the soils on the side of the hill will differ from those at the top. Soils at the foot of the hill receive both moisture and sediment from the hillside, and they will be still different, even though developed

TABLE 8.1
Zonal, intrazonal, and azonal soils

Order	Suborder	Great soil group
Zonal soils	1. Soils of the cold zone	Tundra Soils
	2. Soils of arid regions	Sierozem Brown Soils Reddish-Brown Soils Desert Soils Red Desert Soils
	3. Soil of semi-arid, subhumid, and humid grasslands	Chestnut Soils Reddish Chestnut Soils Chernozem Soils Prairie or Brunizem Soils Reddish Prairie Soils
	4. Soils of the forest-grassland transition	Degraded Chernozem Noncalcic Brown
	5. Podzolized soils of the timbered regions	Podzol Soils Gray Wooded, or Gray Podzolic Soils Brown Podzol Soils Gray-Brown Podzolic Soils Sol Brun Acide Red-Yellow Podzolic Soils
	6. Lateritic soils of forested warm-temperate and tropical regions	Reddish-Brown Lateritic Soils Yellowish-Brown Lateritic Soils Laterite Soils
Intrazonal soils	1. Halomorphic (saline and alkali) soils of imperfectly drained arid regions and littoral deposits	Solonchak, or Saline soils Solonetz soils (partly leached Solonchak) Soloth Soils
	2. Hydromorphic soils of marshes, swamps, seep areas, and flats	Humic Gley Soils Alpine Meadow Soils Bog and Half Bog Soils Low-Humic Gley Soils Planosols Ground-Water Podzol Soils Ground-Water Laterite Soils
	3. Calcimorphic soils	Brown Forest Soils Rendzina Soils
Azonal soils		Lithosols Regosols Alluvial Soils

on the same parent material. The contrast in soil catena relationships in humid and arid regions is illustrated in Figure 8.6.

Other gradational systems are caused by mixing of the soil layers by frost action, especially in arctic regions, and by burrowing animals, especially in grasslands. In bog ground organic matter may accumulate to form peat many feet thick, a special form of organic layer. In arid and semiarid regions ground water may be near enough to the surface to evaporate in the ground and contribute salts to it. But, despite such gradational types, the open-system acid soils generally are distinct from the closed-system alkaline ones, the former being without a limy layer of deposition, and the latter characteristically having one.

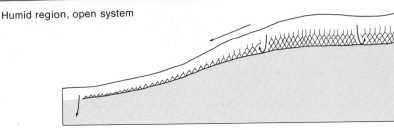

Humid region, open system

Arid region, closed system

FIGURE 8.6

Examples of closely related soils forming catenas in humid and arid regions. In humid regions with open-system soils, layering on well-drained uplands is better developed than on hillsides. The layering is different on the poorly drained ground at the foot of the hill, where moisture collects. In arid regions with closed-system soils, the layering on uplands and hillsides may be that of a closed system, but the runoff that collects at the foot of the hill may provide enough water to develop an open system there.

TABLE 8.2
Modern soils

I. Organic rich soils

A. *Tundra Soils; Tundra* and associated *Arctic Brown Soils.*
Dark brown peaty layer over gray layers mottled yellow or brown subsoil permanently frozen. Arctic Brown Soils on well drained uplands; Tundra Soils on poorly drained ground. Thickness 1 to 2 feet. *Climate:* frigid, humid. *Vegetation:* lichens, mosses, herbs, shrubs. *Weathering process:* decomposition of organic matter slowed by freezing and in Tundra Soils, excessive wetting; frost-heaving mixes organic and mineral matter. *Age:* Holocene.

B. *Alpine Meadow Soils*
Soils on mountain tops above timberline; somewhat like Tundra Soils but the subsoil is not permanently frozen and the ground generally is not so wet.

C. *Peat and Bog Soils.*
Brown, dark brown, or black peaty material over buried peat (*Bog Soil*) or over gray and rust mottled soils of mineral matter (*Half Bog Soil*). May be many feet thick. *Climate:* arctic, temperate, or tropic; standing water. *Vegetation:* swamp forest, sedge, or grass. *Weathering process:* decomposition of plant matter slowed by excessive wetting, which also causes development of organic rich, sticky, compact clayey layer (gley). *Age:* mostly Holocene, rarely late Pleistocene.

← Acid ground, wet →

II. Forest soils

    A. *Light soils; conspicuous light leached layer; Podzol Soils.*

    Surface layer is litter of needles, twigs, and cones over a humus-rich layer containing mineral matter which overlies a conspicuous whitish-gray, leached layer that generally is sandy; layer of deposition clayey and stained brown. Total thickness generally less than 24 inches. Associated with extensive Peat and Bog Soils. *Climate:* cold, humid; annual precipitation mostly between 20 and 40 inches. *Vegetation:* mostly spruce-fir forest. *Weathering process:* acid leaching removes iron and aluminum from the leached layer and redeposits them in the layer of deposition; clay worked downward partly by washing and probably partly by the vigorous freeze and thaw which contributes to the sandiness of the leached layer. *Age:* Holocene.

    B. *Dark soils; brown to gray-brown leached layer; Gray-Brown Podzolic Soils, Sol Brun Acide; related soils modified by cultivation are Brown Podzolic Soils.*

    Surface layer is mostly litter from broadleaf trees, some conifers; under this is humus-rich layer containing mineral matter and overlying a brown or gray-brown leached layer containing more or less clay (compare above); layer of deposition stained darker brown and contains more clay than does the leached layer; locally it is an acid, clay-rich hardpan. Total thickness commonly about 30 inches. Occurrence generally south of the soils having conspicuous light leached layer and only locally associated with Peat and Bog Soils. *Climate:* temperate, humid; annual precipitation 30 to 40 inches. *Vegetation:* in eastern United States mostly deciduous broadleaf forest, some conifer forest on western mountains. *Weathering process:* the difference in leaching between these soils and those farther north in eastern United States and Canada is attributable to several factors: greater carbonate content of the more southerly ground south of the Precambrian Canadian Shield, greater extent of broadleaf forest, less sandy and less permeable leached layers, and (along the Appalachians) greater extent of steep slopes favoring rapid runoff. *Age:* Late Pleistocene and Holocene.

    C. *Dark soils without a noticeable leached layer; Brown Forest Soils.*

    Leaf litter over dark brown friable surface soil grading downward through lighter colored soil to calcareous parent material; the layering is obscure because the leaching is slowed by the carbonate matter. *Climate:* cool temperate; annual precipitation about 30 inches. *Vegetation:* Mostly northern hardwood forest, some mixed broadleaf and conifer forest. *Weathering process:* same as in the other northeastern forest areas but leaching retarded by the high calcium carbonate content of the parent material. *Age:* Late Pleistocene and Holocene

    D. *Red and yellow soils, clayey; Red and Yellow Podzolic Soils.*

    Thin, dark colored organic layer at surface over yellow-gray or gray-brown leached layer over a darker clayey layer of deposition which grades downward into deeply weathered parent material brightly colored and generally mottled red, yellow, purple, brown, white, and gray. The deeply weathered

Acid ground; mostly open system (i.e. alkalis and alkaline earths removed)

TABLE 8.2 (continued)

—— Acid ground (continued) ——

parent material may be scores of feet deep; the organic layer, leached layer, and layer of deposition commonly total 3 to 4 feet in thickness. *Climate:* warm temperate to tropical humid. *Vegetation:* southeastern pine forest or mixed broadleaf and pine forest. *Weathering process:* like the more northerly soils but these soil layers are dominated by the deeply weathered, clayey, residual parent material. Depending on the kind of parent residuum, there are 5 principal varieties of these red and yellow soils: 1, residual deposits on the crystalline rocks of the Piedmont Province; 2, residual deposits on the weakly consolidated sedimentary formations of the Coastal Plain; 3, weathered loess of Pre-Wisconsinan age, especially along the east side of the Mississippi River valley; 4, residual deposits in central and eastern United States developed on the Paleozoic formations, particularly the limestones (*Terra Rossa*); and 5, residual deposits on various kinds of rocks in the rainy Northwest. *Age:* the surface layers are late Pleistocene and Holocene; the residual, deeply weathered parent material is the result of much older weathering, early Pleistocene, Tertiary, and locally even older (see Chapter 7).

E. *Sandy soils over shallow water table; Ground Water Podzol Soils.*
Thin organic layer over thick (2 to 3 feet) light-colored leached layer, usually sandy, over dark brown layer enriched with organic matter and collected at the top of the water table, which fluctuates. *Climate:* cool to tropical, humid. *Vegetation:* mostly southeastern pine forest. *Weathering process:* intensive acid leaching and transfer of organic matter to water table at bottom of the profile. *Age:* Late Pleistocene and Holocene.

III. Grassland soils

—— Transitional ground; mostly open system soil profiles; nearly neutral, or partly acid and partly alkaline ——

A. *Parent materials non-calcareous or only slightly so; no hardpan; Prairie (Brunizem) Soils.*
Dark brown to nearly black, mildly acid, surface soils over brown, well oxidized subsoils; reddish brown towards the south. Grades downward to lighter colored parent material with no layer of lime carbonate accumulation. Total thickness about 3 feet. *Climate:* cool to warm temperate, humid; annual precipitation 25 to 30 inches. *Vegetation:* tall grass. *Weathering Process:* acid leaching, but weak. *Age:* Late Pleistocene and Holocene.

B. *Parent materials non-calcareous or only slightly so; with hardpan; Planosols.*
Leached surface soils over parent material containing clay and developing a clayey hardpan; ground nearly level. *Climate:* cool to warm temperate, humid; annual precipitation 25 to 30 inches. *Vegetation:* tall grasses, some forest. *Weathering process:* acid leaching with gleyization. *Age:* late Pleistocene and Holocene.

C. *Parent materials calcareous; Rendzina Soils.*
Dark gray or black, organic rich, surface layers over soft, light gray or white calcareous material derived from chalk, soft limestone, or marl; associated with swelling clays, Thickness highly variable. *Climate:* variable. *Vegetation:* mostly tall grasses. *Weathering process:* acid leaching but process slowed by the carbonate in the parent material (cf. II,C, forest soil without noticeable leached layer). *Age:* mostly late Pleistocene and Holocene.

Alkaline ground; mostly closed system (i.e. alkalis and alkaline earths retained)

IV. *Thin, woodland soils (e.g. chaparral, pinyon-pine, juniper, oak brush); Noncalcic Brown Soils.*
Brown to red surface layers over redder and more clayey subsoil; alkaline to about neutral. *Climate:* semiarid; seasonally dry. *Weathering process:* weak acid leaching; little or no calcification; some clay and silica accumulated in layer of deposition.

V. *Grassland soils; Chernozem, Chestnut, Brown Soils.*
Black to gray-brown, friable soil to depth of 3 or 4 feet under tall grasses and 1 to 3 feet under short grasses; grades downward through lighter colored layer to a layer where lime carbonate has accumulated. *Climate and vegetation:* subhumid, temperate to cool, annual precipitation 20 to 25 inches in belt of tall grasses (Chernozem Soils); 15 to 20 inches in belt of short grasses (Chestnut and Brown Soils). *Weathering process: calcification,* i.e. accumulation of carbonates in the lower layers. *Age:* late Pleistocene and Holocene.

VI. *Reddish calcareous soils under grasses or shrubs; Reddish Chestnut, Reddish Brown Soils.*
Rather like IV but surface layers are reddish and lime carbonate layer generally thicker and more conspicuous; occur farther south. *Climate and vegetation:* semiarid; annual precipitation 10 to 20 inches; cool to hot; grasses and shrubs. *Weathering process:* calcification. *Age:* mixed ages with late Pleistocene and Holocene weathering profiles superimposed on older Pleistocene ones (See Chapter 9).

VII. *Western forest soils; Chestnut, Brown Soils.*
Rather like IV but developed under woodland of oak, pinyon and juniper, or pine forest.

VIII. *Desert soils under shrubs; Desert, Sierozem, and Red Desert Soils.*
Light gray or brown in north, reddish in south; organic layer thin and may be discontinuous where shrubs are widely spaced; carbonate layer generally within a foot of the surface. *Climate and vegetation:* arid; annual precipitation less than 10 inches; cool to hot. Shrubs, occasional trees; some woodland. *Age:* Late Pleistocene and Holocene soils (shallow, weakly developed profiles); older Pleistocene soils (several feet thick, generally reddish, and with well-developed layers).

IX. *Excessively alkaline and saline soils; Solonetz, Solonchak Soils.*
Accumulated salts on or near the ground surface due to imperfect drainage; salts more than 0.2 percent. Alkaline ground ("Black Alkali"; Solonetz) usually formed by the less soluble salts (esp. sodium carbonate and sulfate); saline ground ("White alkali"; Solonchak) contains more soluble salts (esp. the chlorides). *Climate and vegetation:* arid, semiarid climates but may occur on saline ground in humid regions; vegetation limited to salt-tolerant species. *Age:* Holocene.

TABLE 8.2 (continued)

↑ Acid or alkaline ground	X. *Surface deposits with weakly developed weathering profiles; Skeletal Soils, Lithosols.* (Whether categories X and XI are acid or alkaline depends on parent material.) See Chapter 7. XI. *Disturbed ground.* May be plowed layers on agricultural lands, where the organic and mineral layers would be mixed; may be weathered or unweathered surface deposits heaped in embankments or spread across a surface, or land cleared of top soil.

THE SEVENTH APPROXIMATION

More recently the U.S. Department of Agriculture has developed a new comprehensive classification based on general morphology. Known as the Seventh Approximation, the classification is said to have considerable merit, but its merit is lost in its incredibly horrendous nomenclature, which bars the use of any simple English word. The situation is reminiscent of the stage in geology when sand had to be described as psammite, gravel as rudyte, and a limy sand had to be a calc-arenyte. During this stage of the history of geology, basalt would be classified

> *Class,* Dosaline; *Order,* Perfelic. Germanare; *Rang,* Alkalicalci Andase *Subrang,* Dosodic. Andose. (U.S. Geol. Survey Prof. Paper 99, p. 522).

Geology has survived; thankfully the nomenclature has been simplified. Names of soils, like the names of rocks, can be defined precisely and quantitatively, but the facts of field life are such that they cannot be applied precisely. Moreover, such terms do not aid communication; few will understand the following quotations from *Soil Classification, Seventh Approximation,* U.S. Dept. Agriculture, 1960:

"The Umbraqueptic Cryaquents are comparable to the Cryaqueptic Cryaquents in color values, and to the Orthic Cyraquent in chromas" (p. 107). Or "The Entisols are those soils, exclusive of Vertisols, that have a plaggen horizon or that have a diagnostic horizon other than an ochric or anthropic epipedon, and albic horizon, and agric horizon, or, if the N value exceeds 0.5 in all horizons between 20 and 50 cm. (8 and 20 inches), a histic epipedon" (p. 105).

America is first in technology; the Seventh Approximation would keep America first in terminology!

Such nomenclature fails its purpose of communication, by deadening the senses of the person trying to understand the complexity of the processes and relationships. The name becomes the end! Geologists' efforts to develop a quantitative nomenclature have largely abated, though some still persist in such efforts, as is evidenced by use of the currently fashionable terms "tephra," "tholeiite," and "ignimbrite." The terms "hill" and "mountain" have served us well without being quantified. Let us hope that the nomenclature developed for the Seventh Approximation will also eventually be abandoned in favor of simple terms that can be understood by those who are not specialists in the taxonomy of soils. Given a reasonable nomenclature, the basis for the new classification might survive.

The Seventh Approximation distinguishes 10 orders of soils as follows:

1. Entisols. Soils without layering, except perhaps a plowed layer. These are the Azonal Soils of earlier classifications; as such, they are surface deposits and might as well be so identified. The term Entisol is unnecessary, as are the names of the suborders—for example, Aquents for wet deposits and Psamments for sandy deposits.

2. Vertisols. Soils with upper layers mixed or even inverted because they are developed in clays that swell and in environments that are alternately dry enough to cause the clays to crack and wet enough to cause them to swell. The group is very important for engineering purposes and would become better known sooner if identified simply as "Soils with clays that crack and swell." The varieties are many, depending on the hydrological environment, vegetative cover, land use, etc., and could be so described in understandable terms rather than being referred to as Aquerts, Usterts, etc.

3. Inceptisols. Very young soils having one or more of the diagnostic layers that can develop readily, but without much leaching, deposition in the subsoil, or mineral alteration. This grouping seems to represent those soils having profiles too weakly developed to be truly zonal, yet layered enough not to be typically azonal. It is an old classification problem of what to do with the transition members of a continuum. If Entisols could become known as "Surface deposits without weathering profiles," the Inceptisols could be known as "Surface deposits with weakly developed profiles." And they could be described as wet without being named Aquepts, or as volcanic ash without being named Andepts!

4. Aridosols. Soils of the deserts and semiarid regions, and associated saline or alkaline soils. The different kinds are distinguished chiefly on the basis of mineralogy, but few will find it helpful to read that these soils include Orthids, Camborthids, Durorthids, Calcorthids, Salorthids, Argids, Haplargids, Duragids, Natrargids, and Nadurargids.

5. Mollisols. Grassland soils, mostly lime-rich (the Brown, Chestnut, Chernozem, Prairie (Brunizem), Red Prairie, and Rendzina Soils); also Brown Forest Soils, which are the forest soils developed on lime-rich parent materials. The characteristic feature is the thick, organic-rich surface layer.

6. Spodosols. Mostly the Podzol, Brown Podzol, and Ground Water Podzol Soils. Emphasis is on the layer of deposition, which contains free sesquioxides, organic matter, and clay leached from an overlying, light, ashy gray layer.

7. Alfisols. Includes most of the other acid soils and characterized by clay-enriched subsoils, especially the Gray-Brown Podzols, Planosols, and some Half Bog Soils.

8. Ultisols. Red and Yellow Podzol Soils and certain lateritic soils; weathering more advanced than in Alfisols.

9. Oxisols. Most lateritic soils; weathering more advanced than in Ultisols.

10. Histosols. Bog Soils and some of the Half Bog Soils.

Proponents of the Seventh Approximation need to be reminded of the admonition from Arthur Holmes, the author of "Nomenclature of Petrology," who wrote (1920):

> "There is undoubtedly an attraction in the creation of new names . . . (but) . . . brevity of expression is by no means an unmixed blessing, and the one word may require a whole paragraph of explanation."

With this admonition in mind, I refrain from the temptation of offering science the new word *phreatosol*, for the *oxyhypopedon* formed by ground-water weathering (p. 203).

SOIL DISTRIBUTION IN NORTH AMERICA

The common modern soils in North America are listed and briefly described in Table 8.1; Figure 8.4 shows their distribution in the United States. Figure 8.7 illustrates the changes in kinds of modern soils and their profiles reflecting differences in climate, moisture supply, and vegetation along a transect from the Hudson Bay region to the southwestern United States. In terms of climate, the northeast end of the transect is cold and wet, and the soils are acid; the southwest end is warm and dry and the soils there are alkaline. The vegetation changes from tundra and conifer forest at the northeast, to deciduous forest, and then to grassland on the subhumid and semiarid plains, and to shrub land in the deserts at the southwest. The changes in profile and kind of soil correlate with the changes in climate and vegetation. However, there is another variable too, because the soils average progressively younger toward the northeast end of the transect. The principal differences in the soils, though, reflect the differences in climate and vegetation.

For example, in northerly glaciated latitudes and at high latitudes, where drainage is interrupted, the organic (O) layer may be thick enough to accumulate as peat, and, depending on the kind of vegetation, the peat may be woody or fibrous. Peats also are common in wet tropical or subtropical areas where drainage is imperfect. Because of the high water table at bogs and marshes, most of the ground water drains laterally rather than downward, and consequently the lower layers of the soil profile—the leached layer and the layer of deposition—may be poorly developed or nonexistent. Where the organic matter in peaty ground is less than about 3 feet thick and underlain by impermeable or nearly impermeable ground, the peaty soil is referred to as *Half Bog Soil;* peaty ground underlain by other peat at depths greater than 3 feet is referred to as *Bog Soil.*

At the other extreme, on desert soils, vegetation is scanty and widely spaced because of the slight rainfall. In deserts the organic layer may be absent except on the part of the ground surface directly under shrubs; between shrubs, the mineral layers extend to the surface. The soil layers are weakly developed because of the low rainfall and sparse vegetation. There is not enough moisture to penetrate deeply into the ground, and not much carbonic acid is generated.

Soils in arctic regions also have weakly developed layers, partly because the ground is frozen much of the year, which virtually stops biochemical processes, and partly because the layers are repeatedly mixed by freezing and thawing.

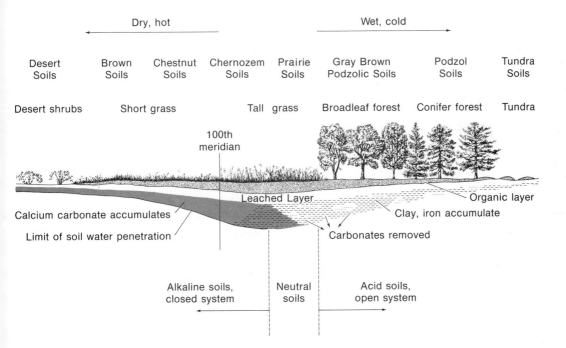

FIGURE 8.7
Transect illustrating changes in soil profiles that accompany changes in vegetation and climate between the tundra in northern Canada and the deserts in southwestern United States. At the 100th meridian the annual precipitation averages about 20 inches; there and to the west the soils are alkaline. The easternmost grassland soils are about neutral; farther east the soils are acid.

Between these extremes are the soils that form under temperate-zone conifer forest, deciduous forest, and grass. In the eastern United States the layers containing organic matter are thicker under deciduous forest than under coniferous forest, and thickest under grasses (Fig. 8.7). Under coniferous forests there is a surface mat of needles, twigs, and cones—the accumulation of several year's fallen litter. Because conifers are evergreen, the annual increment of fallen organic matter is smaller than it is under deciduous forest. Below the layer of litter is partly decomposed organic matter, duff, and below this is a layer that, in the upper part, may be blackened with completely decomposed organic matter, or humus, that has washed into it from above. The organic-rich layers are only about 2 to 4 inches thick, and are highly acid. In the rainy northwestern United States, however, the organic layers may be measured in feet. Conifers require little mineral matter and can occupy poor sandy ground that is deficient in the mineral nutrients required by most plants, hence the frequency of stands of pine without much understory.

The layering of the soils under deciduous forest is similar to that under coniferous forest, but because deciduous trees shed all of their leaves each year, organic-rich layers are commonly about 4 to 6 inches thick.

In general, deciduous trees require more mineral matter than the conifers, and the bases, especially calcium, magnesium, potassium, and sodium, are returned to

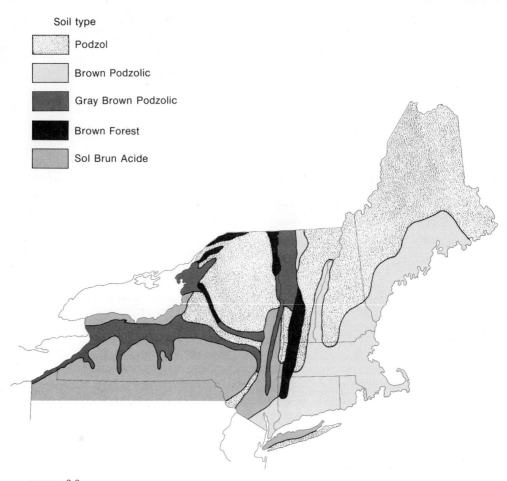

Soil type

Podzol

Brown Podzolic

Gray Brown Podzolic

Brown Forest

Sol Brun Acide

FIGURE 8.8
Map of the acid forest soils in the northeastern United States (compiled from maps by U.S.D.A. and N.Y. State College of Agriculture).

the soil when the fallen leaves decay. For this reason, acidity of soils under deciduous forest averages less than under conifer forest.

Grassland soils have a thin layer of litter. However, the layer of humus-rich mineral matter is as much as 2 feet thick. Grasslands in the United States are mostly in subhumid areas, where annual precipitation is less than 30 inches. The annual precipitation may be no less than in northern coniferous forests, but the ground averages drier because of the more southerly latitude, more completely integrated drainage system, and exposure to wind. Grassland soils are thickest under the tall, deep-rooted grasses and thinnest under the short, shallow-rooted grasses (Fig. 8.7). The short grasses extend into semiarid lands, and the organic layers of even these soils are a few inches to a foot thick. Such grassland soils may be alkaline.

Except in bog soils, the organic layers are underlain by a light-colored layer in which there has been maximum leaching.

Light-Colored Acid Soils

The light-colored acid soils (Podzol Soils), which have a conspicuous, light leached layer are extensive in the northeastern United States (Fig. 8.8) and extend across central Canada and Alaska. They approximately coincide with the spruce-fir forest but are not restricted to it. The soils are characterized by a surface mat of partly decayed needles, twigs, and cones that grades downward into a strongly acid, dark layer of mineral soils that may contain 5 percent of finely divided organic matter (*humus*). This humus-rich, partly leached layer overlies a strongly leached, very light gray (also acid) layer a few inches thick. Soluble salts of the alkalis and alkaline earths are dissolved by the acidic soil water and removed. Carbonates, in carbonate-rich ground, are leached to a depth of 2 to 3 feet. Iron, aluminum, clay, and organic matter also are transported downward, partly in solution and partly in suspension, but they are redeposited in the layer of deposition, which is brown, more clayey than the leached layer, and grades downward to yellow brown subsoil that in turn grades downward to the parent material. These soils rarely are thicker than 3 feet. The parent materials are mostly very late Pleistocene and Holocene glacial drift, and because of the slight weathering of the mineral matter, the composition of these soils depends largely on that of the parent material, which is mostly glacial drift.

In central New York State, systematic relationships have been found between the kinds of parent materials and the occurrence of the different kinds of acid forest soils (Pearson and Cline, 1958). Soils without a noticeable leached layer (Brown Forest Soils) were found in lime-rich parent materials; soils with a brown leached layer (Gray Brown Podzolic Soils and Sol Brun Acide) were found on less limy parent materials, and soils with conspicuous light leached layer (Podzol Soils) were found on nonlimy parent materials (Fig. 8.9; see also Fig. 7.3). Depending on ground drainage, all gradations are found also between these several kinds of acid forest soils and the organic-rich bog and peat soils. Similar relationships are found farther west in the Great Lakes states.

Arctic Soils

North of the belt of Podzol Soils and conifer forest are Tundra and related Arctic Soils. This ground is generally damp or wet; in some localities it is slightly acid, but in most places it is slightly alkaline. The several kinds of Arctic Soils, each developed under a different combination of latitude (temperature) and drainage, all have weakly developed soil profiles, partly because they are young (no more than a few thousand years old), partly because of the long period of freezing and the short period of thaw, and partly because freezing and thawing mixes the layers.

On well-drained ground with shrub vegetation there are Arctic Brown Soils, which are dark brown above and grade to a lighter brown below. On well-drained ground without vegetation there are lithosols. Wet meadows and marshes have Tundra Soil that grades to the Bog and Half Bog Soils. Southward with increasing temperature, these become Gley Soils. At the far north the soils are rather like those of the hot deserts, and are consequently called Polar Desert Soils; they develop salt crusts, alkali flats, and layers of carbonate accumulation.

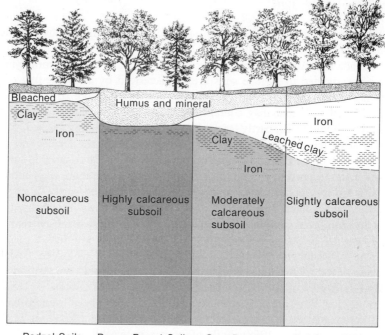

FIGURE 8.9

Differences among soil profiles in New York are in large part due to differences in the lime content of the parent materials. Soils under coniferous forest on ground that is not limy or only slightly so (left) have a surface layer of raw acid humus 2 to 4 inches thick over a bleached layer that is also 2 to 4 inches thick. Below this is a yellowish-brown layer in which iron and clay have accumulated. The layer of accumulation grades downward to lime-free, acid parent material at about 2 feet. Soils on highly calcareous ground, mostly under conifer or northern hardwood forest, are without a noticeable leached layer. The top layer is rich in organic matter; lime is leached to depths of 18 to 24 inches, and a little clay has begun to form immediately below it. The subsoil below this contains free lime. In soils developed on moderately calcareous ground, the leaching is correspondingly deeper, down to about 30 inches. The clay layer is deeper and better developed than where the subsoil is highly calcareous, and above it is a grayish-brown leached layer. Other dark soils (Sol Brun Acide) develop under hardwood forest on ground that is only slightly calcareous; the depth of leaching is more than 3 feet. The upper part of the leached layer is stained yellow with iron; the lower part, enriched with clay, is a kind of hardpan. [After Pearson and Cline, 1958.]

Dark Forest Soils

South of the Valders Moraine (p. 133), the extensive deposits of the Wisconsinan glaciation are the parent materials chiefly for acid forest soils having a brown or gray brown leached layer—the Gray Brown Podzolic Soils. Outliers of soils having a conspicuous, light colored, sandy leached layer are included, and toward the west are some acid grassland soils (Brunizem, or Prairie Soil). These soils developed during late Wisconsinan and Holocene time; they are older than the Podzol Soils farther north, perhaps twice as old.

The Gray Brown Podzolic Soils are developed mostly under broadleaf forests. Typically, they have an organic layer consisting of leaf litter an inch or two thick overlying another inch or two of dark, grayish brown granular humus. The organic layer is thicker than under the more northerly soils and may contain 5 to 10 percent organic matter, but it also contains more bases (alkalis and alkaline earths returned by the deciduous leaves) and generally is less acid. Below these organic layers is a grayish brown leached layer, which may extend to a depth of 8 or 10 inches. The layer of deposition is more clayey than the leached layer, and generally more clayey than the layer of deposition in the more northerly and younger soils that developed on comparable parent materials. The layer of deposition may be yellowish brown or light reddish brown; it grades downward to the parent material at 3 or 4 feet in depth. Carbonates are leached from all the layers, commonly to a depth of 6 feet, which is 2 or 3 times as deep as the average depth of leaching of carbonates in the more northerly and younger soils. This of course partly depends on the amount of calcium carbonate in the parent material. See, for example, Figures 7.3, 8.9.

The differences in kind and degree of leaching in these soils and in the Podzol Soils farther north may be attributed to several factors: differences in age; in parent material (less carbonate on the Canadian Shield); in climatic setting and vegetative cover; and in land use. The organic layers are very different, the northerly soils being under coniferous forest and the more southerly ones under broadleaf forest. Differences in the mineral fractions of the leached layers—the greater sandiness in the north—may be due partly to physical sorting of the mineral grains by the action of freezing and thawing; freezing pushes coarse material upward, and thawing allows the fine material to settle downward. Changes in land use may convert a light-colored leached layer into a brownish one or cause mixing so the layers become indistinct.

Acid forest soils, both on the western mountains and in the northeastern United States, have somber colors and are developed on glacial, loessial, lacustrine, alluvial, and colluvial deposits that are no older than Wisconsinan. The chemical and mineral compositions of these soils reflect the compositions of the deposits on which they have formed.

Grassland Soils

Continuing the transect westward across the United States, we find that west of the forest soils the organic-rich soil layers thicken under deep-rooted tall grass on the prairies (Fig. 8.7). The easternmost grassland soils (*Prairie* or *Brunizem Soils*) are acid. Like other grassland soils, they have a deep, black organic layer, but unlike other grassland soils farther west, they are without a layer of lime accumulation. Annual precipitation is in the range of 25 to 30 inches—sufficient to maintain an open system and flush soluble constituents from the bottom of the profile. The soils are leached of carbonates to a depth of 3 to 6 feet; unlike the acid forest soils, however, they have been leached only slightly of iron and aluminum. South of the youngest drift border, these soils are developed mostly on Wisconsinan loess.

Where forests have advanced over such ground, the upper soil layers have become more acid, and iron and aluminum have been leached from them; the ground takes on some aspects of a forest soil, and is therefore referred to as *Degraded*. Another kind of acid grassland soil has a layer of deposition that is a clay hardpan. These soils, *Planosols*, form mostly on poorly drained, flat lands on clay or silty surface deposits, mostly till, loess, or alluvium. They are controlled more by the topography and clayey kind of ground than by the climatic zonation.

West of the acid soils, in a belt where the annual precipitation is between 20 and 25 inches, still under deep-rooted tall grass, are alkaline grassland soils, *Chernozem Soils.* In these the organic-rich layer may be as much as $2\frac{1}{2}$ feet deep and consists of mineral matter blackened with 5 to 20 percent of organic matter. Below this is a layer in which lime has accumulated; this layer, about 3 feet thick, contains 15 to 30 percent lime carbonate. These and the acid grassland soils to the east are among our richest soils; although they have lost monovalent ions, sodium and potassium, they contain plenty of calcium available for plant use. Surface layers of Chernozem Soils are neutral or even slightly acid; the lower layers are alkaline. The parent material for these soils in the Central United States is mostly loess.

On the Great Plains, at about the 100th meridian (Figs. 8.4, 8.7), the annual precipitation averages about 20 inches. Farther west, the precipitation averages less than 20 inches. In this belt, annual precipitation averages 15 to 20 inches, and the vegetation consists mostly of short, shallow-rooted grasses. The soils are alkaline grassland soils—Chestnut and Brown Soils—but the layers are thinner than to the east. West of the Rocky Mountains, on the Colorado Plateau and Great Basin, precipitation averages 10 inches or less annually, the vegetation is mostly shrub with little grass, and the soils, desert soils, are very thin.

The alkaline soils under short grasses have dark brown organic layers 8 to 15 inches thick. The content of organic matter is less than under the tall grasses, generally no more than 5 percent. Desert soils generally are without an organic layer except on ground immediately underlying shrubs or bunches of grass, and even there the organic layer is rarely more than an inch thick. The leached layers in the grassland soils are 12 to 20 inches thick and lie above a lime-enriched layer about 1 to 2 feet thick. In desert soils the leached layer generally is less than 6 inches thick and the lime-enriched layer 6 to 10 inches thick.

Soils on the Western Mountains

On the western mountains, soils are zoned in much the same way as they are across the plains (Fig. 8.10). With increasing altitude, temperatures decrease and precipitation increases. Foothills of the mountains are semiarid, with an annual precipitation of 10 to 15 inches; on the high parts the precipitation is 25 to 30 inches. Below what is called the *arid timberline,* which averages around 6,500 feet in altitude, the vegetation consists of shrubs, and the soils are Brown or Desert Soils. Above this is a succession of different kinds of forests culminating in spruce-fir forest at the upper timberline, like the northernmost forests on the continent. Below the arid timberline, the evaporation rate exceeds the average annual precipitation by a factor of as much as ten; about 2,000 feet higher on the mountains, however, the evaporation rate roughly equals precipitation. Moving upward on the mountains one finds, in succession, soils like the alkaline grassland soils, acid grassland soils, and, under the upper forests, acid forest soils. The transition from alkaline to acid soils commonly is between 8,000 or 9,000 feet altitude. The summits are rough and rocky and have little soil, but in places there is a tundra-like soil called Alpine Meadow Soil.

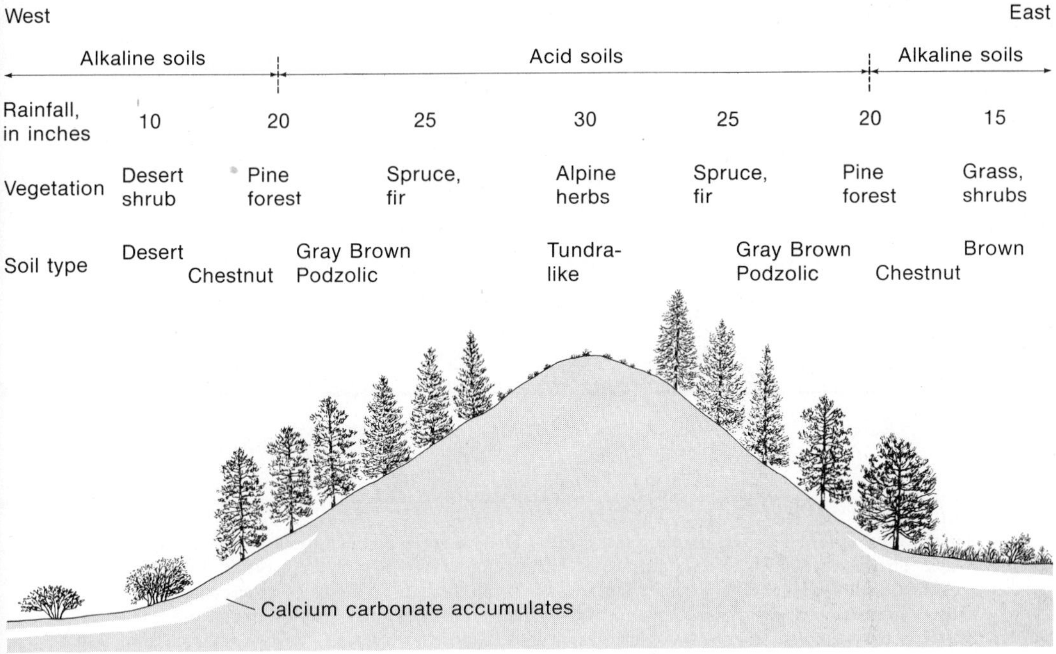

West   East

	Alkaline soils			Acid soils			Alkaline soils	
Rainfall, in inches	10	20	25	30	25	20	15
Vegetation	Desert shrub	Pine forest	Spruce, fir	Alpine herbs	Spruce, fir	Pine forest	Grass, shrubs
Soil type	Desert Chestnut	Gray Brown Podzolic		Tundra-like		Gray Brown Podzolic Chestnut	Brown

Calcium carbonate accumulates

FIGURE 8.10
Generalized relationship between annual precipitation (in inches), vegetation, and soils on mountains in the western United States. At the summits, the growing season is less than 60 days; at the base of the mountains the growing season is 4 to 5 months.

This vertical zoning of soils on the western mountains accords with the climatic zoning and with the zoning of the vegetation, but it also accords with a geologic zoning by age, like the north-south zoning on the plains and farther east. The summits of the western ranges are young surfaces, for they were covered by snow or ice fields in late Pleistocene time. The soils there are young (Holocene), like the tundra soils north of the spruce-fir forest. The soils on the flanks of the mountains average somewhat older, but most of them are no older than the Wisconsinan glaciation because at that time there was vigorous mass-wasting on the mountain slopes below the limit of ice. The pediments and fans that slope away from the foot of the mountains preserve extensive surfaces and soils dating from pre-Wisconsinan time—the red soils.

The Sierra Nevada duplicates the relations on the Rocky Mountains. The summits are bare. Well down on the west flank is a belt of dark acid forest soils, and below these, down to the lower timberline is a wide belt of pre-Wisconsinan soils. These three belts have different climates, but their surfaces are of quite different ages too. The summit surfaces and canyon floors are Holocene; some intermediate slopes are covered by soils of Wisconsinan age; and the foothill belt and inter-canyon soils are of pre-Wisconsinan age.

Another well-developed series of zoned soils occurs on the Columbia Plateau; this series is illustrated in Figure 8.11.

190

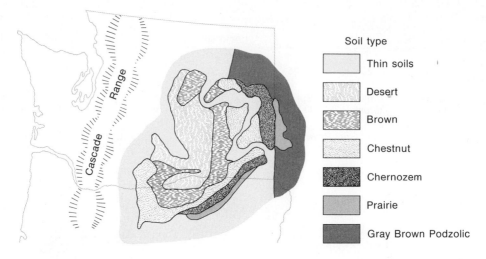

FIGURE 8.11

Soils in southeastern Washington and north central Oregon are zoned like those in the central United States. The region for which soils are mapped lies in the rainshadow of the Cascade Range. Annual precipitation is less than 10 inches in the area of Desert Soils, and increases progressively eastward across the belts of Brown Soils, Chestnut Soils, Chernozem Soils, and Prairie Soils to more than 25 inches in the Northern Rocky Mountains, which have Gray Brown Podzol Soils. A transect from east to west in eastern Washington would duplicate the changes in soils illustrated in Figure 8.7. [After U.S.D.A.]

Reddish Soils in the Southwest

In the Osage Plains, southern part of the Great Plains, and southern sections of the Basin and Range Province, the alkaline grassland and shrubland soils are reddish and are classed as Reddish Prairie, Reddish Chestnut, Reddish Brown, and Red Desert Soils. These soils have thick lime-enriched layers—caliche—and generally have a pinkish caste due to admixed red clay. Overlying the caliche is pink, red, or red brown clay which in turn is topped by a rich organic layer, which thins westward, as it does in the more northerly soils. The clay and caliche layers, however, thicken and thin irregularly, for they are ancient deposits developed to variable thicknesses and eroded irregularly. In places, the caliche layer is at the surface and the clay layer absent. The organic layer and the leached layer immediately under it are of late Pleistocene or Holocene age; the reddish clay and caliche are of pre-Wisconsinan age, although the Red Prairie Soils inherit their color from Permian red beds. The reddish soils are largely ancient multiple soils having one kind of profile superimposed on another, a subject treated more fully in the next chapter.

The boundary between the reddish soils and the younger ones to the north is a sharp geologic contact where the deposits of Wisconsinan or Holocene age overlap the pre-Wisconsinan soils. Examples are given in the next chapter. This contact is as sharp as any transgressive overlap contact in the geologic column; the one kind of soil does not grade laterally into the other. Transitional soils are found only where the pre-Wisconsinan soils have been reworked and mixed with younger materials to form the parent material for a mixed, or transitional, soil (p. 216). Where the reworked pre-Wisconsinan soil predominates, the younger soil is classed as one of the red ones; where the younger surface deposit predominates the young soil is classed with the nonred soils. The transition, however, is entirely because of sedimentary mixing and not because of climatic zoning.

Red and Yellow Soils

South of the limit of glacial drift in the eastern United States (Fig. 2.2) are acid soils that are bright red or yellow and have developed on ancient, deep, residual deposits. These soils, the Red and Yellow Podzolic Soils (Fig. 8.12), have mineral and chemical characteristics largely inherited from the deeply weathered parent material. There is no gradation between the northerly somber-colored forest soils and the more southerly red and yellow ones, as there is between the forest and grassland soils illustrated in Figure 8.7. The occurrence of red and yellow soils depends on the preservation of the ancient clayey residuum, or saprolite. North of the limits reached by the glaciers, the residuum was either eroded by the ice or buried by the glacial drift, and it is found only in exceptional circumstances. On the other hand, south of the drift border, soils like those on the Wisconsinan drift—notably the Gray Brown Podzolic Soils—are developed on alluvial, loessial, and other deposits of Wisconsinan or younger age. These may overlie residuum or rest on bedrock where the residuum had been removed by erosion. They may develop on the residuum, and bear some resemblance to the red and yellow soils where mass-wasting on hillsides has caused creep and mixing of the upper layers of the ground. Because of mass-wasting and other erosion, the red and yellow soils are only spottily preserved in the mountainous areas.

The red and yellow soils change southward with increases in moisture and temperatures, which accelerate the biochemical and other processes. Leaching is more intense in warmer latitudes than in the colder latitudes. In some soils the leaching is so intense that, despite lush plant growth, the organic layers are very thin, and the leached zone may extend all the way to the water table.

Lateritic Soils

The ancient residual deposits that are parent material for the red and yellow soils exemplify an important distinctive process of soil development called *lateritization*. *Laterite*, in the restrictive use of the term (see Pendleton, 1936, Pendleton and Sharasuvana, 1946), refers to the oxides and hydroxides, chiefly of iron and aluminum, which usually form a layer of nodules or a continuous layer precipitated from soil solutions. Such deposits are extensive in wet tropical regions, and the soils in which they form have been depleted of alkalis, alkaline earths, and silica; whatever silica remains is largely a constituent of the clay mineral kaolinite. The total soil is enriched in iron and aluminum, sufficiently so that some deposits may be mined as ore.

The term "Lateritic Soil" has sometimes been used with reference to the ancient residual deposits on the Piedmont Province of eastern United States. Only in a broad sense, however, are these soils related to true Lateritic Soils. Although depleted in

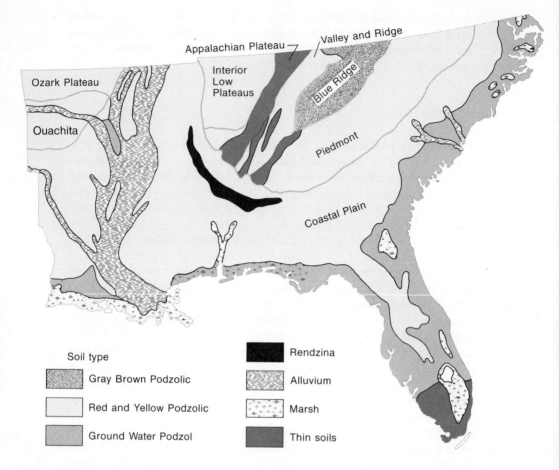

FIGURE 8.12

Soils of southeastern United States. Most of the region has red and yellow soils, which are developed on clayey, deeply weathered ancient soils. On the Piedmont Province the parent materials for the ancient soils are crystalline rocks, in large part igneous and metamorphic. On the Coastal Plain the ancient soils are developed on Cretaceous and Tertiary sedimentary formations. In the Valley and Ridge Province, Interior Low Plateaus, Ozark Plateau, and Ouachita Province, the ancient soils are developed on Paleozoic sedimentary rocks. Bordering the Mississippi River alluvial section, especially to the east, are thick loess deposits that are parent materials for the pre-Wisconsinan soils there. Brownish acid forest soils have formed on the slopes of the Blue Ridge; on these steep slopes the ancient soils are largely reworked by mass wasting and washing to form the parent materials for the brownish soils. Acid grassland soils (Rendzina) have developed on the marly Cretaceous formations that crop out along the Black Belt of Alabama and Mississippi. Along the coast are marshes. Surfaces that are slightly elevated but have shallow water table have a thin mat of organic matter over a reddish sandy layer 2 or 3 feet thick, at the base of which is a layer stained brown with organic matter collected just above the water table (Ground Water Podzol Soils). [After U.S.D.A.]

alkalis, alkaline earths, and silica and enriched in iron, aluminum, and clay, they contain no true laterite (Pendleton and Sharasuvana, 1946). Most of the iron is in the form of oxides and most of the aluminum in the form of silicate clay minerals. The modern red and yellow soils developed on these residual deposits are sometimes referred to as *latosols*, in reference to their similarity to the Lateritic Soils.

The processes by which true laterite and the Lateritic Soils are formed are not entirely understood. Iron and aluminum are precipitated in alkaline environments and dissolved in acid ones, yet so far as we know the environments in which the laterite was precipitated from solution were acid, and highly so. True laterite supposedly forms in a layer where the water table fluctuates from high to low. Where true laterite exists at the surface, it is generally attributed to exposure by erosion of the overlying soil layers.

Lateritic soils in Hawaii (Cline and others, 1955) are characterized by a surface layer containing organic matter and ranging from about 3 to about 12 inches thick. The organic matter ranges from as little as 3 percent to 20 percent or more. Soils having little organic matter commonly have a concentration of manganese dioxide in the surface layer. Below this is a layer 3 or 4 feet thick in which sesquioxides (Fe_2O_3 and Al_2O_3) have been concentrated. Below this is deeply weathered parent material (in large part derived from basaltic lavas) consisting chiefly of secondary minerals. There is no layer of deposition; concentrations of particular constituents have resulted from the removal of others. In particular, all layers have lost silica, alkalis, and alkaline earths.

The soils without much organic matter in the surface layers are only slightly acid and retain more silica, alkalis, and alkaline earths than do the strongly acid soils containing much organic matter. The strongly acid soils generally are in the areas of high rainfall and lush vegetation. In the less acid lateritic soils, the aluminum is combined with silica to form a clay mineral, kaolinite, as it is in the red and yellow soils. The iron occurs as oxides and hydroxides; in the strongly acid lateritic soils it may total nearly 50 percent. In these soils much or most of the aluminum also occurs as oxides and hydroxides and may total more than 30 percent. The silica content may be reduced to less than 10 percent, and the alkalis and alkaline earths reduced to less than 2 percent. These alterations by weathering are impressive when compared to the average composition of unaltered Hawaiian lavas from which most of the soils were derived—roughly 50 percent silica, 14 percent iron oxide, 14 percent aluminum, and 20 percent of alkalis and alkaline earths.

To what extent true laterite is forming today in the tropics is not clear. The literature seems to indicate that most true laterite and the extensive Lateritic Soils in the loose sense of the term are ancient. They have been so described in the West Indies, Australia, Queensland, India, and China (Goldich and Bergquist, 1948; Kaye, 1951; Prescott and Pendleton, 1952; Whitehouse, 1940; for refs., see Chapter 9); Africa (Finck, 1963, p. 48, Fig. 3.14), East Indies (Van der Voort, 1950), and Indonesia (Gourou, 1945, quoted by Mohr and Van Baren, 1954, p. 358).

Ground Water Podzol

In sandy ground in humid climates, leaching may be intensive enough to extend all the way to the water table. Where the water table is shallow and fluctuates, the organic matter washing downward may accumulate and be preserved in a zone kept moist by capillarity above the fluctuating water level. Such soils undergo podzolic leaching and are referred to as Ground Water Podzols. They are common on flat sandy stretches of the coastal plain in the southeastern United States and in poorly drained ground in the glaciated northeastern and north-central United States.

BIBLIOGRAPHY

Baldwin, M., Kellogg, C. E., and Thorp, J., 1938, Soil classification: *in* Soils and Men, U.S. Dept. Agriculture Yearbook, 1938, pp. 979–1001.

Beatty, M. T., and others, 1964, Soils of Wisconsin: Wisconsin Blue Book, University of Wisconsin, pp. 149–170.

Berger, K. C., 1965, Introductory soils: MacMillan, New York, 371 pp.

Bunting, B. T., 1965, The geography of soil: Aldine Publ. Co., Chicago, 213 pp.

Cline, M. G., 1949, Basic principles of soil classification: Soil Sci., v. 67, pp. 81–91.

———, 1963, Soils and soil associations of New York: Cornell Ext. Bull. 930, 64 pp.

Cline, M. G., and others, 1955, Soil Survey of the Territory of Hawaii: U.S. Dept. Agriculture Soil Survey Ser. 1939, no. 25, pp. 67–95.

Donahue, R. L., 1965, Soils—An introduction to soils and plant growth (2nd ed.): Prentice-Hall, Englewood Cliffs, N.J., 363 pp.

Drew, J. V. (editor), 1967, Selected papers in soil formation and classification: Soil Sci. Soc. America, Special Publ. No. 1, 428 pp.

Finck, A., 1963, Tropische Böden: Verlag Paul Parey, Hamburg and Berlin, 188 pp.

Glinka, K. D., 1927, The great soil groups of the world, and their development: Ann Arbor, Mich., 235 pp. Trans. by C. F. Marbut.

Jennings, J. N., and Mabbutt, J. A., 1967, Landform studies from Australia and New Guinea: Cambridge Univ. Press, 434 pp. (See chapters on soils by John Hays, M. J. Mulcahy, and B. Butler.)

Jenny, H., 1941, Factors of soil formation: McGraw-Hill, New York.

Joffe, J. S., 1949, Pedology (2nd ed.), Rutgers Univ. Press.

Kelley, W. P., 1951, Alkali soils: Reinhold, New York.

Kubiana, W. L., 1953, Soils of Europe, diagnosis and systematics: Murby, London.

Lyon, T. L., and Buckman, H. O., 1943, The nature and properties of soils: Macmillan, New York.

Marbut, C. F., 1935, Atlas of American Agriculture: pt. III, Soils of the United States, U.S. Govt. Printing Office, Washington, D.C.

———, 1951, Soils, their genesis and classification: Memorial Volume, Soil Sci. Soc. America, 134 pp.

Miller, C. E., Turk, L. H., and Foth, H. D., 1966, Fundamentals of soil science: Wiley, New York, 491 pp.

Mohr, E. C. J., and Van Baren, F. A., 1954, Tropical Soils: Interscience, New York, 498 pp.

Nikiforoff, C. C., 1949, Weathering and soil evolution: Soil Sci., v. 67, no. 3, pp. 219–230.

———, 1959, Reappraisal of the soil: Science, v. 129, pp. 186–196.

Pearson, C. S., and Cline, M. G., 1958, Soil survey of Ontario and Yates Counties, New York: U.S. Dept. Agriculture Soil Survey Series 1949, no. 5, 126 pp.

Pendleton, R. L., 1936, On the use of the term laterite: Am. Soil Surv. Assoc., Bull. 17, pp. 102–108.

———, 1949, The classification and mapping of tropical soils: Comm. Bur. Soil Sci. Tech. Comm. 46, pp. 93–97.

Prescott, J. A., 1931, The soils of Australia in relation to vegetation and climate: Bull. Council Sci. Ind. Res. Australia, no. 52, p. 82.

Riecken, F. F., and Smith, G. D., 1949, Lower categories of soil classification: family, series, type, phase: Soil Sci. v. 67, pp. 107–115.

Robinson, G. W., 1949, 1951, Soils, their origin, constitution and classification (3rd ed.): Murby, London, 573 pp.

Ruhe, R. V., and Scholtes, W. H., 1956, Ages and development of soil landscapes with relation to climatic and vegetational changes in Iowa: Proc. Soil Sci. Soc. America (May).

Smith, G. D. (editor), 1960, Soil classification—a comprehensive system, 7th Approximation: Soil Survey Staff, Soil Conservation Service, U.S. Dept. Agriculture.

Soil Survey Staff, 1951, Soil Survey Manual, U.S. Dept. Agriculture Handbook No. 18, 503 pp.

Stephens, C. G., 1949-50: Comparative morphology and genetic relationships of certain Australian, North American, and European soils: Jour. Soil Sci., pp. 123-249.

Tedrow, J. C. F., and Cantlon, J. E., 1958, Concepts of soil formation and classification in arctic regions: Arctic, v. 11, pp. 166-179.

Tedrow, J. C. F., and Harries, H., 1960, Tundra soil in relation to vegetation, permafrost, and glaciation: Oikos, v. II no. 2, pp. 237-249.

Terrill, S. E., 1956, Laterite and the materials of similar appearance in S. W. Australia: Jour. Royal Soc. W. Australia, v. 40, pp. 4-14.

Thorp, J., and Smith, G. D., 1949, Higher categories of soil classifications: Order, Suborder, and Great Soil Groups: Soil Sci., v. 67, pp. 117-126.

Van der Voort, M., 1950, The lateritic soils of Indonesia: Trans. 4th Internat. Cong. Soil Sci., v. 1, pp. 277-281.

Wilde, S. A., 1958, Forest soils: Ronald Press, New York.

Yaalon, Dan H., 1963, On the origin and accumulation of salts in groundwater and in soils of Israel: Bull. Research Council of Israel, v. 11G, no. 3, pp. 105-131.

In ancient soils, hard rocks commonly are disintegrated, and may even be decomposed to clay. Rocks altered to clay usually retain the appearance of the original rock and its structure; a granite so altered may look like a granite but may be cut easily with a pen knife—because it is clay. The weathering is not only very much more advanced than in modern soils: it extends very much deeper, in places to depths of more than 100 feet. [Photograph by John Stacy, U.S.G.S.]

9/ Ancient Soils and Superimposed Profiles

Soils as a continuum in geologic history; layering in ancient soils; diagnostic features; contrast in weathering and erosion with Wisconsinan and Holocene soils; superposition of weathering profiles; determining superposition; stratigraphically datable superimposed profiles at the drift border in northeastern United States; superimposed profiles south of the drift border in eastern United States; superimposed profiles in central United States; superimposed profiles of alkaline soils on the Great Plains; stratigraphy of superimposed profiles in the Lake Bonneville Basin; some implications of superimposed profiles; the problem of exceptions; the most ancient soils; bibliography.

SOILS AS A CONTINUUM IN GEOLOGIC HISTORY

The preceding chapter described some of the many varieties of modern soils; the number of varieties becomes even greater when soils, both the ancient and the modern, are considered as a geologic continuum. Soils, after all, are not static; they undergo constant, though gradual change. When climates change, as has happened repeatedly during the latter part of the Cenozoic Era, the biota, the physical-chemical environment, and the processes of soil formation all change.

By ancient soil is meant the saprolite and similar deep residual deposits attributable to weathering in pre-Wisconsinan time, as described in Chapter 7. Some of the ancient soils developed immediately before the Wisconsinan glaciation; others developed as long ago as late Mesozoic. The ancient soils, formed by deep weathering of pre-

Wisconsinan deposits, consist of layers with quite different kinds of alteration. As described by Leighton and MacClintock (1930) these are:

Layer 1: modern soil (see Chapter 8).

Layer 2: chemically decomposed parent material; considerable clay developed by alteration of hard rocks.

Layer 3: leached of carbonates and other soluble salts and oxidized; otherwise little altered.

Layer 4: oxidized, but not leached and not otherwise altered; may contain some clay washed from above.

Layer 5: unaltered parent material.

The diagnostic feature of the ancient soils is the chemical decomposition of hard rocks and their alteration to clay (*argillic alteration*) in Layer 2. Boulders or cobbles of granite may be so altered that they can be sliced by a knife. In contrast, deposits of the Wisconsinan glaciation and younger deposits are little weathered. The kind of alteration attributable to Wisconsinan and later weathering is the kind seen in Layer 3 or Layer 4 of the ancient soils—that is, leaching of soluble salts and oxidation, but little or no argillic alteration. Moreover, the intensity and depth of the leaching and oxidation in Wisconsinan and younger soils are generally less than in Layers 3 or 4 of the ancient soils. The depth to which leaching and other alteration extends depends on the abundance and strength of the soil solutions, the resistance of the minerals to alteration, the ground permeability, and the length of time the alteration has progressed.

On moraines and other deposits of late Wisconsinan age, most stones, even in the upper layers, are firm and have only a very thin weathering rind. They are darkly stained with desert varnish. On early Wisconsinan deposits most stones are firm, but many have an oxidized weathering rind an eighth or quarter inch thick. Of the stones on the surface or in the upper layers of the ancient soils, however, a high percentage are clayey, crumbly, or otherwise unsound—a fact of economic as well as of stratigraphic interest, for it limits the usefulness of the gravel.

Along with this contrast in weathering and soil development on Wisconsinan and pre-Wisconsinan deposits is a difference in the degree of erosion of the deposits. Whether in the eastern, central, or western parts of the United States, the moraines and other glacial deposits of Wisconsinan age retain the constructional geomorphic form they had when they were originally laid down by the ice or its meltwaters. The landforms have not been greatly modified by erosion. The pre-Wisconsinan deposits, however, are not only deeply weathered but are so eroded that only small remnants exist; the original geomorphic forms have been destroyed.

The modern soils developed on these ancient ones are mostly acid red and yellow soils (see Chapter 8), generally 3 to 4 feet deep. The chief characteristic of those soils, the red and yellow clay, is largely inherited from the ancient soil, which is the parent material for the modern one.

The ancient soils are missing north of the Wisconsinan drift border except in a few isolated localities where excavations penetrate the drift or where the drift is absent and the weathered rock is exposed. Along the drift border, the Wisconsinan glacial deposits overlap the old soils. Similar ancient soils are also preserved in isolated patches in the Rocky Mountain region and in the Pacific Northwest; these too are overlapped by Wisconsinan deposits.

SUPERPOSITION OF WEATHERING PROFILES

From the preceding it follows that a pre-Wisconsinan soil that is, say, 15 feet deep, may be the parent material for a superimposed Wisconsinan soil perhaps 4 feet thick. The Wisconsinan soil may, in turn have superimposed on its upper layers a Holocene soil that is perhaps 2 feet thick. Where one soil profile has been superimposed on another, the two may be distinguished by tracing the soil layers laterally to a place where the older profile is clearly separable from the younger one—for example, to a place where a surface deposit intervenes between the layers of the old profile and the layers on the superimposed profile. An actual example is shown in Figure 9.1. Superimposed on the hill of Cretaceous shale and sandstone are soils of three ages. Laterally the pre-Wisconsinan soil becomes separated from the younger soils by a deposit of early Wisconsinan eolian, silty sand, which can be dated paleontologically and by tracing it laterally to Wisconsinan glacial outwash with which it merges. The pre-Wisconsinan soil has an upper clayey layer above a layer of caliche about 3 feet thick. Where this soil is buried, its weathering profile is just as well developed and just as thick as it is where the soil is exposed at the surface. We can conclude, therefore, that the argillic alteration was fully developed before it was buried by the silty sand.

The Wisconsinan soil has undergone no alteration to clay. Below the organic layer is a leached layer that is 10 to 15 inches thick and contains no free lime, but below this is a darker layer of about the same thickness that is whitened with nodules, veins, or finely disseminated calcium carbonate. The leaching and deposition of the caliche is just as well developed and just as deep where this soil is buried as it is where it is exposed at the surface. Evidently that weathering profile had developed before it was buried by the Holocene sand that overlaps it. That sand is dated archeologically, and on it is a Holocene weathering profile that extends continuously across the Wisconsinan and pre-Wisconsinan soils. The Holocene soil consists of an organic layer about 3 inches thick. Under it is a lime-free, leached layer about 6 inches thick that grades downward to another layer about 6 inches thick in which a little lime has accumulated.

Figure 9.1 shows that the Wisconsinan soil on the hill is the parent material for the Holocene soil, and that the pre-Wisconsinan soil is the parent material for the Wisconsinan soil. It must be stressed further that, not only are the three soils separable, but each of the older soils was fully formed *before* the overlapping sand was deposited. Weathering during Wisconsinan and Holocene time has contributed little to the development of the pre-Wisconsinan profile, and weathering during Holocene time has contributed little to the development of the Wisconsinan profile. The differences in the degree and kind of weathering represented by these soils are not due to the factor of *time* per se, but are attributable to differences in the processes that operated during the development of each. For example, in many places the depth of the old soils is very much greater than the depth to which moisture now penetrates.

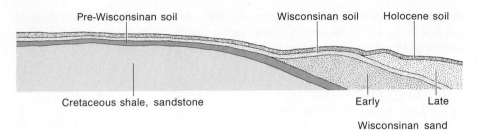

FIGURE 9.1
Section on the Great Plains near Denver, Colorado, illustrating how superimposed soil profiles can be traced laterally to where unweathered parent materials separate the soils of different ages. At the right, three soil profiles of different ages are separated by unweathered deposits datable by fossils and artifacts. Toward the left, the late Wisconsinan sand thins and the Holocene soil is superimposed on the Wisconsinan soil. Farther left, the early Wisconsinan sand thins and the soils are superimposed on a pre-Wisconsinan soil. [After Hunt 1954, p. 127.]

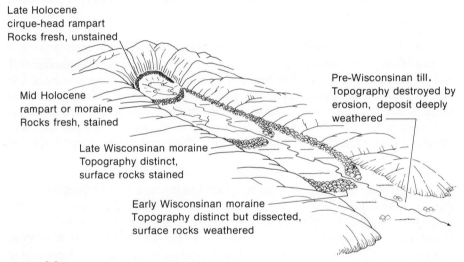

FIGURE 9.2
The arrangement and weathering characteristics of glacial deposits in a western mountain valley suggest a series of glacial and interglacial episodes. The record, though, is incomplete, and each of the glacial advances represented could have reworked and destroyed the deposits left by earlier, lesser glacial advances.

Such discontinuities in soil stratigraphy suggest episodes of soil development, but this is not so, for soil development is a continuous process. Soil stratigraphy is like glacial stratigraphy in that only maxima are completely preserved. Glacial deposits are arranged in the order of increasing age away from the source. This is true whether the deposits were laid down by continental glaciers (Fig. 2.2) or by mountain glaciers (Fig. 9.2). Between any two glacial maxima, there may have been lesser glacial advances whose records have been largely destroyed by a later, greater advance. Because only the maxima are preserved, the stratigraphic record gives the impression of a series of glacial episodes rather than a continuum with fluctuations.

The same is true of soils. A climatic change that causes weathering to a depth of 10 feet could destroy all record of a previous environment in which the ground was weathered to only a depth of 5 feet. Moreover, a climatic change may increase precipitation sufficiently to change an alkaline soil, with its closed-system mode of weathering, toward an open-system, acid soil; the alkaline soil may even become acid if the soluble constituents stored in the carbonate layer are flushed from the system. On the other hand, if a climatic change were to reduce precipitation so that an acid soil became altered by closed-system weathering, the layers of the acid soil would not accumulate much alkaline material because the readily soluble constituents would already have been largely removed. Unless loessial material is added to the surface, a likely result of a reduction in precipitation, the only source of carbonate would be the organic matter, for the calcium would already have been depleted. The complexities of climatic change seem to favor the preservation of acid soils and the alteration of alkaline soils to acid ones.

These kinds of relationships are found in every state, and have been reported and described in most of them. The remainder of this chapter presents actual examples of superimposed soil profiles.

STRATIGRAPHICALLY DATABLE SUPERIMPOSED PROFILES AT THE DRIFT BORDER IN NORTHEASTERN UNITED STATES

In northern New Jersey, the Wisconsinan drift is composed of firm, hardrock cobbles and boulders, and is parent material for acid Brown Forest Soils. A pre-Wisconsinan drift, the Jerseyan drift, contains boulders and cobbles that are altered to clay, and this deeply weathered drift is parent material for red and yellow soils. Outwash from the Wisconsinan drift overlaps the Jerseyan drift.

Another example is found in northern Pennsylvania, where Wisconsinan drift is parent material for Gray Brown Podzolic Soils and Sol Brun Acide; the drift overlaps red, deeply weathered, older drift in which the boulders and cobbles are altered to clay (Fig. 9.3). The pre-Wisconsinan soil is parent material for a Red and Yellow Podzol Soil (Sweden Soil); southward from the drift border, this soil grades into yellowish-brown, strongly acid, friable loam about 3 feet thick overlying reddish-brown, yellowish-red, or red silt-clay 3 to 7 feet thick containing pebbles that have been altered to clay.

SUPERIMPOSED PROFILES SOUTH OF THE DRIFT BORDER IN EASTERN UNITED STATES

South of the glacial drift border, the exact ages of the surface deposits and their soils are uncertain; only the relative ages of the deposits can be determined, by superposition. But the kinds of soils are similar to those that can be dated strati-graphically at the drift border. At Turkey Hill, on the Delaware River (Fig. 9.4, A), the Cape May Formation, containing outwash of Wisconsinan drift, is only

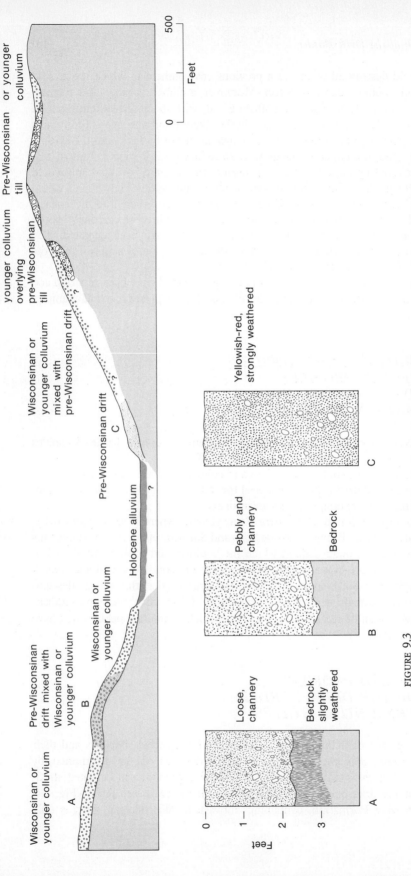

FIGURE 9.3

Surface deposits and soils in an area south of the Wisconsinan drift border in northern Pennsylvania. Idealized section shows dissected uplands mantled by pre-Wisconsinan drift, Wisconsinan or younger colluvium, and mixtures of these deposits. The soils are chiefly Sols Bruns Acides (A), Gray-Brown Podzolic Soils and Red and Yellow Soils (B), and Red and Yellow Podzolic Soils (C). The Holocene alluvium on the flood plain probably overlies Wisconsinan outwash. Thickness of surface deposits exaggerated. [After Lyford, 1963.]

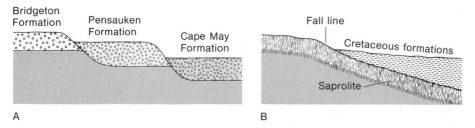

FIGURE 9.4

(A) Cross section at Turkey Hill, by the Delaware River, showing relations of the little-weathered Cape May Formation and the deeply weathered older ones. [After Lockwood and Messler, 1960]. The Bridgeton Formation caps the highest hills. The Pensauken Formation (pre-Wisconsinan, possibly Illinoian) consists of deeply weathered mixed glacial outwash and fluvial deposits from Cretaceous and Tertiary formations to the south and east. Near the surface is lag gravel with ventifacts; this is capped by loess. The Cape May Formation consists of outwash from Wisconsinan glacial drift; it is unweathered. Both the Bridgeton and Pensauken formations contain warm-climate plant fossils. (B) Cross section of the Fall Line at Baltimore, showing Cretaceous formations overlapping saprolite.

slightly weathered, whereas older deposits, the Pensauken and Bridgeton formations, are deeply weathered. Both the Pensauken and the Bridgeton formations contain plant remains indicative of a warm climate.

Still farther from the drift border, along the Fall Line in Maryland and even as far south as southern Virginia, Cretaceous and Tertiary formations of the Coastal Plain unconformably overlap saprolite developed on the metamorphic rocks of the Piedmont Province (Fig. 9.4, B). Some of the alteration that formed the saprolite may be younger than the overlying Cretaceous and may have been caused by ground water seeping along the base of the saprolite. But most of the weathering must be older than the Coastal Plain formations, for these formations contain masses of reworked saprolite. The contact between the saprolite and the sediments derived from it is gradational, and the sediment is difficult to distinguish from the parent soil.

Figure 9.5 shows examples of saprolite developed on various kinds of formations in the Appalachian Highlands. At some places in this region, there is evidence that ground-water weathering considerably deepened the saprolite. For example, along the Gunpowder Falls River, in the Baltimore area, saprolite is developed to a depth about 50 feet lower than the stream bed (Fig. 9.5, C), and the base of the saprolite is 30 feet lower than the stream channel where it crosses quartzite.

In the Tennessee and Shenandoah valleys (Fig. 9.5, A, B), saprolite is also developed much deeper than the river channels. Such deepening of the saprolite must be due to weathering by ground water, but when did the weathering take place? The relationships along the Fall Line, and the lack of comparable subsurface weathering north of the border of the Wisconsinan glacial drift, suggest that most of the ground water weathering dates from pre-Wisconsinan time.

In the southern Appalachians, depressions that probably originated as sinkholes contain slumped beds of bauxite and kaolinitic clay, carbonaceous clay, silt, and sand; some of these deposits have yielded early Tertiary plant remains. The available evidence indicates that some of the saprolite is as old as Cretaceous and that some is as old as early Tertiary (like the bauxite deposits). Little if any can be attributed to the late Pleistocene and Holocene, when our modern soils were forming.

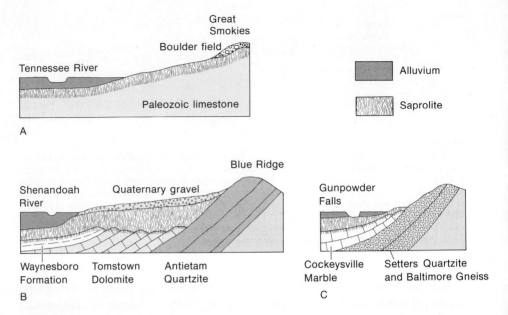

FIGURE 9.5
Examples of alluvium overlapping saprolite in the valley of the Tennessee River (A), Shenandoah Valley (B), and Gunpowder Falls near Baltimore (C). At each of these valleys the alluvium overlying the saprolite seems to be late Pleistocene in age. In the Great Smokies the saprolite can be traced onto the mountain sides under the boulder fields, that also are regarded as late Pleistocene.

Genetically related old soils are found on part of the Coastal Plain—in the phosphate district near Plant City, Florida. These old soils have developed partly on Miocene, partly on Pliocene, and partly on deposits thought to be early Pleistocene. The late Pleistocene deposits are unweathered; near the coast a late Pleistocene shell bed 25 feet above sea level overlaps the old soil. Most of the weathering took place during early and middle Pleistocene, but some may date from the late Pliocene and even Miocene time. The stratigraphic relationships of the ancient and modern soils in the district are illustrated in Figure 9.6, A.

Superimposed on the surface layers of the old soils are younger soils, the youngest and least developed of which occur on pottery-bearing Indian shell mounds along the coast. Similar mounds with similar weakly developed soils are found as far north from there as coastal New England. The archeological evidence shows that these soils are no more than a few hundred years old. In that time, leaching has been slight; the soils consist of a humus-enriched layer over a parent material that is so little weathered that carbonate shells are still intact immediately under the organic layer (Fig. 9.7).

Other modern soils in that part of Florida (the Leon, Norfolk, Lakeland, Eustis, Blanton) are pre-pottery in age yet clearly younger than the old phosphatic soil. They are probably Holocene, because projectile points found buried deep in the soils are of the kind made before pottery came into use, and they were probably buried before the soil profile developed.

Relationships quite similar to those in Florida have been reported in South

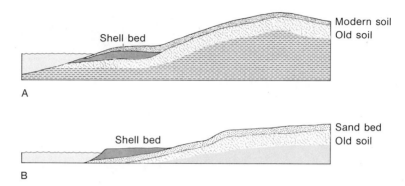

FIGURE 9.6
Cross sections illustrating stratigraphic relations of old soils in west Florida phosphate district (A) and in the South Carolina phosphate district (B). In Florida the old soil (nodular phosphate with sand and clay) developed on top of the phosphatic Miocene and Pliocene formations, and is overlapped by a late Pleistocene shell bed. [After Hunt and Hunt, 1957.] In South Carolina, a similar old phosphatic soil is overlain by a sand bed that extends under a late Pleistocene shell bed. [After Malde, 1959.]

Carolina (Fig. 6.11, B). Farther north, in Virginia, the rather unstable mineral hornblende is preserved in deposits that are Sangamon (?) or younger, but is scarce in older deposits. Modern soil at the surface is 1 to 2 feet thick; below this are the Sangamon (?) deposits, which are oxidized and enriched with clay to a depth of 3 to 6 feet. In older Pleistocene deposits the depth of oxidation and clay enrichment is as much as 14 feet.

These unconformable surface deposits and soil profiles illustrate the general physical differences between pre-Wisconsinan and younger soils. Pre-Wisconsinan soils are characterized by alteration of hard rocks to clay and by reddening with iron oxides. With only minor exceptions, the hard rocks in soils no older than the Wisconsinan stage remain firm, and there is little reddening. The clayey alteration is especially noticeable in the easternmost part of the United States and in the far western states because the weathered deposits in those areas are stoney. The alteration by weathering of a granite cobble to a clay pseudomorph is impressive indeed!

The old soils that once existed in the glaciated areas were destroyed by the ice; the few remnants are now covered by drift or other glacial deposits. That the old soils are overlapped by the Wisconsinan drift is shown in many outcrops and by borings taken at numerous places along the southern border of the drift. How far north the old soils originally extended is uncertain, but remnants have been found in some of the northernmost parts of the United States, such as in northern Vermont and in the Boston area (see Chapter 7). The clays of the old soils may have contributed to the mobility of the glacial ice.

The differences between pre-Wisconsinan and later alteration provide a satisfactory guide for distinguishing soils of these ages throughout the northern hemisphere. This is not to say that the occurrence of a single cobble or boulder altered to clay indicates a pre-Wisconsinan age for the deposit, but finding a bed of them very nearly does.

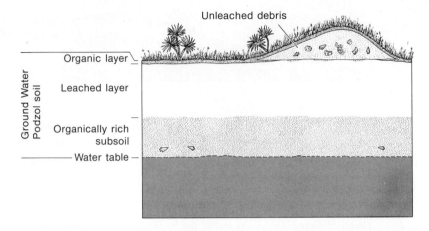

Unleached debris

Ground Water Podzol soil

Organic layer

Leached layer

Organically rich subsoil

Water table

FIGURE 9.7

Diagrammatic section of modern Ground Water Podzol Soil in Florida. Indian shell mounds that consist of only slightly leached debris and contain pottery and arrow points overlie the Ground Water Podzols. These shell mounds range in age from historic time back to about A.D. 500. The Ground Water Podzols are older than the mounds, but seem to be younger than some earlier, pre-pottery archeological remains that occur in the subsoil down to the water table.

Parallels are found in paleontology, where an entire fauna is regarded as significant but an individual fossil as possibly misleading, and in archeology, where an entire array of evidence is regarded as significant but an individual artifact as possibly misleading. In the study of soils, there is no substitute for good judgment.

SUPERIMPOSED PROFILES
IN CENTRAL UNITED STATES

In southern Ohio, Indiana, and Illinois, south of the border of the Wisconsinan drift, are flat, poorly drained uplands covered partly by loess and partly by Illinoian till. The weathering is deep, commonly more than 10 feet, even on steep hillsides. That this weathering is pre-Wisconsinan is shown by the northward extension of the deeply weathered layers beneath the Wisconsinan drift. The comparatively thin modern soils on the surface of the old weathered deposits are Planosols with light-colored surface layers and deeper layers mottled brown and reddish brown. The mottled layers are enriched in clay and may be tough and not very permeable—a clay pan.

On the Wisconsinan glacial drift the parent materials are unweathered within 3 or 4 feet of the surface. The Gray Brown Podzolic Soils developed on them have undergone very little alteration to clay; the weathering has for the most part only leached calcareous materials to a depth of about 30 inches.

A large number of measurements across the contact between the Wisconsinan and Illinoian drifts in Indiana show an average depth of leaching of 40 to 60 inches in the Wisconsinan drift, and an average of twice to three times that depth in the Illinoian. In no pair of test holes was the difference less than 2 to 1. Depth of leaching can

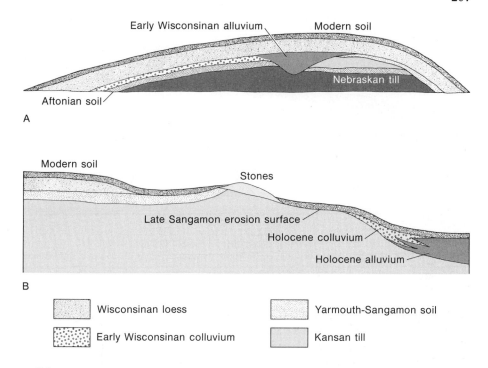

FIGURE 9.8
Buried soils and superimposed soil profiles in Iowa (generalized from Ruhe, Daniels, and Cady, 1967). (A) In the upper section a soil of the Aftonian interglaciation is developed on Nebraskan till and is buried at the right by Kansan till on which is developed a soil of the Yarmouth or Sangamon interglaciation. In places this Yarmouth or Sangamon soil extends across the Aftonian soil and is developed on Nebraskan till that had been exposed by erosion. These old soils are buried under Wisconsinan loess on which a modern soil has developed. (B) In the lower section, a Yarmouth-Sangamon soil developed on Kansan till is partly buried by Wisconsinan loess on which a modern soil has developed. The old soil and till are truncated by a pediment marked by a lag layer of stones. At the right is Holocene colluvium and alluvium. In this section the modern soils have developed on five quite different kinds of parent materials.

provide useful evidence bearing on relative ages of weathering profiles, but it must be used with caution. The original carbonate content of the parent materials must be demonstrated to have been similar, and the ground being compared must be equally susceptible to leaching.

In Iowa, deeply weathered pre-Wisconsinan till, called gumbotil, is overlapped by loess and till of Wisconsinan age (Fig. 9.8). The gumbotils are Kansan and Nebraskan, and they are much more deeply weathered than is the Illinoian till farther east. The decomposition of rock by weathering becomes progressively more advanced toward the surface. In the fresh, comparatively unaltered till, nearly half the stones are composed of granite and other igneous rock; in the gumbotil only about 10 percent of these igneous rocks have survived the weathering. In the transition zone between the unweathered till and the gumbotil, stones that are almost completely altered to clay can be identified only by their outlines. The unweathered till may contain 10 percent of carbonates whereas the gumbotil contains none.

In all these areas of acid soils, south of the Wisconsinan glacial drift and ranging from Iowa eastward to the Atlantic Coast, are found deep ancient soils with shallow

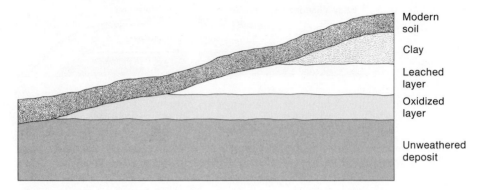

Modern
soil

Clay

Leached
layer

Oxidized
layer

Unweathered
deposit

FIGURE 9.9
Superimposed soil profiles. Depending on accidents of erosion, the modern soil may be developed on unweathered parent material, on an overlying layer that is oxidized but not leached, on a still higher layer that is both oxidized and leached, or on the upper most layer, where there has been alteration to clay.

modern soils developed on them. Because accidents of erosion have removed parts of the upper layers of the ancient soils, there is a complex array of modern soils. The modern soils may be developed on any of the layers in the ancient soils: the upper layer, in which clay alteration is advanced; the next lower layer, which is leached and oxidized; the layer that is oxidized and not leached; or unweathered original deposits (Fig. 9.9).

SUPERIMPOSED PROFILES OF ALKALINE SOILS ON THE GREAT PLAINS

The surface deposits and soils on the Great Plains, notably in the Denver area, Colorado, further reveal the variations provided by superimposed soil profiles. Soil profiles of three different ages of alkaline soils have been distinguished there—one dating from pre-Wisconsinan time, a second dating from early Wisconsinan time, and the third dating from the late Wisconsinan or Holocene. The stratigraphic basis for separating these soils is illustrated in Figure 9.1.

Table 9.1 summarizes the soils in the Denver area and their geologic relationships. Holocene soils (Laurel and Cass loam, sandy loam, and clay loam) are those that have formed on various Holocene alluvial and dunal deposits. Alluvial deposits dated archeologically as having been deposited about 2,000 years ago have very feebly developed profiles. With acid, one can detect slight leaching of lime in the upper 3 to 5 inches, and slight accumulation in the zone 5 to 10 inches below the surface. The alluvium predates the use of pottery and the bow and arrow, but it contains a modern fauna.

Among the soils that date from the early Holocene are some very sandy ones (such as the Greeley loamy fine sand) that are reddened and slightly cemented. They developed on stabilized dune sand. In many places this sand has been reworked by

TABLE 9.1
Correlation between the geology and some of the soil types in the Denver area

Soil types and description (from Harper, Acott, and Frahm, 1932, p. 9, 10)	Parent material and geologic history of the soils
Laurel fine sand and very fine sandy loam. Light-colored soils of the first bottoms, developed over recently deposited material; stratified; sand noncalcareous, sandy loam mildly calcareous and micaceous.	*Protohistoric or historic alluvium.* On modern flood plains (about A.D. 1600 or later).
Cass loam and fine sandy loam. Nearly black, dark-brown, or dark reddish-brown surface soils; very slightly lighter brown subsoils, in many places, heavy. Waterworn gravel, fine sand, and finely divided mica are distributed throughout the surface soils and subsoils; essentially noncalcareous. Underlain by a gravelly layer.	*Protohistoric alluvium.* In the major valleys only (about A.D. 1400). These soils have been subject to flooding.
Laurel clay and clay loam. Gray-brown clay to depth of 10 inches (in the clay-loam type this layer contains more fine sand). Below this layer to depth of 6 feet or more is a gray slightly compact layer; in most places mildly calcareous at all depths.	*Piney Creek alluvium.* Older than A.D. 1000; probably deposited about the beginning of the Christian Era. After this alluvium was deposited it was subject to flooding; but this condition ended when streams became incised into the alluvium soon after A.D. 1000.
Greeley loamy fine sand. To depth of 15 inches grayish-brown, noncalcareous mellow fine sand; contains many plant roots, slightly dark staining caused by accumulations of humus. Between depths of 15 to 30 inches rich-brown noncalcareous loam or heavy loam, high content of fine sand; firm, slightly cemented, tendency towards massive prismatic structure; commonly mildly calcareous. A few lime-carbonate nodules at depths of 4 to 6 feet. *Greeley fine sandy loam.* Similar in profile to loamy fine sand, but each horizon is sandy loam or loam.	*Late Pleistocene or early Holocene eolian sand.* The parent material dates from the late Pleistocene or early Holocene. The climatic changes since then include 2 relatively moist and 2 relatively dry periods. The history of these soils is further complicated because their upper few inches must have been repeatedly reworked by wind action during the dry periods.
Weld fine sandy loam. The 2½-inch surface layer is light fine sandy loam matted with grass roots; dark brown, noncalcareous, slightly laminated. Between depths of 2½ to 10 inches is light-brown noncalcareous firm but friable fine sandy loam or loam. Between depths of 10 to 18 inches is rich-brown noncalcareous slightly columnar clay loam or clay. Below 18 inches is olive-brown friable or loose fine sandy loam or loamy fine sand, moderately or highly calcareous.	The history and parent material of the top 18 inches or so are the same as those of the Greeley loamy fine sand and fine sandy loam. At about 18 inches from the top is an unconformity below which are eolian deposits of Wisconsinan age. Part of the weathering in this subsoil therefore dates from late Pleistocene time. Locally this same condition may be found in the Greeley soils.
Gilcrest gravelly sandy loam and gravelly loam. Developed on the terraces of the large stream valleys; the surface soils are slightly laminated, sandy, and more or less mixed with gravel; in many places the heavier subsoil is prismatic and has a well-defined layer of lime accumulation. The deep subsoil is gravel.	The deep subsoil is the Wisconsinan gravel fill that forms the Broadway terrace. The surface soils comprise the alluvial layer capping that gravel and are early Holocene in age. These soils were subject to flooding immediately following deposition of the Broadway terrace, but this condition ended when the South Platte River became incised into the gravel fill at the beginning of the Holocene. The climatic changes since then include the dry period at the beginning of the Holocene and subsequently 2 relatively moist and 2 relatively dry periods.

TABLE 9.1 (*continued*)

Soil types and description (from Harper, Acott, and Frahm, 1932, p. 9, 10)	Parent material and geologic history of the soils
Fort Collins clay loam. The 8-inch surface layer is slightly laminated grayish-brown mellow fine sandy loam. At average depth of 8 inches, color changes to brown, dark brown, or reddish brown; texture changes to clay; and structure becomes decidedly prismatic. The upper 22 inches has been leached of lime and carbonate but contains an accumulation of decomposed organic residue. Between 22 and 38 inches is grayish-brown calcareous clay with prismatic structure. Between 38 and 54 inches lime content is high, between 54 and 96 inches it is slightly less. The lime occurs principally in streaks and pockets.	*Wisconsinan eolian deposits.* If these deposits date from early Wisconsinan their climatic history includes the periglacial climates of the later substages of the Wisconsinan, the mild interstadials of the Wisconsinan, and the Holocene climates. At various stages in this history the upper layers in some places were eroded and at others a few inches of younger sediment were added.
Fort Collins clay. Heavier texture than Fort Collins clay loam but otherwise similar.	The upper layers of this soil represent a fine grain facies of Wisconsinan eolian deposits that have become admixed with the upper part of a pre-Wisconsinan soil. The deeply weathered subsoil is pre-Wisconsinan in age.
Fort Collins loam. The surface soil is gritty loam and contains both well-rounded and angular sand. The lower part of the surface soil is lighter in texture than that of the Fort Collins clay loam. The columnar clay layer beneath contains a few more angular fragments than the clay loam.	This soil has the same parent material and history as the Fort Collins clay but in addition has had some late Pleistocene or early Holocene eolian sand added to its upper layers.
Larimer gravelly clay loam. Developed from old weathered materials. Top 8 inches dark-brown noncalcareous clay loam containing a few pebbles. Between 8 and 15 inches is reddish-brown noncalcareous clay having a vertical breakage. Below 15 or 20 inches is grayish-brown highly calcareous clay loam. Between 3 and 6 feet is horizon of still higher lime content. Below this is weathered parent materials of various kinds.	This soil dates from pre-Wisconsinan time. Its climatic history includes the mild climate of the Sangamon stage, the alternating periglacial and mild interstadial climates of the Wisconsinan, and the Holocene climates. At various stages in this history the upper layers in some places were eroded and at others a few inches of younger sediment were added.

Source: From Hunt (1954).

later wind erosion, and an unconformity separates the modern loose sand from the older, reddish, slightly cemented sand. Another soil, which dates from late Pleistocene (Wisconsinan) time but is primarily a product of Holocene weathering, includes gravelly loam (Gilcrest) on the late Pleistocene terraces, a gravel fill containing the remains of mammoth, camel, and other extinct mammals. The Gilcrest soil has a $1\frac{1}{2}$ foot layer of brown silt-cemented gravel at the surface underlain by $1\frac{1}{2}$ feet of gravel that is stained with iron and manganese oxides and may or may not have a well-developed carbonate layer. Even though the Greeley loamy fine sand and the Gilcrest gravelly loam underwent weathering during much or all of Holocene time—at least 10,000 years and perhaps much more—they are feebly developed.

Soils that date from Wisconsinan time include the Fort Collins clay loam that developed on Wisconsinan loess. This well-developed soil has been widely recognized on Wisconsinan loess in the Great Plains, where it is called "Brady soil." The organic-rich layers at the surface are about 6 to 8 inches thick; beneath them is a clayey layer that is brown, dark brown, or reddish brown and has a distinct prismatic structure. Lime carbonate has been leached from the upper 20 to 24 inches and deposited as veins and nodules in a zone 2 to 4 feet below the surface.

Although there is some question about just how late in the Wisconsinan the loess was deposited, there is little doubt, on either stratigraphic or physical grounds, that the soils that developed on it are at least twice as old as the oldest of the Holocene soils.

Pre-Wisconsinan soils, which are common in the Rocky Mountain region, are overlapped by loess or other Wisconsinan age deposits. The deposits that underlie the soils contain the fossil remains of pre-Wisconsinan types of animals, such as the musk ox, *Symbos* (Table 2.5).

In the Denver area, the pre-Wisconsinan soils include the Larimer gravelly clay loam, a northern outlier of the Reddish Brown Soils of the Southwest. This soil probably originally consisted of several feet of reddish clay above a layer 10 or more feet thick composed of weathered parent material strongly impregnated with lime carbonate. The upper layers of this old soil have been eroded, and younger soils have developed on the eroded surface. In some places, however, a thin cap of younger loess or gravel lies on the eroded surface of the older soil, separating it from the superimposed younger soil.

Unless a soil that formed under an ancient climate became buried and stayed buried beneath younger deposits, that soil would become the parent material for a younger soil. Such parent material may be either the leached or the depositional layer of the older soil, depending on the depth to which the old soil was eroded before the younger developed. In the Denver area, soils of three different ages—pre-Wisconsinan, Wisconsinan, and Holocene—provide nine distinctive kinds of superimposed profiles and nine different soils.

In the simplest example (Fig. 9.10, A) the older soils are buried and the Holocene soil is developed on unweathered parent material. Other examples, however, are more complex:

1. The parent material of the Holocene soil may be either the leached layer or layer of deposition of a Wisconsinan soil that developed on material that was unweathered when the Wisconsinan soil started to form (Fig. 9.10, B).

2. The parent material for the Holocene soil may be either the leached layer or the depositional layer of a Wisconsinan soil that developed on the leached layer of a pre-Wisconsinan soil (Fig. 9.10, C).

3. The parent material of the Holocene soil may be either the leached layer or the depositional layer of a Wisconsinan soil that developed on the depositional layer of a pre-Wisconsinan soil (Fig. 9.10, D).

4. The deposits and soils of Wisconsinan age may have been eroded, and the Holocene soil may have as parent material either the leached layer or depositional layer of a pre-Wisconsinan soil (Fig. 9.10, E).

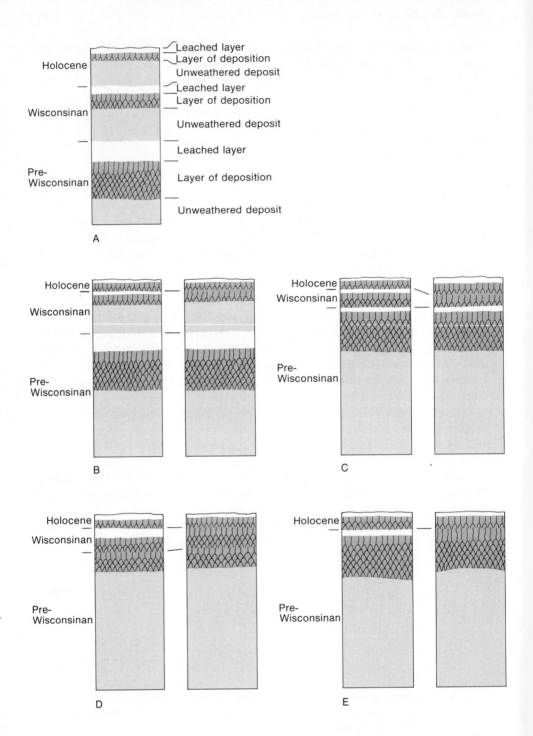

FIGURE 9.10

Diagrammatic stratigraphic section illustrating relationship among pre-Wisconsinan, Wisconsinan, and Holocene soils where the older soils are buried under younger deposits. In this example each of the three soils developed from materials that were unweathered when those soils began to form. [After Hunt, 1954.]

STRATIGRAPHY OF SUPERIMPOSED PROFILES IN THE LAKE BONNEVILLE BASIN

Stratigraphic relations like those between surface deposits and soils in the Denver area are found also in the Lake Bonneville basin, where a pre-Wisconsinan soil is overlapped by the Lake Bonneville formations of Wisconsinan age (Fig. 9.11). The old soil consists of several feet of reddish clay underlain by several feet of lime-enriched, weathered parent material. As in other pre-Wisconsinan soils, most cobbles and boulders in the soil are altered to clay. The Lake Bonneville formations can be correlated with the Wisconsinan moraines in the Wasatch Mountains. Pre-Wisconsinan moraines in the mountains are deeply weathered, and the soil on them correlates with the soil overlapped by the Lake Bonneville deposits.

Not only is the old soil in the Lake Bonneville basin overlapped by the Lake Bonneville formations, but the oldest of those formations (Alpine Formation) contains a much higher proportion of clay and silt than do the younger ones (Bonneville and Provo Formations). The Alpine Formation in Utah Valley has a volume of about 2.6 cubic miles and more than half of this is silt and clay. The Bonneville and Provo Formations there aggregate only a half cubic mile of sediment, and probably less than a fourth is silt and clay. In terms of sedimentation, therefore, Lake Bonneville changed from a lake in which much clay was deposited along with the sand and gravel from the glacial meltwaters to a lake in which very little clay was deposited. This change in sedimentation probably was due to removal of most of the clayey pre-Wisconsinan soil that had blanketed the Wasatch Range before the Wisconsinan glaciation.

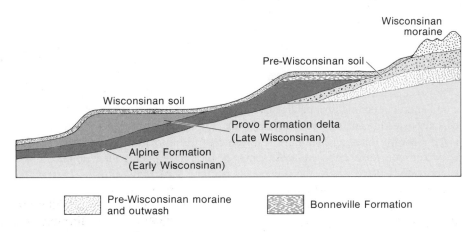

FIGURE 9.11
Diagram illustrating relationships between soils and lake deposits in the Bonneville Basin, A pre-Wisconsinan moraine with a pre-Wisconsinan soil is overlapped by the lake beds and by the Wisconsinan moraine. A Wisconsinan soil is developed on the lake beds and superimposed on the pre-Wisconsinan soil where it is at the surface. At the Stansbury shoreline (to left of area shown) and below it, both the deposits and the soils developed on them are Holocene.

Soils that developed on the surfaces of the deltas formed of the Provo Formation have gravelly loam top soils over gravelly subsoils. This difference in texture between the surface and subsurface layers is not due to weathering; it is stratification. The surface layer of silt and sand, probably in part eolian, was deposited on the topset beds of the gravelly delta. At a depth of 18 to 24 inches, the gravel is partly cemented with lime carbonate, but in part this lime carbonate represents caliche deposited in the capillary fringe when the lake stood just below the level of the deltas and the water table extended through the porous gravel. These soils, which are Holocene and late Pleistocene in age have distinct profiles, but the layering is due partly to stratification and partly to evaporation of ground water. The mineral alteration is slight, as it is elsewhere in the United States on late Wisconsinan and younger deposits.

In all these western soils the lime-cemented layers (caliche) are thickest in the pre-Wisconsinan soils and commonly are many feet thick. They may be nearly as hard as concrete or may be friable. The lime occurs in layers or as nodules. Where the lime zone is gravelly, the lime coats the pebbles and enters those made porous by weathering. The interstitial silt and clay is thoroughly impregnated with lime carbonate. Where the lime zone is silty or clayey, the whole layer may be uniformly impregnated with lime carbonate that becomes less abundant and divides downward in veinlets.

SOME IMPLICATIONS OF SUPERIMPOSED PROFILES

Superimposed soil profiles illustrate some complexities of soil genesis and of soil stratigraphy arising from geologic history, and why the geology must be understood to understand soil genesis. Superimposed soil profiles are the rule rather than the exception, particularly beyond the limits of the younger glacial drifts. Soils in stratigraphy provide limits quite opposite to those provided by fossils. A deposit can be no older than the fossils it contains, and may be much younger (the fossils could be reworked from an older deposit). On the other hand, the deposit can be no younger than the soil developed on it, and may be much older.

Superimposed profiles of different ages may be suggested by discontinuities in a soil profile, such as abrupt changes in texture, abrupt reversals or other abrupt changes in acidity, alkalinity, color, or (in alkaline areas) by the repetition of lime-free layers beneath lime-enriched ones. But the compelling evidence that a soil consists of two or more superimposed profiles must be sought by tracing the weathered layers laterally to a place where the older soils become separated from the base of the younger ones by intervening unweathered surface deposits.

The unconformities between surface deposits and buried ancient soils also illustrate an important principle bearing on soil genesis. Too often in geologic literature and in that of agronomic soil science, the greater weathering of old soils is attributed to their great age—that is, to longer duration of the process of the weathering. But this is a gross oversimplification, and one that can lead to error, because the advanced weathering typical of pre-Wisconsinan soils *had occurred before their burial.* Very little of the extensive and intensive mineral alteration can be attributed to Wisconsinan and Holocene time.

This is well shown by the closed-system, alkaline soils. In the western United States, Wisconsinan soils are several times deeper than Holocene soils, and pre-Wisconsinan weathering profiles are several times deeper than the Wisconsinan ones. These old soils could not have developed under present conditions because there simply is not enough moisture available to wet the ground to the depths of those old soils.

In areas of acid, open-system soils, the burial of the old soils by younger deposits is stratigraphic evidence that the old soils had developed and the boulders in them altered before they were buried. That the clays are commonly reworked into the overlapping deposit is additional evidence. Conceivably, the intensive weathering of the old soils happened rapidly.

We cannot be sure how long the processes of weathering and soil development operated on any ancient soil until we know what the environment was like and what caused the alteration of rock to clay and other products of weathering. There are gaps in our knowledge; the intensive alteration may have been caused by an environment and process that still need to be entered on some of the many blank pages in our record book of geologic history.

THE PROBLEM OF EXCEPTIONS

In at least one locality, Baisman Run, near Baltimore (Fig. 9.12), strongly acid ground water is known to have altered rock to clay, apparently during late Pleistocene or even during Holocene time; there are no doubt other localities. At Baisman Run, a marsh is maintained by seepage from a spring near the foot of an alluvial fan, and the water is perched on a flat bedrock surface on which has developed an inch or two of structured saprolite. Over this is 12 to 18 inches of organic-rich, clayey muck—gley—which in places contains quartz gravel.

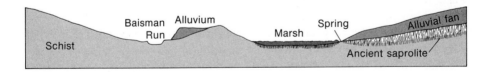

FIGURE 9.12
Relationship between saprolite, alluvium, and ground water at Baisman Run, near Baltimore. Baisman Run is in a rocky channel on schist; remnants of an alluvial deposit that once must have been widespread cover knobs of the bedrock, and these knobs are not deeply weathered. Springs discharge from the foot of an alluvial fan that overlaps ancient reddish saprolite on the hillside, and the springs maintain a marsh on a flat part of the old valley bottom. The water in this marsh is highly acid; the flat consists of 12 to 18 inches of gley over an inch or two of saprolite darkly stained with organic matter. This thin layer of saprolite is attributed to accelerated weathering in the highly acid flat ground and is thought to be much younger than the saprolite on the hillside.

Baisman Run once flowed across this surface, eroded the bedrock platform, and deposited alluvium on it. Later, most of the alluvium was removed by erosion. The saprolite layer and the gleyed layer above it are saturated with water; both are strongly acid. The weathering is attributed to the highly acid ground water seeping along the bedrock surface at the base of the alluvium. Dry hillocks of bedrock protrude through the saprolite and are not much weathered.

Ancient soils may grade into late Pleistocene and Holocene soils where mass wasting has reworked an ancient soil and mixed its materials with unweathered materials. Such a mixed deposit may be the parent material for a Gray Brown Podzolic Soil or Sol Brun Acide. On the Piedmont Province the saprolite and the red and yellow soils on it are best developed on the uplands, where slopes are gentle. On steep hillsides, mass wasting has mixed the saprolite with fresh materials moved downslope; soils on these mixed materials are young. Similar mixing of subsurface and surface soil layers, even the mixing of Holocene soil layers with ancient soil materials, occurs near the border of the Wisconsinan glacial drift. The toppling of trees, frost heaving, earth slides, and the activities of burrowing animals contribute to the downhill creep of the colluvium and soil that causes the mixing. Such mixing results in the development of a complete gradation between the old and the young soils, but the gradation is of course due to mixing by transportation and not a result of gradation in the weathering processes.

Very possibly there are sites, perhaps on flat uplands in the southeastern states, where conditions have been stable for so long and the weathering so advanced that the ground has become essentially inert. At such locations, present-day changes would be minimal, and soil development could simply further accent the ancient layering if for no other reason than textural differences between the layers. While such a state can be visualized, I do not know of its existence; in my experience, the old soils antedate the geomorphic surfaces and are cut discordantly by them. Modern weathering profiles conform to the topographic surface and similarly cut discordantly across the older profiles. Such pedologic unconformity certainly is the general case, although admittedly there may be exceptions.

THE MOST ANCIENT SOILS

Taking a much longer view of the continuum of weathering and soil development throughout geologic time, we may infer four major stages of the kinds of soil developed on surface deposits correlating with four major stages of the evolution of vegetation (see also Yaalon, 1962). In earliest Precambrian time, while the earth's crust was developing during the first billion or two years of earth history, the ground must have been like that of the moon. There was no terrestrial life, the atmosphere was thin, and the regolith resulting from rock disintegration must have been due to physical and physical-chemical processes.

Later in the Precambrian, when the land became occupied by small primitive life forms (algae? lichens?), biochemical weathering became increasingly important. By mid-Paleozoic time, forests of tree-like ferns grew in wet places. Some effects of this biochemical weathering are recorded in the clays associated with Paleozoic coal beds and by some of the variegated Triassic and Jurassic formations (see for example Van Houten, 1961). Flowering plants first appeared during the Mesozoic, and since then weathering profiles probably have been similar to the kinds we know.

Yaalon (1962) has referred to these four stages of soil development as: protosoil during the early Precambrian; primitive soil during late Precambrian and early Paleozoic; rudimentary pedosphere, Devonian into Cenozoic; and fully developed pedosphere, Mesozoic to the present.

BIBLIOGRAPHY

Bryan, K., 1927, The "Palouse" soil problem, with an account of elephant remains in wind-borne soil on the Columbia Plateau of Washington: U.S. Geol. Survey Bull. 790, pp. 21–45.

Bryan, K., and Albritton, C. C., Jr., 1943, Soil phenomena as evidence of climatic changes: Am. Jour. Sci. v. 241, pp. 469–490.

Bryan, W. H., 1947, The geological approach to the study of soils: Rept. 25th Meeting Australian and New Zealand Assoc. Adv. Science, Sect. C-Geology, Adelaide Mtg., Aug. 1946.

Cady, J. G., and Daniels, R. B., 1968, Genesis of some very old soils—the Paleudults: Trans. 9th Internat. Cong. Soil Science, v. IV, paper 11, pp. 103–112.

Cleaves, E. T., 1968, Piedmont and Coastal Plain geology along the Susquehanna aqueduct, Baltimore to Aberdeen, Maryland: Md. Geol. Survey Rept. Inves. 8, 45 pp.

Cline, M. G., and others, 1955, Soil survey of the territory of Hawaii: U.S. Dept. Agriculture Soil Survey Ser. 1939, no. 25, 644 pp.

Coch, N. K., 1966, Post-Miocene stratigraphy and morphology, inner Coastal Plain, southeastern Virginia: Tech. Rept. 6, Task Order 388–064, Geog. Br., U.S. Office Naval Research.

Crocker, R. L., 1946, Post-Miocene climatic and geologic history and its significance in relation to the genesis of the major soil types of South Australia: Australia Council for Sci. and Ind. Res., Bull. 193, 56 pp.

Denny, C. S., 1956, Surficial geology and geomorphology of Potter County, Pennsylvania: U.S. Geol. Survey Prof. Paper 288, 72 pp.

———, and Lyford, W. H., 1963, Surficial geology and soils of the Elmira-Williamsport region, New York and Pennsylvania: U.S. Geol. Survey Prof. Paper 379, 60 pp.

Dickson, B. A., and Crocker, R. L., 1953 and 1955, A chronosequence of soils and vegetation near Mt. Shasta, California: Jour. Soil Sci., pt. 1, v. 4, pp. 123–141; pt. 2, v. 4, pp. 142–154; pt. 3, v. 5, pp. 173–191.

Frye, J. C., 1949, Use of fossil soils in Kansas Pleistocene stratigraphy: Trans. Kansas Acad. Sci., v. 52, no. 4, pp. 478–482.

Goldich, S. S., and Bergquist, H. R., 1948, Aluminous lateritic soil of the Republic of Haiti, W. I.: U.S. Geol. Survey Bull. 954-C, pp. 63–109.

Goodman, K. V., and others, 1958, Soil Survey of Potter County, Pa.: U.S. Dept. Agriculture Soil Survey, Series 1953, no. 2, 101 pp.

Hack, J. T., 1965, Geomorphology of the Shenandoah Valley, Virginia and West Virginia, and origin of the residual ore deposits: U.S. Geol. Survey Prof. Paper 484, 84 pp.

Harper, W. G., Acott, L., and Frahm, E., 1935, Soil Survey of the Brighton area, Colorado: U.S. Bur. Chemistry and Soils, ser. 1932, no. 1.

Hunt, Chas. B., 1954, Pleistocene and Recent deposits in the Denver area, Colorado: U.S. Geol. Survey Bull. 966-C, pp. 91–140.

Hunt, Chas. B., and Hunt, Alice P., 1957, Stratigraphy and archeology of some Florida soils: Geol. Soc. America Bull., v. 68, no. 7, pp. 797–806.

Hunt, Chas. B., and Sokoloff, V. P., 1950, Pre-Wisconsin soil in the Rocky Mountain region: U.S. Geol. Survey Prof. Paper 221-G, pp. 109–121.

Kay, G. F., and Pearce, J. N., 1920, Origin of gumbotil: Jour. Geology, v. 28, no. 2, pp. 89–125.

Kaye, C. A., 1951, Some paleosols of Puerto Rico: Soil Science, v. 71, pp. 329–336.

King, P. B., 1949, The floor of the Shenandoah Valley: Am. Jour. Sci., v. 247, pp. 73–93.

Leighton, M. M., and MacClintock, Paul, 1930, Weathered zones of the drift sheets of Illinois: Jour. Geology, v. 38, pp. 28–53.

Leverett, F., 1898, Weathered zones and soils (Yarmouth and Sangamon) between drift sheets: Bull. Geol. Soc. America, v. 65, no. 5, pp. 369–383.

Lockwood, W. N., and Meisler, H., 1960, Illinoian outwash in southeastern Pennsylvania: U.S. Geol. Survey Bull. 1121-B, pp. B1–B9.

Malde, H. E., 1959, Geology of the Charleston phosphate area, South Carolina: U.S. Geol. Survey Bull. 1079, 105 pp.

Mohr, E. C. J., and Van Baren, F. A., 1954, Tropical soils: Interscience, New York, 498 pp.

Nikiforoff, C. C., 1943, Introduction to paleopedology: Am. Jour. Sci., v. 241, pp. 194–200.

Ruhe, R. V., Daniels, R. B., and Cady, J. G., 1967, Landscape evolution and soil formation in southwestern Iowa: U.S. Dept. Agriculture Tech. Bull. 1349, 242 pp.

Ruhe, R. V., 1969, Principles for dating pedogenic events in the Quaternary: Soil Sci., v. 107, pp. 398–402.

Simonson, R. W., 1954, Identification and interpretation of buried soils: Am. Jour. Sci., v. 252, pp. 705–732.

Thorp, J., Johnson, W. M., and Reed, E. C., 1950, Some post-Pliocene buried soils of central United States: Jour. Soil Sci., v. 2, pt. 1, pp. 1-19.

Van Houten, F. B., 1961, Climatic significance of red beds, *in* Nairn, A.E.M. (editor), Descriptive pataeoclimatology: Interscience, New York, pp. 89–139.

Whitehouse, F. W., 1940, Studies in the late geological history of Queensland. I. The lateritic soils of western Queensland: Univ. Queensl. Paper, Dept. Geol., v. 2, no. 1, pp. 2–22.

Williams, B. H., 1945, Sequence of soil profiles in loess: Am. Jour. Sci., v. 243, no. 5, pp. 271–277.

Yaalon, Dan H., 1962, Weathering and soil development through geologic time: Bull. Research Council Israel, Section G, v. 11G, Proc. Israel Geol. Soc. in 4th Congr. Israel Assoc. Adv. Sci., 1961.

———. (editor), 1971, Paleopedology—origin, nature and dating of paleosols: Internat. Soc. Soil Sci. and Israel Universities Press, 350 pp.

Salt marsh and mud flats, southern part of San Francisco Bay, California. Only 60 percent of the Bay's marshes remain today, because they have been filled to make new land for developments and for garbage disposal. Among the physical properties of this kind of ground, commonly called "young bay mud," are low bearing strength and high compressiblity. The main engineering problem when building on mud is to squeeze the water out, but natural subsidence must also be taken into account. Marshes abound in food for a great variety of animals. Cordgrasses (Spartina), rich oxygen producers, are at the bottom of estuarine food chains. During high tide, food drifts from the marshes and becomes available to fish. When the tide recedes, snails eat the decaying cordgrass; the snails in turn become food for other animals. Shorebirds feed upon the numerous worms, clams, insects, and other forms that either live on the mud or drift there from the marshes. Ducks—a half million or more—rest and feed in deeper waters nearby. Only in recent years have our estuaries—veritable wildlife nurseries—come to be recognized as a valuable resource to Man. [Photograph courtesy of U. S. Dept. of Interior, Bureau of Sport Fisheries and Wildlife.]

10 / Physical Properties of the Ground

The physical properties; texture—grain sizes; particle shapes; stony ground; fabric and structure; density and consistency; soil water and ground water; permafrost; organic matter; color; bibliography.

Among the physical properties of the ground that can be noted and described are texture, structure, moisture content, density, toughness, permeability, homogeneity, surface roughness, and mechanical stability. These properties plus such features as color, the atmosphere in and near the ground, and its moisture content and temperature depend partly on the mineral and chemical compositions of the deposit, partly on the primary structures (such as bedding), partly on secondary structures caused by weathering and soil development, and partly on other environmental influences, such as worm burrows and decay of plant roots.

TEXTURE

The term "texture" commonly refers to the sizes of the particles that make up a surface deposit or soil. But the term also refers to the size and shapes of the particles and their arrangement, or packing. The three features determine the porosity of a surface deposit or soil. The simplest approach to the study of porosity consists in examining the possible arrangements of uniform spheres, which may be loosely packed or tightly packed. As Figure 10.1 shows, the pore space around each sphere is controlled by the arrangement of the spheres and is independent of their size so long as they are uniform. This, of course, is a highly theoretical, situation, because in nature different sizes and different shapes of particles are mixed (Figs. 3.1, 3.3).

222

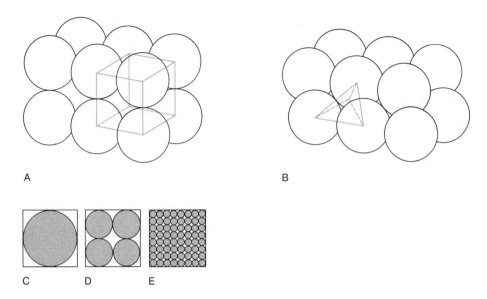

A   B

C   D   E

FIGURE 10.1
Diagram illustrating porosity. Porosity around loosely or tightly packed uniform spheres (B) is independent of grain size. In C, D, and E the amount of pore space is the same; in each the voids account
for slightly more than 45 percent of the whole volume. This is the most open packing. In the tightest
possible packing of uniform spheres (B) the void space amounts to about 25 percent of the total volume.

Figure 10.2 illustrates a field chart for estimating texture and gives commonly used
definitions of different sizes and shapes of particles.

Porosity is a measure of the volume of void space between particles, whereas
permeability is a measure of the rate at which water or air can move between the
particles. Permeability is controlled by the size of the individual voids and their
connecting passages, and not by the total void space. In Figure 10.1, the total pore
space is the same in D and in E, but clearly D is the more permeable. Actually,
some clayey ground is highly porous, yet it may be nearly impermeable because the
individual voids are so minute. Shape of the particles also affects permeability; in
general, in sand-size material, the more angular the grains, the greater the permeability—and by as much as a factor of 2.

Texture can be exceedingly difficult to determine, and commonly is, because clays,
salts, oxides, hydroxides, and organic colloids resulting from weathering cause
aggregation (cementing together) of individual grains of primary minerals or rocks.
This aggregation can be overcome by washing or other treatments, but this causes
hydration or other changes that affect texture or other properties. Oxides, hydroxides,
and salts coat the surfaces of primary and secondary minerals, and these together
with the very fine, clay-size materials (diameters less than 0.002 mm) almost defy
size determination. Because of such difficulties, the adequacy of textural descriptions
of the ground depend to considerable degree on the age of the ground and the stage
of the weathering. Textural descriptions of Wisconsinan and younger soils generally
are satisfactory because not much clay, oxide, hydroxide, or salts have formed in
them; textural descriptions of pre-Wisconsinan soils are in some instances very
misleading.

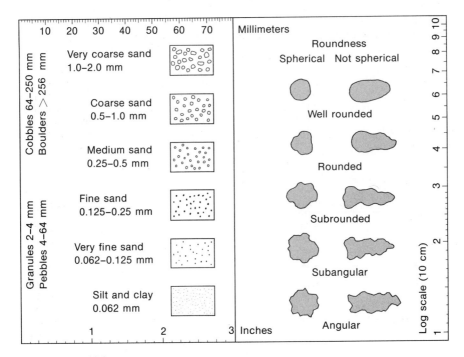

FIGURE 10.2
Field chart for estimating textures. Depressions in thick cardboard contains grains of proper size glued into position.

Grain Size

There are numerous scales for describing grain sizes of surface deposits and soils. Geologists have favored a classification that is graded geometrically, known as the Wentworth Scale (Table 10.1). The sieve scale of the American Society for Testing Materials (Table 10.1) also is a geometric scale. Agriculturalists use the Department of Agriculture or Atterberg scale. In all systems the determinations are made on organic free, oven-dried samples. Most mineral particles are silt size or larger, but those newly formed in the ground (e.g., carbonates and uncombined oxides) occur in both the coarse and fine fraction.

Engineers use the same terms for grain size as agriculturalists and geologists but have different definitions (Fig. 10.3). This usage extends the meaning of the term "clay," and gives correspondingly greater emphasis on the clayey fraction of the sands, loams, and silts. Highway engineers classify soils in 8 size groups according to their suitability as subgrade materials (Table 10.2). Engineers also use a numerical expression, *sorting coefficient,* for indicating the dimensional spread of particle sizes in a sample. The sorting coefficient (S_o) equals

$$\sqrt{\frac{\text{Max. diameter of the smallest 75\% by weight (3rd quartile)}}{\text{Max. diameter of the smallest 25\% by weight (1st quartile)}}}$$

TABLE 10.1

Four scales for describing grain sizes of surface deposits and soils

Wentworth scale

Boulder	Above 256 mm	Medium sand	0.5–0.25 mm
Cobble	256–64 mm	Fine sand	0.25–0.125 mm
Pebble	64–4 mm	Very fine sand	0.125–0.062 mm
Granule	4–2 mm	Silt	0.062–0.0039 mm
Very coarse sand	2–1 mm	Clay	less than 0.0039 mm
Coarse sand	1–0.5 mm		

Sieve scale, American Society for Testing Materials
(Meshes and openings corresponding to the Wentworth scale are underlined).

Mesh	Opening (mm)	Mesh	Opening (mm)
5	4.00	40	0.42
6	3.36	45	0.35
7	2.83	50	0.297
8	2.38	60	0.250
10	2.00	70	0.210
12	1.68	80	0.177
14	1.41	100	0.149
16	1.19	120	0.125
18	1.00	140	0.105
20	0.84	170	0.088
25	0.71	200	0.074
30	0.59	230	0.062
35	0.50	270	0.053
		325	0.044

U.S. Dept. of Agriculture		The Atterberg System	
Name of fraction	Diameters (millimeters)	Name of fraction	Diameters (millimeters)
Very coarse sand	2.0–1.0	Blocks	2,000–200
Coarse sand	1.0–0.5	Cobbles	200–20
Medium sand	0.5–0.25	Pebbles	20–2
Fine sand	0.25–0.10	Coarse sand	2–0.2
Very fine sand	0.10–0.05	Fine sand	0.2–0.02
Silt	0.05–0.002	Silt	0.02–0.002
Clay	below 0.002	Clay	below 0.002

Note: Special terms are used for describing fragmental volcanic rocks. Collectively they are referred to as *pyroclastics*. Fragments less than 0.25 mm in diameter are classed as *volcanic dust;* those between 0.25 and 4 mm in diameter are classed as *volcanic ash;* those larger than 4 mm in diameter are *lapilli.*

The fine-grained materials in surface deposits and soils include not only the clay-size mineral particles but also colloidal organic matter. If the fine-grained fraction is dispersed in water, the larger particles settle to the bottom while the very fine ones remain suspended. In general, the particles that settle are larger than about 1 micron (10^{-3} mm) and can be seen with polarizing microscopes. The particles that remain dispersed—that is, those smaller than 1 micron in diameter—are considered colloidal. The coarser of these colloidal particles, down to about 1 millimicron in

TABLE 10.2
Field identification of grain sizes

Field test	Sandy loam	Silty loam	Loam	Clay loam	Clay
Cast	Cast bears careful handling without breaking	Cohensionless silty loam bears careful handling without breaking. Better graded silty loam casts may be handled freely without breaking	Cast may be handled freely without breaking	Cast bears much handling without breaking	Cast can be molded to various shapes without breaking
Thread	Thick, crumbly, easily broken	Thick, soft, easily broken	Can be pointed as fine as pencil lead that is easily broken	Strong thread can be rolled to a pin point	Strong, plastic thread that can be rolled to a pin point
Ribbon	Will not form ribbon	Will not form ribbon	Forms short, thick ribbon that breaks under its own weight	Forms thin ribbon that breaks under its own weight	Long, thin, flexible ribbon that does not break under its own weight

Source: From U.S. Public Road Adm. (1943).

diameter (10^{-6} mm), can be seen under the electron microscope. Still finer particles (to about 10^{-8} mm) may be identified by X-ray diffraction. Such fine materials constitute a considerable fraction of most surface deposits and soils, and many of them are still unidentified. Units used in measuring the sizes of the very fine-grained materials are:

$$\text{micron, } \mu, = 10^{-3} \text{ millimeter,}$$
$$\text{millimicron, } m\mu, = 10^{-6} \text{ millimeter,}$$
$$\text{angstrom, } \text{Å} = 10^{-7} \text{ millimeter,}$$
$$\text{micromicron, } \mu\mu, = 10^{-9} \text{ millimeter.}$$

In the field, grain size can be estimated by making casts, threads, and ribbons of material wetted to give it a moisture content at which it is plastic. A cast is formed by squeezing a lump in the hand, and the type of material is judged by its toughness or resistance to breaking or crumbling. A thread of plastic material is made by rolling it between the hands and noting the toughness of the thread. A ribbon is formed by drawing the index finger under the thumb and pinching the material to form a ribbon. Responses of different kinds of materials to these field tests are given in Table 10.2. Sand feels gritty. Silt feels rough but not gritty, and clay feels greasy. Silt dries quickly and does not stick to the fingers; clay dries slowly and does stick. In applying such tests, though, one must be aware that there are differences in the plasticity and stickiness of different minerals. Micaceous grains in silt may give it a greasy feel like that of clay. Nonclay minerals in the clay-size fraction may cause the thread or ribbon suggestive of a silt.

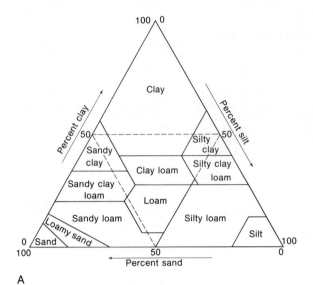

A

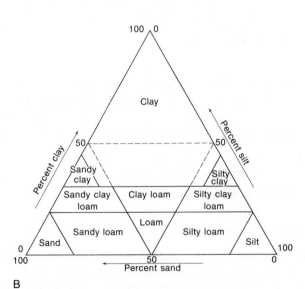

B

FIGURE 10.3
Textural classes of soils. (A) Agricultural definitions
(from U.S.D.A.). (B) Engineering definitions (from
U.S. Bur. Public Roads).

Particle Shapes

In shape, minerals or rock particles may be angular or rounded (Fig. 10.2), spherical, platy, cylindrical, or even fibrous. As already noted, the closest packing of uniform spherical particles gives a void space of about 25 percent of the whole; the loosest packing gives a void space of more than 45 percent. Admixed fine material reduces the pore space but angularity of the particles tends to increase it.

In general, weathering of rocks produces angular particles; transportion of the particles as sediment rounds them. The angularity or roundness of particles is a function of the hardness, or resistance to abrasion, of the particles, the velocity with which they are transported, and the distance they have travelled. Resistance of minerals to abrasion or rounding is partly a function of their shape and weight, and partly a function of their hardness and cleavage. Some of the common minerals are listed below, in order of decreasing resistance (after Friese, 1931; Thiel, 1945):

1. quartz	8. epidote
2. tourmaline	9. hornblende
3. potash feldspar	10. apatite
4. titanite (= sphene)	11. augite
5. magnetite	12. hematite
6. garnet	13. kyanite
7. ilmenite	14. siderite

Similarly, particles of rock resist abrasion unequally. Some common rock types found in soils and surface deposits are listed below in order of decreasing resistance (after Kuenen, 1956):

1. chert	7. sandstone
2. quartzite	8. scoriacous lavas
3. granitic rocks	9. gneiss
4. basaltic rocks	10. schist
5. dolomite	11. glassy volcanic rocks
6. limestone	

Pebbles of even the moderately resistant rocks, like granite and dense basalt, become rounded in a few miles of transportation by streams. Pebbles of diorite porphyry from the laccolithic mountains on the Colorado Plateau and Great Plains are subround after being transported only 2 or 3 miles (Fig. 10.4). Ten miles from the foot of the mountains, most pebbles are rounded; 20 miles from the mountains, most of them are well rounded. On the other hand, sand-size materials of the moderately resistant minerals (such as hornblende, apatite, augite, and feldspar) require much more transportation before they become rounded. Twenty miles from

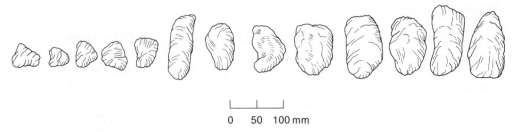

0   50   100 mm

FIGURE 10.4
Rounding of pebbles and cobbles of porphyry as a result of stream transportation in a distance of about 10 miles. The examples are from the Colorado Plateau.

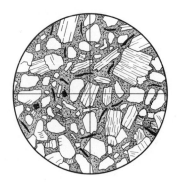

FIGURE 10.5
Mineral proportions, rounding, and other properties of soils and surface deposits can be estimated by counting grains along cross hairs (or any scale) seen through a microscope. The same principle can be used in the field by counting stones along stream beds or other surfaces that are touched by an extended 50 or 100 foot tape.

the mountains, sand-sized particles of hornblende and oligoclase contained in alluvium still retain their crystal faces and crystal edges; only the corners are rounded. Hundreds of miles of transportation would be needed for them to become well rounded. According to Pettijohn (1957), resistant minerals like quartz require thousands of miles of stream transportation to become well rounded.

Transportation by wind, however, is very much more effective in rounding even the resistant minerals, because the impact upon collision of two particles is so much greater in air than in water. The impact is a function of the difference between the density of the particle and the fluid. Wind abrasion may glaze or frost the surfaces of the most resistant minerals.

Surface deposits and soils generally are mixtures of grains of differing sizes and shapes. The identification of individual minerals requires the use of a petrographic microscope, which also can be used for estimating the proportions of particles that have particular sizes and shapes. A simple but satisfactory technique consists in covering a glass slide with a powder made from the sample and counting the grains of the various kinds, sizes, and shapes "touched" by the cross hairs of the microscope (Fig. 10.5); a similar technique can be used in the field by extending a 50-foot or 100-foot tape and counting the pebbles and cobbles of the various kinds, sizes, and shapes touched by the tape. The silt and clay fractions may be analyzed by methods based on the settling rates of particles in water (Stokes Law); many such methods have been devised, including centrifuging, but all are, at best, approximations because the particles are irregular in density and shape and many are altered by reaction with water. Moreover, in many theoretical calculations about the physical properties of fine-grained particles, the particles are assumed to be spherical, but the clay minerals and many or most colloidal particles are irregular, tabular, and even fibrous in shape.

STONY GROUND

Coarse materials, those of pebble size and larger, may constitute a significant fraction of a surface deposit. Distinctions need to be made about the kind of stones, their size and their roundness or angularity. Stones at the surface of pebbly, cobbly, or bouldery deposits older than the Wisconsinan glaciation are commonly found to be in the process of disintegrating—forming a new crop of angular rock fragments. Geologists distinguish angular fragments from rounded or subrounded ones of equivalent size by the following names (Woodford, 1925):

Rounded or subrounded	Angular or subangular
boulder	block
cobble	slab
pebble	flake

Engineers use the terms "boulder" and "gravel" for rounded, water-worn materials; angular fragments are referred to as "rock" and "stone."

An equally important property of gravels, in addition to size and roundness, is the soundness of the stones—that is, the degree to which they are weathered. Gravel, cobble, or boulder deposits of pre-Wisconsinan age are likely to contain a large percentage of crumbly decayed rock. Indeed, in many such deposits there are only clay relicts of the pebbles, cobbles, and boulders. Deposits of early Wisconsinan age generally contain a large proportion of rocks that are fresh and firm beneath an unsound weathered rind. Deposits of late Wisconsinan and Holocene age are the least weathered and generally are the best sources of sound gravel.

The amount of stones on the surface and in the ground is of practical importance as well as of scientific interest. Stones obstruct excavating machinery and make it difficult to collect accurate samples of the ground. Where there is a weathered zone on the bedrock under a surface deposit, blocks of bedrock may be detached and rotated, and easily weathered layers may be turned up. Boulders and stones in the ground act as a dilutant by reducing the volume of material that can be penetrated by roots and burrowing animals. The volume of water that can be held in the ground is also reduced, and the weathering is concentrated in the finer parts. Table 10.3 provides a scale for estimating the percentage and volume of stones on the ground.

Many or most laboratory analyses of soils are reported (on a weight basis) on the fraction that is less than 2 mm in diameter. In stony ground such analyses can

TABLE 10.3
Estimating the percent and volume of stones on the ground

Diameter of stones (feet)	Spacing of stones center to center (feet)	Percent of area covered with stones	Volume of stones per acre-foot assuming no sorting (cubic yards)
2	11	3	97
	5	15	485
	2.7	50	1,616
1	5.5	3	48
	2.5	15	242
	1.3	50	808
0.5	2.7	3	24
	1.2	15	121
	.7	50	404

Source: Soil Survey Manual (1951).

The percent area covered with stone can also be estimated by visual comparison with a chart (e.g., Yaalon, 1966).

be misleading by giving an erroneous indication of mineral content and moisture content, both of which are important to plant growth, to engineering properties, and weathering processes.

The U.S. Department of Agriculture classes ground as non-stony if less than 0.01 percent of the surface is covered by stones. Ground is classed as stony if the stones cover between 0.01 and 0.1 percent of the surface. More than 0.1 percent cover is regarded as very stony. A similar scale is used for indicating the extent of bare rock surface. If outcroppings of bedrock make up 2 to 10 percent of the surface, the land is classed as rocky; if they make up 10 to 50 percent of the surface, the land is classed as very rocky; if more than 50 percent, the land is classed as rock land or rock outcrop.

Depth of stoniness is equally important, particularly if one is concerned with foundations, excavations, or drainage problems or is seeking a source of gravel. Identification of the kind of deposit can greatly assist in making preliminary estimates of depth.

A special kind of stony ground, known as *desert pavement,* is widely distributed on gravel fans and gravel terraces in arid and semiarid parts of the country. Desert pavement consists of a single layer of closely spaced stones, which may be angular or rounded (Fig. 10.6), and which overlies a generally vesicular layer of sand and silt; the sand and silt, in turn, usually overlies gravel. The stones collect at the surface by a sorting action, apparently due to frost and/or salt heaving, in much the same way as grit or small stones in a jar or sand work their way to the top when the jar is shaken. The sand and silt beneath the layer of desert pavement probably is in part of eolian origin, because in many places the parent gravel deposit lacks comparable fine-grained material. In general, within a particular region, the thickness of the silt layer underlying the pavement increases with increasing age of the surface, and the angularity of the stones forming the pavement also increases with increasing age of the surface because of advanced weathering and disintegration (Fig. 3.5).

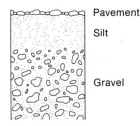

FIGURE 10.6
Most desert pavement consists of a surface layer of closely spaced stones above a stone-free layer of silt that grades downward to gravel. Thickness of the silt ranges from an inch or two to about a foot; in any particular region, the older the surface, the greater is the thickness of silt. Its origin is not understood. In places it seems to be loessial, and not like the fine fraction in the underlying gravel. Perhaps it was deposited as loess in small increments, and the stones were moved upward by frost action or by swelling and shrinking of the clay as a result of alternate wetting and drying.

Another form of desert pavement results where wind removes sand and silt, causing the stones that remain to settle until eventually they touch one another and form a protective pavement on the ground.

STRUCTURE

The term *soil structure* in this book is used to refer to the forms assumed by aggregates of particles, whether in soils or surface deposits, and to the partings that separate the aggregates. The various natural structural forms are intermediate between two nonstructural extremes—that is, loose, single grains without any cohesiveness, and tight massiveness due to uniform cohesion of particles (Nikiforoff, 1941, p. 193). Soil structure is to a considerable degree controlled by the kind and amount of clay and its moisture content. There are eight principal kinds of soil structure; some of the terms given here are borrowed from the agriculturalists (see U.S.D.A., Soil Survey Manual 1951, p. 225–230) but are used in more general ways.

1. Loose single grains without cohesiveness.

2. Crumb: small, soft, porous aggregates.

3. Platy: particles or partings arranged in planes, generally at or near the ground surface and parallel to it: plates generally less than 10 mm thick.

4. Prismatic or columnar: particles or partings arranged in columns; partings planar or curved; columns generally 2 to 5 cm in diameter and normal to the surface. In many kinds of ground the prisms or columns end upward at the base of a layer having platy structure.

5. Blocky: aggregated, closely packed clumps, commonly 2 to 3 cm in diameter and roughly equidimensional but having very irregular surface; corners round or angular.

6. Granular: similar to 5 but aggregates small, mostly less than 5 mm.

7. Nodular: like 4 and 5 but nodules not closely packed and commonly widely spaced; nodules generally differ in composition from the matrix.

8. Tubular: irregular tubelike fillings; casts of foreign material filling former cavities such as those developed from decaying roots or burrows of rodents or insects.

9. Massive, or structureless.

These structures may be weakly, moderately, or strongly developed, and the aggregates or structural units may be small, medium, or large.

The soil structure may develop what sometimes is referred to as *fabric*, a term used with various meanings. In this book, "fabric" is used to mean the arrangement of the individual particles and voids, which may be oriented either parallel to one another (*lineation*), or may have their long axes in the same plane and not necessarily parallel (*foliation*). Or, the three-dimensional arrangement may be random. The fabric may be inherited from the parent material or be developed as a result of the depositional or weathering processes. Fabric may be megascopic or submicroscopic, as when newly formed clay minerals develop a planar arrangement. Bedding planes are familiar examples of fabric inherited in soils; another is the parallelism of particles developed as a result of downhill creep. Familiar examples of fabric include desert pavement in arid regions and the sorting of stones resulting from frost action or salt-heaving. Fabric in hardpans (p. 234) may be random or planar.

A special category of soil structure is represented by the different kinds of *patterned ground*. The term refers to ground whose surface layers have developed orderly patterns, such as polygons, circles, steplike forms (terracettes), and stripes. Polygonal mudcracks due to drying and shrinking of the surface layer of an area that has been flooded are a familiar example of patterned ground. Patterned ground occurs in most regions but is particularly well developed where frost action is intense, as in polar, subpolar, and alpine regions. It also tends to develop wherever marked changes in volume of surface or near-surface layers of the ground occur, as by desiccation, thermal change, and hydration or other chemical change. Patterned ground is also well developed in many arid regions. Figure 10.7 illustrates some common kinds of ground patterns.

The varieties of soil structures and of patterned ground are almost infinite, and there is gradation between the various kinds. Mud cracks, for example, consist of both prisms and plates, and their internal structure may be granular or massive. Where composed of layers of sediment, the mud cracks tend to curl in the direction of the finer-grained layer, because the fine-grained sediments shrink most.

The common occurrence of platy structure in the surface layer of the ground above columnar or prismatic structure in the layer below is analogous to the platy joints that form at the surface above columnar joints in a basaltic lava. Whatever its cause, the shrinkage that develops the partings at the surface develops maximum tension normal to the ground surface; below this the maximum tension is parallel to the ground surface and the fractures form around columns. The diameter of the prisms and columns is a function of the toughness of the soil and the rate of shrinking.

The walls of the opened partings become coated with films of colloidal and other materials (*cutans*), which may be of clay with or without organic matter, oxides and hydroxides of iron and manganese, or salts. The partings are maintained despite alternate swelling and contraction of the ground caused by wetting and drying; as the ground contracts, the textural discontinuities represented by the cutans evidently favor reopening of the old partings rather than the opening of new ones.

Loess and fine-grained alluvial deposits provide the best examples of soil structures developed at depth—that is, below the soil profile. In some alluvial and other clayey deposits, the openings resulting from the structures become conduits for water that

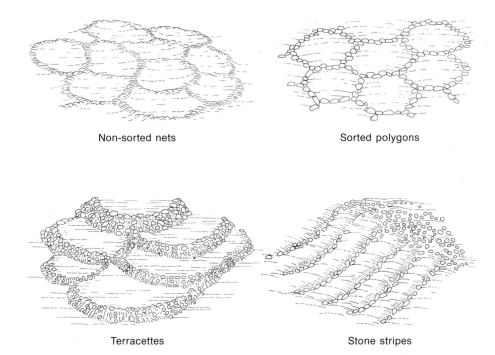

Non-sorted nets   Sorted polygons

Terracettes   Stone stripes

FIGURE 10.7

Some common kinds of patterned ground. Heaving of the ground by frost in cold regions or by salts in arid regions produces various kinds of ground patterns. These may be nearly circular and form nets, or be polygonal, generally hexagonal. Stones that are present collect in the troughs, between heaved patches of ground, forming sorted polygons. The stones may be in the central part of the polygons or along their edges. Mass wasting on hillsides may develop terracettes or stone stripes. In terracettes, the stones collect at the risers; the fines form the treads. Stone stripes form where stones collect in parallel linear troughs; the ground between the troughs is heaved.

develops underground channels not unlike those in some limestone formations. Such underground discharge of water is referred to as *piping*.

For the best development of soil structure, there must be an optimal moisture content. Too much wetting leads to puddling of the ground and collapse of the structures; excess drying, in some ground, may lead to crumbling.

The agricultural potential of a piece of ground depends strongly on soil structure because the structure affects pore space. Massive silt or clay is dense, tough to penetrate, and almost impermeable. Silt or clay having prismatic, blocky, or granular structure may be densely aggregated, but the structural partings provide openings. Other silt or clay may have prismatic, blocky, or granular structure and be loosely aggregated and porous. The structures are of course altered by cultivation; in general the porosity is increased, aeration is improved, water can infiltrate more readily, and roots can spread farther and more easily. Porosity in the surface layers also can be increased by the addition of organic matter.

Other ground features, such as surface crusts and hardpans, could be considered as structural. In arid and semiarid regions, evaporation may cause a salt crust to be deposited over the surface. Expansion due to crystallization of the salts causes the crust to buckle and become separated from the underlying layers. *Hardpans* are tough, subsurface layers resulting from deposition of silica, clay, oxides and hydroxides, or carbonate or other salts in the layer of deposition. In humid areas the hardpans are mostly of clay. In the western United States caliche hardpans are extensive, and in places their structure has been shown to vary in an orderly way with age of the deposit (p. 157). Some hardpans are formed by accumulations of organic matter, as in the well cemented, lower layers (ortstein) of a Ground Water Podzol Soil.

DENSITY AND CONSISTENCE

Rock and mineral particles have an average specific gravity or density of about 2.6 to 2.7, but a third or half of most ground is pore space, and the bulk density of the mass is only about 1.3 to 1.8. A cubic foot of loam, when dry, weighs about 80 to 95 pounds. Peats may weigh less than an equal volume of water; some calcareous loams weigh more than 100 pounds per cubic foot. In modern soils, density increases downward. In dry ground averaging about 85 pounds per cubic foot, the organic and leached layers may weigh only about 70 pounds per cubic foot (sp. g. about 1.1) while the layer of deposition weighs 100 pounds per cubic foot (sp. g. about 1.5). On farmed land, the dry weight of the disturbed, plowed layers (depth 6.5 to 7 inches) averages roughly 2,000,000 pounds per acre.

The porous surface layers of the ground, which help collect and retain moisture, are highly beneficial agriculturally. But these layers generally are unsatisfactory for road foundation or other engineering uses because of their low density, and need to be removed. A feature that is beneficial for one kind of use is detrimental for another.

Specific gravity—the weight of a known volume of soil or surface deposit (including its pore space) divided by the weight of an equal volume of water—can be closely approximated in the field by simple methods. A spadeful of the layer to be tested is removed from the ground and weighed immediately; the cavity is then lined with light plastic and filled with water. The specific gravity is the weight of the sample (in grams) divided by the weight of the water that fills the cavity. In the field the weight of the water is measured by pouring it from a calibrated flask; the volume in cubic centimeters equals the weight in grams. The sample can then be dried in the sun and weighed again to determine what its moisture content was before drying.

Consistence refers to the degree and kind of cohesion and adhesion of a soil or surface deposit—that is, its resistance to deformation or rupture. When the material is wet, the property of stickiness is measured by adhesion to the hands, and plasticity is measured by rolling it into a thread (Table 10.2). When the material is moist, it may be loose, friable, or firm, and when dry may be loose or hard. Consistence also includes the property of cementation, which may be continuous or discontinuous, and may be weak or strong. Some kinds of cementation are not affected by prolonged wetting.

Specific gravity and consistence of the ground depend greatly on the moisture content. A cubic foot of loam that weighs 85 pounds when dry might weigh 110 or 115 pounds when wet, a difference of 30 percent. For optimal plant growth, ground must not be too firm or too loose, and for maximum strength as a foundation material there is an optimal moisture content and consistence, which may be obtained by compaction. For cultivation with heavy implements the moisture content is optimal when the soil is near or at the plastic limit.

The following terminology (from Denny, 1956) is useful for describing consistence of surface deposits or soils.

1. Consistence when wet at, or slightly above, field capacity:

   Stickiness. The quality of adhesion to other objects. For field determination, press material between thumb and finger and observe its tendency to adhere.

   > Nonsticky. After release of pressure, practically no material adheres to fingers.

   > Sticky. After release of pressure, material adheres to fingers and tends to stretch and pull apart rather than pulling free from the fingers.

   Plasticity. The ability to change shape continuously under the influence of an applied stress and to retain the impressed shape after removal of the stress. For field determination, roll material between thumb and finger and observe whether or not a wire can be formed.
   Nonplastic. No wire can be formed.
   Plastic. Wire formed with moderate pressure.

2. Consistence when moist:

   Loose. Noncoherent.

   Friable. Material crushes with gentle to moderate pressure between the fingers and coheres when pressed together.

   Firm. Material crushes under moderate pressure between the fingers but resistance is distinctly noticeable.

   Very firm. Material barely crushable between the fingers.

   Compact. Denotes a combination of firm consistence and close packing or arrangement of particles.

3. Consistence when dry:

   Loose. Noncoherent.

   Soft. Material is very weakly coherent and fragile; breaks to powder under very slight pressure.

   Hard. Moderately resistant to pressure; can be broken in the hands only with difficulty.

SOIL WATER AND GROUND WATER

For healthy growth, plants require water, and anyone who doubts this profundity needs only to visit Death Valley. Good agricultural soil should contain (by volume) about one-half solid material, one-fourth to one-third water, and one-fourth to one-sixth air.

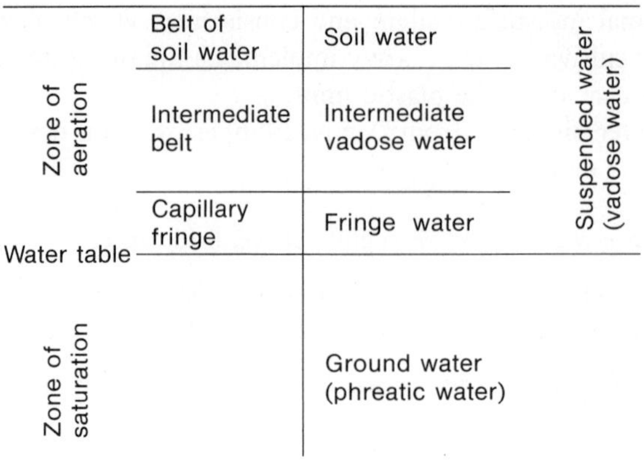

	Belt of soil water	Soil water	
Zone of aeration	Intermediate belt	Intermediate vadose water	Suspended water (vadose water)
	Capillary fringe	Fringe water	
Water table			
Zone of saturation		Ground water (phreatic water)	

FIGURE 10.8
Modes of occurrence and nomenclature of water in the ground.

Water on the surface or in the ground is but a transitory phase of the hydrologic cycle, which is a balanced economy. Over the whole earth, precipitation equals evaporation plus transpiration. About two-thirds of the water that falls on land is returned to the atmosphere by evaporation and transpiration. A third discharges to the oceans in streams and glaciers. Precipitation, evaporation, and transpiration differ greatly in different parts of the country. In arid regions as much as 99 percent of the water may be lost by evaporation; in humid regions less than half may be lost in evaporation and transpiration. In arid regions, as already noted, precipitation is sufficient only to wet the surface layers of the soil.

Only a small part—perhaps 5 percent—enters the ground. Salts transported downward by water entering the ground become deposited when the ground dries; their accumulation contributes to the layer of deposition. In humid regions, precipitation is frequent enough and sufficient in amount so that water escapes from the bottom of the soil profile and carries away soluble constituents. Figure 10.8 illustrates the principal modes of occurrence of water in the ground.

Water in the ground occupies the voids, and its occurrence and movement are controlled largely by the number, shape, and size of the openings—that is, the porosity—and by their shape and continuity. Ground is saturated with water when all its voids are filled (Fig. 10.9), which is the condition below the water table. But saturation also is caused by water perched on a relatively impermeable layer, and by rainwater soaking the surface layers of the ground.

Capillary interstices are small enough to cause water to rise above the water table (Fig. 10.10). In coarse silt (diameter 0.02 mm) the capillary rise may be as great as 10 feet; in sand, with large openings, the rise would be less. Openings that are subcapillary in size, as in clay, also cause less capillary rise. Large openings lack the property of capillarity, and water seeps downward or laterally through them; such ground is permeable. As already noted (Fig. 10.1), ground that is fine-grained (e.g., shale) may have a high porosity yet be nearly impermeable. Moreover, such ground may have considerable moisture, but the moisture, held tenaciously by capillarity in the tiny openings, is not available to plants. In cold latitudes and high altitudes, permeable ground may become impermeable because the water in it freezes.

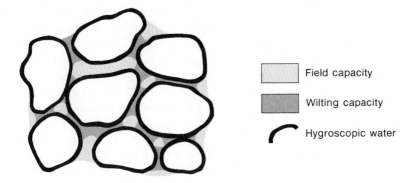

FIGURE 10.9
Occurence of water in pores in the ground. Ground is saturated when the pores are filled. Field capacity is the water held in the ground by capillarity against gravity (stipple pattern). Wilting capacity is water held by surface tension around the points of contact between particles and not available to plants. Hygroscopic water is a thin film on mineral grains; it is not available to plants, nor is the chemically combined water within the grains.

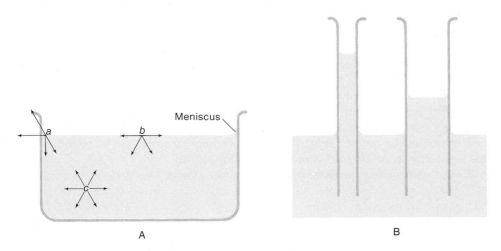

FIGURE 10.10
(A) Surface tension. Water molecules cohere to one another; those below the surface are attracted equally on all sides whereas molecule in the layer at the surface are attracted from the sides and from below but not from above. This creates a bonding that is strong enough to support a needle. The adhesion of water to glass is greater than the cohesive force between water molecules, so that the surface layer curves upward, forming a concave meniscus. (B) Diagram showing inverse relation between size of tube or other opening and capillary lift. The lifting force increases with decrease in diameter of the tube or capillary.

Permeability and capillarity may vary greatly in short distances because of differences in the texture and structure of the ground. Highly permeable beds may alternate with less permeable ones, so that permeability parallel with the bedding is greater than that across the bedding. In cavernous ground, as in some alluvial deposits and some lake beds, the water in the ground may move as freely as surface water. In clean gravel the rate of movement may be about 1,000 feet per day and in clean sand perhaps 100 feet per day. In clayey ground the rate of movement of ground water is practically nil.

The *field capacity*, or *moisture-holding capacity*, of ground is the maximum amount of water that can be held in the pores without seeping downward. The *wilting point* refers to the minimum amount of water contained in ground before plants permanently wilt. Even after plants wilt, the ground still contains water, but it is hygroscopic or chemically combined water and is unavailable to the plants. Field capacity and wilting point depend chiefly on the texture of the ground. In general, wilting occurs when the amount of water in the ground is less than half the moisture-holding capacity. The amount is related to the clay content; the wilting point averages roughly 40 percent of the clay fraction. The moisture content of air-dried soil may average about 10 percent of the clay fraction. Water losses due to evaporation on bared ground may be reduced by cover crops, mulches, or by windbreaks.

The wilting point differs for each plant species; the term has been used chiefly in connection with agricultural crops, but it is applicable to the natural vegetation too. In arid regions the wilting point is, of course, least for the drought-resistant shrubs. In Death Valley, the creosote bush (*Larrea tridentata*) wilts where the desert holly (*Atriplex hymenelytra*) survives. On the Colorado Plateau and Great Basin, sagebrush (*Artemesia tridentata*) wilts where little rabbitbrush (*Chrysothamus puberulus*) survives.

Moisture in surface deposits and soils is affected also by the permeability of the underlying bedrock. Ground at two locations may be alike in slope, texture, and catchment area, yet one may be drier than the other because of differences in texture or structure of the bedrock (Fig. 10.11). The bedrock structure can be important too as a factor in contamination or pollution (Fig. 10.12).

Several factors affect the recharge of water in the ground. One is the texture and

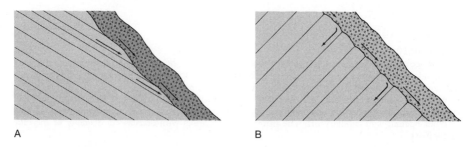

A   B

FIGURE 10.11
The moisture content of similar kinds of ground is affected by the kind and structure of the bedrock. In the example at the left, water (shown by arrows) seeping through the bedrock discharges into the base of the colluvium; in the example at the right, water at the base of the colluvium discharges into the bedrock.

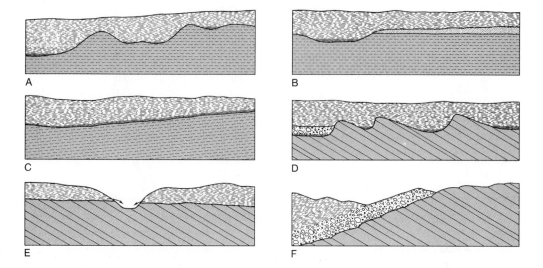

FIGURE 10.12

The configuration of the unconformable surface of the bedrock under surface deposits influences water movement, water supplies, pollution, and underground weathering (adapted from Meinzer, 1923a). The bedrock surface may be rough, with hills and valleys buried (A,D). Ground water is more likely to escape into bedrock that has been folded or tilted (D) than into bedrock that is horizontal (A). On gently dissected bedrock surfaces (B,C) ground water may collect in buried valleys that are shallow (B) or move as a sheet down an inclined plain (C). Other differences will depend on whether the upper beds of the bedrock are more or less permeable than the basal beds of the surface deposit. Where valleys have cut through a surface deposit and into the underlying bedrock (E), the water moving along the unconformity discharges in springs on the hillsides. In certain places, especially in the Basin and Range Province, the surface deposits as well as the bedrock have been folded (F), and in these areas the ground water may move along very complex channelways.

structure of the ground, including the vegetation, which determines the rate of runoff. The second is the slope of the surface, which also affects runoff. A third is the extent of the surface catchment area discharging at a particular place. A fourth is the climate—frequency and amounts of precipitation and rates of evaporation. In most parts of the country, but especially in the arid and semiarid regions, the differences in recharge and moisture availability are expressed in differences in the natural vegetation, because available moisture is the major factor that controls the distribution of many species.

Infiltration is more rapid in sandy than in clayey ground, and sandy ground may hold more water that can continue to drain downward after surface recharge has ended. Examples have been cited (Hilgard, 1910) of water penetrating 3 feet in clay loam under an irrigation ditch in which water ran for 48 hours, whereas in sandy loam the penetration was 5 feet. Twenty-four hours after irrigation had been stopped, the water had penetrated only 3 inches deeper in the clay loam but was 18 inches deeper in the sandy loam. In all kinds of ground the rate of infiltration greatly slows as the depth increases. Where 12 hours may be needed to wet the first foot of depth, 24 hours may be needed to wet the second foot, and 36 hours for the third.

Vadose water (Fig. 10.8) is lost from the ground mostly by evaporation and transpiration (Fig. 10.13). Water evaporates from evaporating pans at a rate of about 50 inches annually in the northeastern United States and at a rate of 100 inches

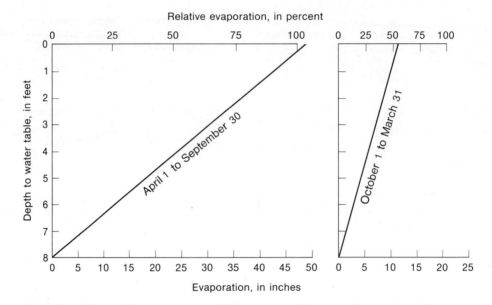

FIGURE 10.13
Diagrams showing evaporativity in Owens Valley, Calif., and actual and relative evaporation from soil surfaces covered by saltgrass, with different depths to water table. [After C. H. Lee, U.S. Geol. Survey Water-Supply Paper 294, Fig. 5.]

or more in the southwestern deserts. The maximum recorded is 150 inches annually at Death Valley, California. Evaporation from soils is much slower. Even in Death Valley more than 2 months may be needed to dry a 12-inch column of wet, saline mud (Hunt, 1966A).

WATER SUPPLIES, DRAINAGE AND IRRIGATION

For water supply and sewage disposal the surface deposits probably are as important as all the underlying rocks taken together, for they have the greatest surface area, supply most wells, and yield the greatest quantities of ground water. Where the annual precipitation exceeds about 20 inches (Fig. 5.9), surface deposits are kept moist and water collects in them where they rest on or contain relatively impermeable layers. Not only is ground water more plentiful in humid than in semiarid regions, it generally is of better quality. Ground water in surface deposits in humid regions generally contains no more than a few hundred parts per million of dissolved solids and may contain as little as 100 parts; in semiarid and arid regions the total dissolved solids in ground water commonly amounts to many hundreds of parts per million.

The rate at which water moves through ground varies greatly depending on its permeability, and this rate of water movement is a major factor in determining whether ground is suitable for septic tanks and other means of sewage disposal. The ground must be open-textured enough to provide adequate drainage and aeration, but not so open as to allow wastes to seep into and pollute the ground water.

Listed below are the surface deposits regarded as most important either as sources of water or for waste disposal (largely from Meinzer, 1923a):

1. Glacial deposits, including both till and stratified outwash, generally supply only small quantities of ground water, but it is adequate for individual homes. These sources commonly are polluted by barnyard and other wastes. The stratified deposits of outwash yield tremendous quantities of good quality water and are a major water source for supplies of towns, villages, and homes in all the states that were glaciated.

2. Quaternary and older volcanic rocks in the Columbia Plateau, Basin and Range Province, Colorado Plateau, and Southern Rocky Mountains (Fig. 7.14) yield tremendous quantities of water that generally is of good quality. On the Snake River Plains, some of the springs issuing from the lavas are large enough to be used for hydroelectric power.

3. Valley fill in the basins of the Basin and Range Province are the major water source for that region, but the supply and quality are variable. Similar valley fill supplies water in central and southern California, and in the Willamette and Puget troughs in Oregon and Washington.

4. Loess deposits yield small supplies of water locally, but in general are not important sources.

5. Lake deposits likewise are not generally important sources for water, but may serve to contain water in underlying valley fill, as in the Lake Bonneville Basin.

6. Alluvial deposits are a major source of water; the deposits in the Mississippi River valley yield tremendous quantities.

7. Coastal deposits generally yield small supplies.

8. Residual deposits are important sources for water in many areas, particularly the suburbs of the Fall Line cities. The water generally is of good quality and occurs on top of the bedrock under the residuum.

Ground in much of the eastern United States has excess water and must be drained (Fig. 10.14). Ground drainage refers to the relative rates of recharge and discharge of water. Poorly drained ground has excessive recharge relative to discharge; well drained ground does not accumulate water. Seven classes of ground drainage distinguished on the basis of slope, surface runoff, infiltration, and permeability are given in Table 10.4.

In contrast to these lands with excess water, the dry, parched ground in the western United States requires addition of water by irrigation in order to grow crops (Fig. 10.14). In irrigation, not only water supply but also water quality is important. Both the ground being irrigated and the water being applied have a high content of alkali and alkaline-earth ions, and these may accumulate as salts in troublesome quantities. Nevertheless, irrigation has many advantages, because the water can be applied in the right amount at the right time to promote optimum growth of a crop. It has advantages too for the home owner who dislikes cutting grass, because the water can be applied in just the right amount to keep the grass from dying but without growing fast enough to require much cutting.

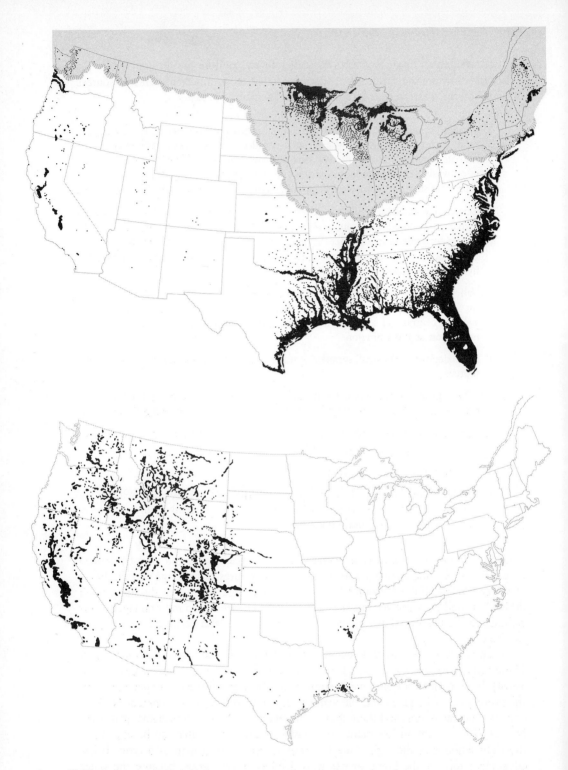

FIGURE 10.14
(Top) Wet lands in the United States occur mostly along the Atlantic and Gulf coasts and in the glaciated part of the country. [After U.S.D.A.] (Bottom) Irrigated lands in the United States. In 1954 the total irrigated area was almost 30 million acres of a total cropland area of more than 400 million acres. [After U.S.D.A.]

TABLE 10.4
Ground drainage classes

Drainage class	Characteristics
Excessively drained	Water is removed from the ground very rapidly. Most excessively drained ground is very porous or occurs on steep slopes, or both, and is free of mottlings.
Somewhat excessively drained	Water is removed from the ground rapidly. Most somewhat excessively drained ground is sandy and very porous, and free of mottling through the profile.
Well-drained	Water is removed from the ground readily, but not rapidly. Most well-drained ground is intermediate in texture and is free of mottling to depths of several feet.
Moderately well-drained	Water is removed from the ground somewhat slowly, and it is wet for a small but significant part of the time. Most moderately well-drained ground has a slowly permeable layer at a depth of 2 to 3 feet, a relatively high water table, additions of water through seepage or combinations thereof. Most moderately well-drained ground is mottled at depths of 2 to 3 feet.
Imperfectly or somewhat poorly drained	Water is removed from the ground slowly enough to keep it wet for significant periods, but not all the time. Most imperfectly or poorly drained ground has a slowly permeable layer at a depth of 1 to 2 feet, a high water table, additions through seepage, or combinations thereof. Most of the imperfectly, or somewhat poorly, drained ground is mottled at depths of 6 to 16 inches.
Poorly drained	Water is removed from the ground so slowly that it remains wet for a large part of the time. In poorly drained ground, the water table commonly is at, or near, the surface during a considerable part of the year and is due to a high water table, to a slowly permeable layer, to seepage, or to some combination thereof. Poorly drained ground is light gray from the surface downward, with or without mottlings.
Very poorly drained	Water is removed from the ground so slowly that the water table remains at, or near, the surface most of the time. Most very poorly drained ground occurs in level, or depressed, sites, frequently ponded. Most very poorly drained ground has dark-gray or black surface layers and is light gray, with or without mottlings, in deeper parts.

Source: U.S.D.A. Soil Survey Manual, 1951.

Land that is to be irrigated must be well drained, for there must be a net outward flow of water to prevent salts from accumulating. Some mistakes have been made on reclamation projects because not enough attention was given to this. Moreover, water that has been through the irrigation cycle becomes ever more salty, which is one of the reasons why Mexico has complained about the quality of water delivered there by the United States from the Rio Grande and the Colorado.

PERMAFROST

Permafrost is the name applied to permanently frozen ground, specifically to ground in which the temperature has been at or below freezing for a long time—from a minimum of 2 years to tens of thousands of years. Ground that freezes during the winter and thaws during the summer is referred to as *seasonally frozen ground.* Ground that is frozen, either permanently or seasonally, may extend into and include bedrock. (The term "frozen" here refers solely to the temperature, and not to the water content (ice) or other factors.) Ground that is below 0°C but contains no ice is *dry permafrost;* it is most common in coarse-grained deposits that drain easily.

Permafrost occurs where the mean annual temperature is below 0°C. Where the annual temperatures average below −5°C the permafrost is *continuous.* The maximum known thickness, more than 2,000 feet, is in Siberia; the maximum thickness known in Alaska, somewhat more than 1,300 feet, is on the arctic slope south of Point Barrow.

Southward the permafrost thins, and where the temperatures average between −1.5° and −5°C, the permafrost is *discontinuous,* with islands of unfrozen ground (*talik*) separating the more extensive areas of permafrost. At its southern limit, *sporadic permafrost* occurs in small, isolated patches.

Permafrost is generally regarded as a relict of the Pleistocene glaciations, but it is not coextensive with the glaciations (Fig. 5.8). The polar deserts (for example, the arctic slope of Alaska) have continuous permafrost, but they were not glaciated because there was not enough precipitation. The southern limit of permafrost about coincides with the treeline. General opinion seems to be that the permafrost is retreating, but the evidence is not conclusive.

In temperate latitudes we are accustomed to thinking in terms of depth of penetration of winter frost; in ground with permafrost, however, one thinks in terms of the depth of summer thaw! The layer of ground above the permafrost that thaws during the summer is referred to as the *active layer.* Its thickness depends on exposure, texture and water content of the ground (wet ground freezes more slowly than dry ground because of the latent heat of fusion of the water, as is shown in Fig. 5.6), and on insulating effects of vegetation, snow, or of water standing on the ground. The active layer may be 10 feet deep in sandy ground, half that deep in clayey ground, and only a foot deep in neighboring ground with peat or swamp (Fig. 10.15). Permafrost may be continuous across the north side of a hill but discontinuous across the south side. Permafrost thins under bodies of surface water—coastal waters, rivers,

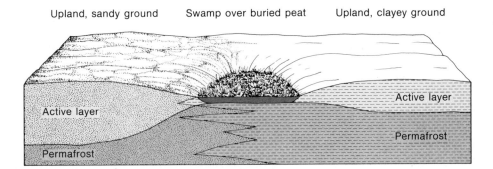

FIGURE 10.15
Diagram illustrating differences in depth of the active layer over permafrost as a result of differences in the kind of ground. Swamps and peat beds are insulators, and the active layer under them is thin. Well-drained ground freezes and thaws more deeply than clayey ground. [After Nikiforoff, 1928, and Muller, 1945.]

and lakes. The top of the permafrost may be depressed 200 feet under a lake a half mile wide.

Islands of ground without permafrost may be surrounded by extensive areas with permafrost. But thawed layers may also occur within the ground, as for example the layer above the top of permafrost and below the depth reached by seasonal freezing. Such a thawed layer may become sealed in by an overlying layer of seasonal ice that survives the following summer thaw. Free water trapped between frozen layers moves because of the hydraulic pressures, and not only the water but the entire saturated layer may move. The ground surface becomes hummocky or otherwise distorted with patterned ground and the wet mass may even erupt onto the surface. In areas of permafrost the active layer is active in more ways than one! Some engineering problems peculiar to such ground are described in Chapter 13.

ORGANIC MATTER

Organic matter affects many of the physical properties of the ground. At the surface, the layer of litter forms an insulating mat that helps retain moisture and protects the ground against compaction and washing. The surface layers are made porous by the burrows of insects and larger animals and by the decay of roots. The content of humus to a considerable degree controls the soil color. Finally, the decayed organic matter washed downward by the soil solutions forms colloids and gels that coat the particles, or the walls of soil columns and prisms, or that serve to bind particles together and cause them to aggregate in granules. Some differences in soil structures are attributable to the organic matter. In general, organic matter tends to make sandy materials more cohesive and clayey ones less plastic, and it increases the water-holding capacity, especially on the weight basis (as compared to the volume basis).

COLOR

Color is one of the most obvious physical properties of soils and surface deposits, yet the causes of many color differences still are imperfectly understood, because things that at first seem obvious prove not to be so. Many black soils contain little organic matter; some yellow and brown soils contain little iron hydroxide, the coloring being caused by organic compounds. The problems perhaps are best illustrated by the occurrence in the tropics of black soils, containing carbonates, alongside highly leached, brilliant red soils that apparently are on the same parent material (Mohr and Van Baren, 1954). In some such black soils the total organic matter may be less than 3 percent, whereas nearby red and yellow soils may have two or three times that amount of organic matter. In such examples the blackness evidently is not due to total organic matter but more likely is attributable to the clay mineralogy and to the stage of humification—that is, the nature of the organic compounds. Finely divided organic matter may coat mineral grains and be absorbed or adsorbed by the clay, especially montmorillonite.

Reds, browns, and yellows commonly are attributable to chemical weathering. Red ground generally is attributed to nonhydrated iron oxides and yellow ground to hydrated iron oxides. But some yellow ground lacks hydrated iron oxides; their yellows may be due to organic matter, as are some browns, or they might be due to mixtures of iron and aluminum oxides (U.S.D.A., Soil Survey Manual, 1951).

In residual deposits, the colors generally vary according to climatic regions. In cold humid latitudes much of the ground is gray; in more temperate latitudes, much of it is gray-brown, brown or chestnut. Farther south it becomes reddish, and in still more southerly and more humid areas the ground commonly is brilliant red and yellow. Red and yellow soils may be mixed with black ones, as noted above. In most Holocene soils that do not contain much organic matter, the color of the parent surface deposit controls that of the soil.

Deposits mottled gray, brown, yellow, or red generally are poorly drained. Light grays to whites may be due to leaching of organic matter and of iron oxides, as in the leached layers of an acid forest soil or of a saprolite, or to deposition of salts, especially calcium carbonate or calcium sulfate, as in the layer of deposition of an alkaline soil. Mottling may be faint, distinct, or prominent; the contrasting mottles may be few, common, or many, and they may be small or large.

Our understanding of what produces these soil colors has lagged behind our ability to measure and describe them. Highly satisfactory means have been developed for determining colors in the field, and there is no reason why color terms should be inexact and used differently by different people. A few percent of people are color blind, and many—like the author of this book—are color dumb. Today, however, color charts are available that give standards of references for naming soil, rock, and ground surface colors.

Color has three attributes: hue (the spectral, or rainbow, color), value (the amount of light reflected), and chroma (strength or purity of the color). A widely accepted system of color identification used in the United States, the Munsell system, represents these attributes by a color solid (Fig. 10.16) that has a neutral gray axis grading

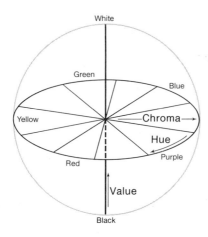

FIGURE 10.16
The different colors (hues) are represented by segments of a sphere radiating from a central axis. The vividness of the color (chroma) increases radially away from the central axis. Lightness (value) increases vertically through the sphere, from black at the base of the axis to white at its top. Midway between black and white is gray.

from black at the bottom through gray to white at the top. This variation in the amount of light is termed *value*. Ground albedo, already referred to, is an effect of value. The color sphere is divided radially into 10 segments joining at the central axis, and each segment represents a *hue*. The five principal ones are red (R), yellow (Y), green (G), blue (B), and purple (P), and they are separated by 5 gradational ones—yellow-red (YR), green-yellow (GY), blue-green (BG), purple-blue (PB), and red-purple (RP). Each of these 10 segments is, in turn, divided into radial subdivisions; 5 marks the middle of a hue and 10 marks the boundary with the next hue clockwise. The radial segments are divided into concentric divisions representing *chroma,* which ranges from gray at the central axis of the sphere to the most vivid color outward.

Every possible color is represented by a point in the sphere, and any particular color can be identified by the coordinates of the point it occupies. A color having a hue along 5 YR, a value of 6, and a chroma of 4 would be recorded as 5YR 6/4.

The sphere is divided into zones identified by color names, which have been standardized by intersociety committees working cooperatively with the U.S. Bureau of Standards. Adjectives *light, medium,* and *dark* designate degrees of value; adjectives *grayish, moderate, strong,* and *vivid* designate increasing degrees of chroma. Hue names are used both as adjectives and nouns. The soil color used as an example in the preceding paragraph would be written: Light reddish brown (5 YR 6/4).

Modern soils have only a limited range of hues, mostly within the range 10R, 5YR, 10YR, and 5Y, although poorly drained soils may be in the range 5GY, 5G,

FIGURE 10.17

Color chart to serve as a guide to soil color names. The color tabs for the correct hue, chroma, and value and neutral gray cardboard for mounting are available at Munsell Color Company, Inc., 2441 Calvert Street, Baltimore, Maryland.

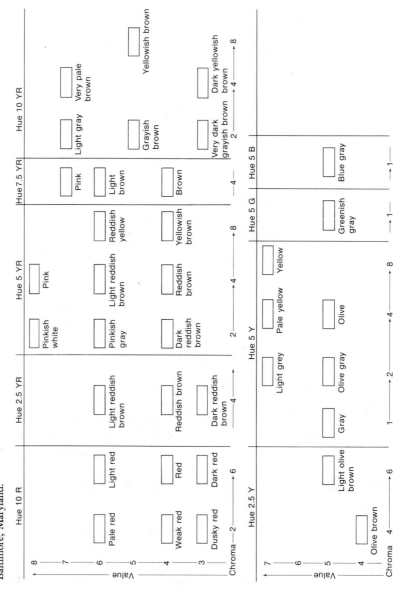

5BG, and 5B. Because of this limited range of hues, only a moderate number of color names are needed for soil descriptions. Hues of rocks and of surface deposits, on the other hand, are more varied and require the use of hues in all ten segments of the color solid. Geologists, though, have reduced rock color names to reasonable numbers by restricting the subdivisions of the hues.

The chief value of the refinement in color determination is the basis it provides for everyone to use standard color names. Agriculturalists have tended to emphasize color description by giving the Munsell coordinates as well as the color name, but this reporting requires an expensive set of color charts and, except for specialized studies, is overly refined. For most work a much simpler approach is adequate. The principal need is for consistent use of color terms, and this can be achieved by using a simplified chart costing only a few dollars that serves to identify the color names rather than the color coordinates (Fig. 10.17).

Opinions differ about the effect of moisture on color. On rock specimens wetting is said merely to decrease the value—that is, to darken the specimen (Goddard, 1948). On the other hand, soil colors which also darken when moistened, by one to three steps in value, may also change as much as two steps in chroma (U.S.D.A., Soil Survey Manual, 1951).

Color descriptions, whether of soils or of rocks, involve determining the color and also describing the color patterns. Color contrasts between crystals and matrix or between mottles may be faint, distinct, or prominent. The contrasting parts may be abundant, common, or few. Sizes of individual parts may be minute or very large.

BIBLIOGRAPHY

Amer. Soc. Agron., 1965, Methods of soil analysis: Monograph 9 of the Society.

Baver, L. D., 1956, Soil water, *in* Soil Physics (3rd ed.): Wiley, New York, pp. 224–303.

Bear, J., Zaslavsky, D., and Irmay, S., 1968, Physical principles of water percolation and seepage: UNESCO, Paris.

Black, R. F., 1954, Permafrost—A review: Geol. Soc. America Bull., v. 65, pp. 839–855.

Black, T. A., Gardner, W. R., and Thurtell, G. W., 1969, The prediction of evaporation, drainage and soil water storage for a bare soil: Proc. Soil Sci. Soc. Amer., v. 33, pp. 655–660.

Bodman, G. B., and Coleman, E. A., 1944, Moisture and energy conditions during downward entry of water into soils: Proc. Soil Sci. Soc. America, v. 8, pp. 116–122.

Bresler, E., and Hanks, R. J., 1969, Numerical method for estimating simultaneous flow of water and salt in unsaturated soil: Proc. Soil Sci. Soc. America, v. 33, pp. 827–832.

Brewer, R., 1964, Fabric and mineral analysis of soils: Wiley, New York, 470 pp.

Buckman, H. O., and Brady, N. C., 1960, The nature and properties of soils: Macmillan, New York.

Colman, E. A., and Bodman, G. B., 1945, Moisture and energy conditions during downward entry of water into moist and layered soils: Proc. Soil Sci. Soc. America, v. 9, pp. 3–11.

Darcy, H., 1856, Les Fontaines Publique de la Ville de Dijon: Dalmont, Paris.

Day, P. R., 1965, Particle fractionation and particle size analysis: Agronomy, v. 9, pp. 545–567.

Drew, J. F., Tedrow, J. C. F., Shanks, R. E., and Koranda, J. J., 1958, Rate and depth of thaw in arctic soils: Trans. Am. Geophys. Union, v. 39, pp. 697–701.

Duley, F. L., 1945, Infiltration into loess soil: Am. Jour. Sci., v. 243, no. 5, pp. 278–282.

Dyal, R. S., and Hendricks, S. B., 1950, Total surface of clays in polar liquids as a characteristic index: Soil Sci., v. 69, pp. 421–432.

Emerson, W. W., 1959, The structure of soil crumbs: J. Soil Sci., v. 10, p. 235.

Feustel, I. C., Dutilly, A., and Anderson, M. S., 1939, Properties of soils from North American arctic regions: Soil Sci., v. 48, pp. 183–189.

Gardner, W. R., 1960, Dynamic aspects of water availability to plants: Soil Sci., v. 89, pp. 63–73.

Hillel, D., 1971, Soil and water—Physical principles and processes: Academic Press, New York, 288 pp.

Holmes, A., 1930, Petrographic methods and calculations: Murby, London, 515 pp.

Johnson, A. I., 1962, Methods of measuring soil moisture in the field: U.S. Geol. Survey Water Supply Paper 1619-U, 24 pp.

Kelly, K. L., and Judd, D. B., 1955, Method of designating colors and a dictionary of color names: Nat. Bur. Standards Circ. 553.

Krumbein, W. C., and Sloss, L. L., 1963, Stratigraphy and sedimentation (2nd ed.): W. H. Freeman and Company, San Francisco.

Kuenen, P. H., 1956, Rolling by current (Pt) 2 of Experimental abrasion of pebbles: Jour. Geol. v. 64, pp. 336–368.

Lull, H. W., and Reinhardt, K. G., 1955, Soil moisture measurements: U.S. Dept. Agri. Southern Forest Exp. Sta. Occasional Paper 140, 56 pp.

Luthin, J. N. (editor), 1957, Drainage of agricultural lands: Amer. Soc. Agron. Monograph 7.

Martin, J. P., 1945, Microorganisms and soil aggregation. I. Origin and nature of some aggregating substances: Soil Sci., v. 50, pp. 163–174.

McCalla, T. M., 1942, Influence of biological products on soil structure and infiltration: Soil Sci. Amer. Proc. v. 7, pp. 209–214.

Marshall, T. J., 1958, A relation between permeability and size distribution of pores: J. Soil Sci., v. 9, pp. 1–8.

Meinzer, O. E., 1923, A, The occurrence of ground water in the United States, with a discussion of principles: U.S. Geol. Survey Water-Supply Paper 489, 321 pp.

———, 1923, B, Outline of ground-water hydrology, with definitions: U.S. Geol. Survey Water-Supply Paper 494, 35 pp.

Muller, S. W., 1945, Permafrost: U.S. Geol. Survey, Strategic Engineering Study No. 62, 231 pp. Reprinted by J. W. Edwards, Inc., Ann Arbor, Mich., 1947.

Nikiforoff, C. C., 1941, Morphological classification of soil structure: Soil Sci., v. 52, pp. 193–207.

Parr, J. F., and Bertrand, A. R., 1960, Water infiltration into soils: Advan. Agron. v. 12, pp. 311–363.

Penman, H. L., 1948, Natural evaporation from open water, bare soil and grass: Proc. Roy. Soc. London, A193, pp. 120–146.

Pettijohn, F. J., 1957, Sedimentary rocks (2nd ed.): Harper, New York.

Rose, C. W., 1966, Agricultural physics: Pergamon Press, Oxford.

Schofield, R. K., 1950, Soil moisture and evaporation: Trans. 4th Internat. Cong. Soil Sci., Amsterdam, v. II, pp. 20–28.

Simonson, R. W., 1951, Description of mottling in soils: Soil Sci., v. 7, pp. 187–192.

Swaby, R. J., 1949, The relationship between micro-organisms and soil aggregation: J. Gen. Microb., v. 3, pp. 236–254.

Swineford, A., and Frye, J. C., 1945, A mechanical analysis of windblown dust compared with analyses of loess: Am. Jour. Sci., v. 243, pp. 249–255.

Taber, S. M., 1929, Frost heaving: Jour. Geology, v. 37, pp. 428–461.

Taylor, S. A., 1955, Field determination of soil moisture: Agri. Eng., v. 36, pp. 654–659.

Washburn, A. L., 1956, Classification of patterned ground and review of suggested origins: Bull. Geol. Soc. America, v. 67, pp. 823–866.

Woodford, A. O., 1925, The San Onofre breccia; its nature and origin: Univ. California Dept. Geol. Sci. Bull., v. 15, no. 7, pp. 159–280.

Yaalon, D. H., 1966, Chart for the quantitative estimation of mottling and of nodules in soil profiles: Soil Sci., v. 102, pp. 212–213.

The geochemistry and mineralogy of soils and surface deposits are major factors that affect the suitability of ground for various uses. Chemical analyses can be made by various means. Shown here is a thin section cut from an igneous rock, a diorite porphyry. Such sections are ground thin enough to transmit light, and the minerals can be identified by their optical properties. The white crystals are feldspar; the large, dark, rectangular ones are hornblende. The grayish spots in the feldspars indicate clayey alteration. The large crystals were once floating in a melt, which is now represented by the fine-grained groundmass. Width of field: 2.5 mm.

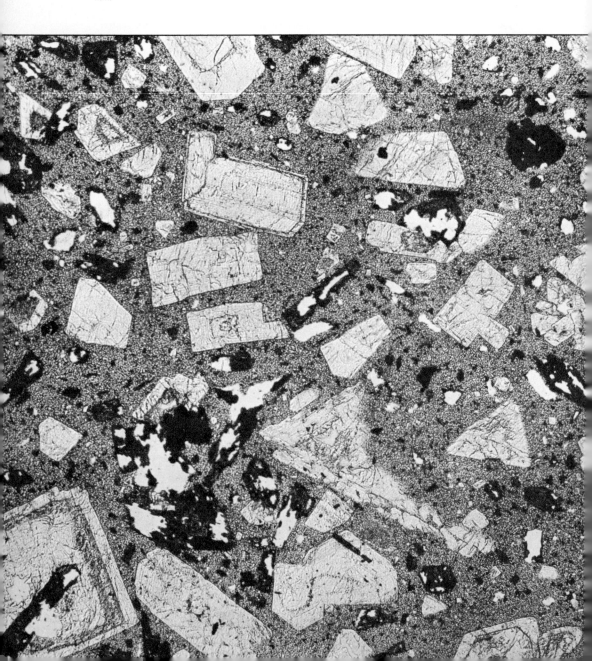

11/ *Mineralogy and Geochemistry*

Composition of rocks; the ground—general features; acidity, alkalinity; water; soil atmosphere; salts; stability of minerals; structure of minerals; clay minerals; solid oxides and hydroxides; organic matter, production, decay, nitrogen compounds; trace elements; ion exchange; chemical weathering processes; bibliography.

The mineralogy and chemistry of the ground depend partly on the composition of the original surface deposit and partly on changes caused by external influences of the environment. The most notable changes are those that take place in the upper layers, in the modern soil, but changes also take place in the deeper parts of the surface deposit on which the soil is developed. The materials that constitute the ground are of 6 principal kinds:

1. Fragments of rocks and of primary rock-forming minerals.

2. Secondary minerals derived by weathering and alteration of the primary ones.

3. Living organisms.

4. Dead organic matter in various stages of decomposition.

5. Water.

6. Air.

All of these are variables. Not only do they differ from one deposit to another, but even within an individual deposit or soil, all but the first of these variables may change in response to changes in external conditions (humidity, temperature, barometric pressure, earth movement). The outstanding feature of the mineralogy and geochemistry of surface deposits and soils is that they undergo virtually continuous change.

Chemical compounds in the ground that interact in the processes of weathering are of four kinds:

1. Acids, in which hydrogen is combined with an acid element or radical (e.g., $H^+ + Cl^- \longrightarrow HCl$, hydrochloric acid).

2. Oxides, in which oxygen is combined with another element (e.g., $2Fe^{+++} + 3O^{--} \longrightarrow Fe_2O_3$, iron oxide, which is ordinary iron rust).

3. Bases, in which the hydroxyl radical $(OH)^-$ is combined with an element (e.g., $Na^+ + (OH)^- \longrightarrow NaOH$, sodium hydroxide, lye).

4. Salts, in which the hydrogen of an acid is displaced by a metallic element (e.g., $Na^+ + Cl^- \longrightarrow NaCl$, sodium chloride, table salt).

The compounds may be organic or inorganic.

The acids in the ground that cause weathering are mostly those containing oxygen; by far the most important one is carbonic acid (H_2CO_3). Other acids that might be expected in the upper layers of the ground are silicic acid, H_4SiO_4; nitric acid, HNO_3; phosphoric acid, H_3PO_4; perchloric acid, $HClO_4$; chloric acid, $HClO_3$; sulfuric acid, H_2SO_4; and sulfurous acid, H_2SO_3.

COMPOSITION OF ROCKS

The rocks that form the earth's crust are composed chiefly of the following eight elements (approximate percentage by weight, generalized from Mason, 1958):

oxygen	46.6	calcium	3.6
silicon	27.7	magnesium	2
aluminum	8	sodium	2.8
iron	5	potassium	2.6

Table 11.1 gives the average composition of igneous rock and of the two common end members, basalt and granite. The chemical composition of the average igneous rock probably is about average for the earth's crust.

Sedimentary rocks and metamorphic rocks differ greatly in composition from igneous rocks. A pure *limestone*, for example, or its metamorphic equivalent, *marble*, is composed almost entirely of calcium carbonate, $CaCO_3$. A pure *sandstone* and its metamorphic equivalent, *quartzite*, consists almost entirely of silicon dioxide, SiO_2. A mixture of limestone and sandstone could be sandy limestone or limy (calcareous)

TABLE 11.1
Chemical composition of basalt, granite,
and the average for all igneous rocks

	Basalt	Granite	Average igneous rock
SiO_2	50	72	59
Al_2O_3	17.5	13.5	15.5
Fe_2O_3	3	1	3
FeO	7	1	3.75
CaO	8	1	5
MgO	6.5	0.25	3.50
Na_2O	3	3.5	3.75
K_2O	2	5	3
H_2O	1	1	1.25
TiO_2	1.25	0.5	1
other	0.75	1.25	1.25

Source: Data rounded off from Clarke, (1924) and
Mason (1958).

sandstone, depending on the proportion of lime and sand. A *shale,* or its metamorphic equivalent, *schist, slate,* or *phyllite,* may approach the composition of pure clay, roughly 47 percent SiO_2, 40 percent Al_2O_3, and 13 percent H_2O. Mixtures of shale and sand form shaley sand or sandy shale, depending on the proportions; mixtures of shale and lime form shaley lime (marl) or limy shale. To the degree that there are other impurities in such rocks, the compositions approach those of the igneous rocks. Some are rich in iron, *ferruginous,* others are rich in magnesium. Most rocks are mixtures of minerals, and the unequal weatherability of the minerals leads to collapse of the support for the resistant ones. For example, micaceous sandstone weathers easily because the mica becomes hydrated and the resistant sand grains are shoved apart.

ACIDITY, ALKALINITY

Acidity and alkalinity are expressed in terms of the hydrogen ion concentration, pH, which is the logarithm of the reciprocal of the concentration. Accordingly, low pH means high H ion concentration, and conversely, high pH means low concentration. There is a tenfold difference in concentration between each unit of pH: pH 7.0 is neutral; pH 6.0 indicates 10 times as great a concentration of H-ions as pH 7.0 and is acid; pH 8.0 indicates one tenth as great a concentration of H ions as pH 7.0 and is alkaline. Most ground is between pH 4 and pH 9.

In general, pH is a measure of the exchangeable ions. Low pH means that the ground is high in exchangeable hydrogen; high pH means high concentration of other exchangeable bases. The pH measures degree of acidity but not total amount. Sandy ground may have the same pH as clayey ground, but the total amount of the available hydrogen is much greater in the clay. For example, to raise the pH of a piece of ground from 5.5 to 6.5 may require half a ton of limestone if the ground is sandy, twice that amount if the ground is loam, and four times that amount (2 tons) if the

256

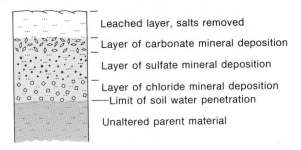

FIGURE 11.1
Distribution of salts near the ground surface in an arid environment reflects frequency of wetting of the ground and the relative
solubilities of the salts. Salts are leached from the uppermost
layers, which are most frequently wetted. Carbonates are deposited below this, but this layer is wet frequently enough to remove
the sulfates and chlorides, which are deposited below it. Chlorides
are moved downward from the occasionally wet sulfate layer to
the infrequently wet bottom layer. Below this the ground remains
dry and is free of secondary salts.

ground is clay loam (U.S.D.A., Soil Survey Manual, 1951). And, of course, many
times those amounts of limestone are needed if the ground pH is much below 5.5.

Within a weathering profile the pH may vary greatly from one layer to another
and from one particle to another, and may vary with moisture content, CO_2 pressure
in the soil air, and with season (growing season and resting season). Many, perhaps
most, surfaces coated with organic colloids are acid, probably even in alkaline ground.
But in alkaline ground the content of organic matter is low, and the pH, which is
an average of all the sites on the particles, is alkaline. Indicator dyes can be used
with reasonable success for field determination of differences in ground pH, but
refined laboratory methods are needed for the absolute determinations on the micro-
scale at particular sites.

The term "salinity" generally refers to the chlorides of the alkalis or alkaline earths;
"alkalinity" generally refers to their sulfates or carbonates. Depending on the salts
present, ground may be saline without being alkaline (e.g., sodium chloride ground)
or it may be alkaline without being saline (e.g., sodium carbonate ground), or it may
be both. The salts may occur as rock fragments, as crystals in the parent material,
as veinlets, nodules, disseminated masses, or as isolated, well-formed crystals (*eu-
hedral crystals*) of secondary minerals developed by the weathering. Moreover, the
salt content may vary greatly in short distances, and from layer to layer within the
ground (Figs. 11.1, 11.2).

WATER, DISSOLVED MATTER

Without water (H_2O), there would be no weathering; nor would there be any
organisms or accumulated matter in the surface layers of the ground. Water is
important in ground chemistry even before it reaches the ground, because water in

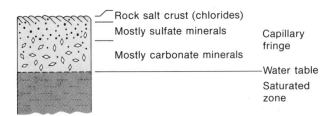

FIGURE 11.2
Distribution of salts deposited by rising water in the capillary fringe
above a shallow water table in an arid environment. Just above the water
table carbonates are deposited; above this are sulfates; chlorides form
a surface crust of rock salt.

the atmosphere contains at least three kinds of *aerosols,* matter suspended in the
atmospheric vapors (Fisher, 1968). *Marine aerosols* derived from the oceans supply
most of the chloride and much of the sodium in the ground; *acidic aerosols,* which
contribute sulfates, are derived from industrial and automobile smog, volcanic
activity, and some sulfate-bearing surface waters. *Alkaline aerosols,* probably derived
from loessial dust, provide much calcium and potassium and some sodium and
magnesium. Those who suffer from hay fever will also think of pollen.

The quantities of these materials supplied to the ground by rain water depend
on proximity to the three principal kinds of sources. Chlorides in rain water increase
oceanward. In England, at Rothamsted, about 15 pounds of chlorine are added per
year per acre; nearer the coast, it is reported that more than 5,000 pounds of chlorine
may be added per acre (Russell, 1956; Carroll, 1962). Such contributions by the
marine aerosols need to be accounted for in estimates of rates of erosion based on
the dissolved solids and suspended loads in streams.

Rain and snow also contribute nitrogen to the ground. According to the U.S.
Department of Agriculture (Yearbook, 1938, p. 364) measurements at a number of
stations in the world suggest that something like 4 to 8 pounds of nitrogen is added
per acre per year, in the form of ammonia compounds and as nitrates. This nitrogen
sustains the natural vegetation, but it is not enough to replenish the losses from
farmed crops.

As soon as water enters the ground, it encounters organic matter in the surface
layers and there dissolves carbon dioxide, carbonic acid, and humic compounds. The
water becomes acid, chiefly because of carbonic acid (H_2CO_3) from the organic
matter. Sulfate and nitrate are other acid ions derived from the organic matter.
Locally, sulfuric acid (H_2SO_4) is developed from decomposition of pyrite (iron
sulfide, FeS_2). The acid dissolves mineral matter as it percolates downward.

In the lower parts of modern soils, and deeper in the ground, the dissolved matter
consists mostly of carbonate, sulfate, calcium, magnesium, sodium, and potassium
ions, with (usually) small amounts of chlorides, iron, and silica. Common dissolved
gases include carbon dioxide (CO_2), nitrogen (N), oxygen (O), methane (CH_4), and
hydrogen sulfide (H_2S). The list on the next page gives the chief sources and
significance of the common mineral constituents dissolved in water below the organic
layers.

Silica (SiO_2). Usually present in amounts between 1 and 30 ppm (parts per million); derived mostly from weathering of silicate minerals; in part redeposited in subsoil as silica cement that may form a hardpan; in part unites with free bases to form secondary silicates; in part removed by ground water.

Iron (Fe^{++}). Usually less than 1 ppm; amounts are greater where there is acid water from mines or deposits containing pyrite; some of the iron leached from the upper soil layers becomes redeposited in the layer of deposition, probably because of decreasing acidity of water in the ground as it seeps downward from the organic layer; in deeper, aerated ground the iron becomes oxidized and precipitated as reddish brown sediment; in water-logged ground, where oxygen is deficient, the iron is in the form of ferrous compounds.

Manganese (Mn^{++}). Like iron but usually present only in trace amounts.

Calcium (Ca^{++}) and Magnesium (Mg^{++}). Dissolved from calcium- and magnesium-bearing silicate minerals and from fragments of limestone, dolomite, and gypsum in the surface deposit or soil; forms soluble salts that are removed from acid ground but accumulate as carbonate or sulfate caliche in the subsoil in alkaline areas; in aerated ground may be deposited as caliche in the capillary fringe above a water table.

Sodium (Na^+) and potassium (K^+). Principally derived from weathering of feldspars in surface deposits and soils; in general, water retains these ions in solution longer than those of calcium and magnesium.

Bicarbonate (HCO_3^-) and carbonate (CO_3^{--}). Derived principally from organic matter in the surface layers of the ground; also derived from mineral matter where there is much limestone and dolomite in the surface deposit or soil; in acid environments, the water may leave the system as carbonate water; in alkaline ground the carbonates are reprecipitated in the subsoil as the ground dries.

Sulfate (SO_4^{++}). Chief source is gypsum ($CaSO_4 \cdot 2H_2O$) or pyrite; in general, sulfates are retained in solution longer than the carbonates; sulfates are removed from acid ground; in alkaline environments they become reprecipitated as the ground dries.

Chloride (Cl^-). Derived from sewage, ancient saline deposits, sea water and rainwater; amount generally small compared to carbonate or sulfate ions and in general more soluble.

Nitrate (NO_3^-). Chief source is organic matter in sewage and fertilizer; amounts generally small; highly soluble; can vary greatly during the year depending on temperature and other factors affecting biological activity.

Oxygen (O_2). Dissolved in water from the air, both above and in the ground.

Many or most of these mineral constituents may have cycled through plant tissue before and after becoming dissolved in the water.

SOIL ATMOSPHERE

The air that occupies voids in the ground (that is, those voids not filled with water) contains the same constituents as the air above ground, and in about the same proportions. The air we breathe contains about 21 percent oxygen and 78 percent nitrogen; air in the near-surface layers of the ground contains about 20 percent oxygen and 78.5 percent nitrogen. The content of inert gases, principally argon, is about the same, slightly less than 1 percent in each. The content of carbon dioxide however, is quite different. The air we breathe contains only 0.03 percent CO_2, whereas air in the surface layers of the ground—in the soil—contains about 0.6 percent CO_2. At depth, the nitrogen and argon content remain about the same, but the oxygen in some soils decreases several percent while the CO_2 content and activity increase correspondingly.

The carbon dioxide in the soil atmosphere is derived chiefly from roots, the metabolism of the microorganisms, and the decay of the organic matter in the ground; the oxygen is consumed by the organisms in building the carbon dioxide. The CO_2 combines with water to form carbonic acid, and in humid regions this is removed by water seeping through the ground.

SALTS

Salts in the ground may be formed in four principal ways:

1. By interaction of an acid and base (e.g., $HCl + NaOH \longrightarrow NaCl + H_2O$).

2. By a metal more active than hydrogen that displaces hydrogen from an acid (e.g., $Na^+ + H_2CO_3 \longrightarrow Na\ HCO_3 + H^+$). Among the common metals that are more active than hydrogen and that can displace it are potassium (most active), barium, strontium, calcium, sodium, magnesium, aluminum, manganese, and iron (least active of these). Moreover, the more active of these elements can displace the less active ones as well as hydrogen (i.e., calcium can displace sodium).

3. By interaction of two salts (e.g., $2NaHCO_3 + MgSO_4 \longrightarrow Mg(HCO_3)_2 + Na_2SO_4$).

4. By direct combination of a metal and a nonmetal (e.g., $Na^+ + Cl^- \longrightarrow NaCl$). Most salts are neutral, but some contain displaceable hydrogen (e.g., sodium bicarbonate, $NaHCO_3$), and in a sense these are acid salts. However, in the example given, sodium bicarbonate dissociates in water to form a strong base (sodium hydroxide) and a weak acid (carbonic acid) with the result that the solution is alkaline.

Some salts are highly soluble. All the nitrates, acetates, and all the common chlorides are soluble (except those of lead, silver, and mercurous mercury). All the sulfates are soluble except those of barium, strontium, and lead. The sulfates of calcium, silver, and mercury are sparingly soluble. All the salts of sodium, potassium, and ammonium are soluble except a few rare ones.

A consequence of the easy solubility of these salts is that they become depleted in ground having an open drainage system, as in humid regions, but in arid and semiarid regions or localities where evaporation exceeds the recharge, salts accumulate. In alkaline soils they accumulate in the lower layers, mostly as carbonates, the carbon being contributed by the organic matter.

The percentage of alkalis and alkaline earths remaining in a soil or soil layer is one measure of the degree of weathering that has occurred. Many ancient pre-Wisconsinan soils are developed on parent rocks containing 15 percent or more of the alkalies and alkaline earths, but in most of these old soils the amount has been reduced to less than 5 percent and in many of them practically to zero. Some layers in late Pleistocene and Holocene soils also may be greatly depleted in salts, as in the leached layers of acid forest soils. Such layers, however, may retain most of the alkalis and alkaline earths if these constituents are contained in primary rock minerals, which, for the most part, are little altered in such young soils.

Salts in the ground develop partly from the alteration of primary minerals, but, as already noted, they may also be introduced by rain and by drainage or irrigation waters. Saline ground is common in arid and semiarid regions, where surface waters may contain moderate or considerable quantities of dissolved solids. The evaporation of the water leaves these as salts. Where water tables are near the surface, subsurface evaporation causes salts to be deposited in the capillary fringe, which may be manifest on the surface. Irrigation systems are especially vulnerable to introduction of salts by the waters that are used.

Because of differences in their solubilities, the salts are not equally persistent. In arid regions, for example, the near-surface layers of the ground are wetted by rain or runoff more frequently than the deeper layers. Near the surface is a layer in which carbonate minerals persist, but from which sulfates and chloride minerals are removed. Below this is a layer chiefly of sulfates from which chlorides are removed, and still lower, at the greatest depth wetted, are chlorides (Fig. 11.1). On the other hand this layering is reversed where water rises in the ground at the capillary fringe above a water table (Fig. 11.2).

Because of their ready solubility and comparative ease of study, salts serve to illustrate some principles of geochemical processes that may bear on some of the more difficult problems posed by the clays, oxides, and hydroxides. One principle is that when two electrolytes that do not yield a common ion are brought into solution, the solubilities of both are increased. For example, a liter of pure water at $25°C$ will dissolve about 2.1 grams of calcium sulfate, but if the water contains 10 grams of NaCl, it will dissolve 3.5 grams of calcium sulfate. If the water contains about 135 grams of NaCl, it will dissolve 7.5 grams of calcium sulfate.

On the other hand, if two electrolytes that both yield a common ion are brought into solution, the solubilities of both are decreased. Continuing the example with calcium sulfate, if the liter of water contains 0.5 grams of Na_2SO_4, it will dissolve only 1.53 grams of the calcium sulfate—about three-quarters as much as the pure water.

Solubilities can be affected in other ways too. For example, carbonates in the ground very likely affect the mobility of the iron. The carbonic acid in a soil solution is partly a function of the CO_2 pressure in the soil atmosphere, and this affects the amount of the iron that can be taken into solution. If the solution moves to aerated ground having lower CO_2 pressure, the iron is precipitated as an oxide or hydroxide.

STABILITY OF MINERALS

The mineral grains in weathered rocks include primary minerals that resist weathering and remain unaltered, others that are partly altered, and still other new, or secondary, minerals formed from chemical elements freed during alteration. Table 11.2 gives the relative stabilities of the minerals common in surface deposits.

TABLE 11.2
Relative stabilities of minute grains (diam. < 1 mm) of common minerals in surface deposits

Oxides and hydroxides	
Magnetite, Fe_3O_4	Not very persistent in most surface deposits; usually alters to the red hematite, brown geothite, or yellow limonite; stable under anaerobic conditions
Hematite, Fe_2O_3	
Goethite, $Fe_2O_3 \cdot H_2O$	
Limonite, a mixture of oxides, approximately $2Fe_2O_3 \cdot 3H_2O$	Most persistent form; a mixture of the iron oxides and hydroxides
Corundum, Al_2O_3	Rare primary mineral that resists abrasion and is not acted upon by acids; persists in deposits derived from some metamorphic rocks
Diaspore and boehmite, $Al_2O_3 \cdot H_2O$, and gibbsite, $Al_2O_3 \cdot 3H_2O$	Secondary minerals in surface deposits and soils where alumina has been freed by alteration of aluminum-bearing silicates
Titanium compounds	
Anatase and rutile, TiO_2	Persistent
Ilmenite, $FeO \cdot TiO_2$	Moderately persistent
Perovskite, $CaO \cdot TiO_2$, and doelterite (leucoxene), $TiO_2 \cdot H_2O$	Alteration products of sphene ($=$ titanite, $CaTiSiO_5$) and ilmenite (resistant) in surface deposits and soils
Sulphides	
Pyrite, FeS_2	Unstable
Phosphates	
Apatite, $Ca_5(PO_4)_3(F,Cl,OH)$	Moderately persistent; primary
Crandallite, hydrous calcium-aluminum phosphate	Alteration product of apatite; alters to wavellite
Wavellite, $Al_3(PO_4)_2(OH)_3 \cdot 5H_2O$	Secondary
Monazite, rare earth phosphate	Persistent; primary
Carbonates (mostly secondary)	
Calcite (limestone, also aragonite), $CaCO_3$	Soluble; primary or secondary
Dolomite, $CaCO_3 \cdot MgCO_3$	Less soluble than calcite; primary or secondary
Sodium carbonate minerals	Several, mostly with combined water; highly soluble; secondary
Sulfates (mostly secondary)	
Gypsum, $CaSO_4 \cdot 2H_2O$	More soluble than calcite
Anhydrite, $CaSO_4$	Like gypsum
Sodium sulfate and mixed sodium-calcium sulfate	Several, mostly with combined water; readily soluble

(continued)

Table 11.2 (*continued*)

Chlorides (secondary)

Halite, NaCl	Highly soluble
Ammonium chloride, NH_4Cl	Highly soluble

Nitrates, nitrites (secondary)

Soda niter, $NaNO_3$	Highly soluble
Saltpeter, KNO_3	Highly soluble

Silica

Quartz, SiO_2	Mostly primary; highly resistant to abrasion and solution
Chalcedony, SiO_2	Microcrystalline; also known as chert, flint, jasper; less stable than quartz; primary or secondary
Opal, $SiO_2 \cdot nH_2O$	Least stable form of silica; mostly secondary in surface deposits and soils

---

Silicates (the principal rock-forming minerals)

Aluminum silicates without iron or magnesium (light minerals)

Feldspars

Potash feldspars: orthoclase, microcline, $KAlSi_3O_8$	Most persistent of the feldspars
Soda-lime feldspars: plagioclase	Gradational series ranging from lime feldspar (anorthite), $CaAl_2Si_2O_8$, to soda feldspars (albite), $NaAlSi_3O_8$; pure end members rare; soda-lime feldspars (oligoclase, andesine) more stable than the lime-soda ones (labradorite, bytownite)
Muscovite (white mica), $KAl_2(Al,Si_3)O_{10}(OH,F)_2$	Primary; stable

Aluminum silicates with or without iron or magnesium

Clay minerals (the principal secondary minerals in surface deposits and soils; also primary; stable)

Kaolinite, $Al_2Si_2O_5(OH)_4$	Most stable clay mineral
Illite	Complex hydrous white mica; alters to kaolinite or montmorillonite
Vermiculite	Complex hydrous biotite; alteration product of chlorite and biotite, and alters to kaolinite or montmorillonite
Chlorite	Complex; mostly a primary mineral in surface deposits and soils (i.e., relict from parent rock); least stable of the clay minerals; alters readily to any or all of the others
Montmorillonite	Complex swelling clay that readily absorbs iron, magnesium, calcium, sodium, hydroxyl, or other ions; alters to kaolinite; alteration product of chlorite, vermiculite, and illite

Aluminum silicates with iron, magnesium, or calcium (dark minerals); all primary in surface deposits and soils

Biotite (dark mica), $K(Mg,Fe)_3(Al,Si_3)O_{10}(OH,F)_2$	Most easily weathered of the dark minerals; readily altered to vermiculite
Hornblende: one of a series of minerals called amphiboles, $(Ca,Na,Fe,Mg,Al)_7(Al,Si)_8O_{22}(OH)_2$	More persistent than augite
Augite: one of a series of minerals called pyroxenes, $Ca(Mg,Fe,Al)(Al,Si)_2O_6$	More persistent than olivine
Olivine, $(MgFe)_2SiO_4$	Least persistent of the dark minerals
Tourmaline, garnet	Persistent

Magnesium silicate

Serpentine, $Mg_6Si_4O_{10}(OH)_8$	Primary or secondary; stable

The persistence of various minerals in sedimentary rocks of different geologic ages provides another index of the stability of rock-forming minerals (Tables 11.2, 11.3). The younger geological formations are more complex mineralogically than the ancient ones, the most complex being the Pleistocene and Holocene deposits. Minerals that occur in bedrock also occur as detrital grains in surface deposits, which contain, in addition, the unstable secondary minerals and organic ones. The relative stabilities of minerals listed in Tables 11.2 and 11.3 are for minerals weathered in situ. The list would differ if abrasion and alteration during transportation, deposition, diagenesis, lithifaction, as well as any subsequent alteration were taken into account.

The common silicate minerals that form igneous rocks (Table 11.3) are susceptible to weathering in about the same order as their sequence of crystallization in igneous melts. Minerals that crystallize late, such as quartz and muscovite, are most resistant to weathering whereas the minerals that crystallize early, such as olivine and calcic plagioclase, are most susceptible. The minerals provide a measure of degree of weathering. Where moderately calcic plagioclase was present in the original material, the preservation of labradorite or andesine indicates only moderate or slight weathering, but if only albite or oligoclase is preserved, the stage of weathering is advanced (Graham, 1950).

The least stable of clay-size particles are the highly soluble nitrates, chlorides, sulfates, carbonates, and phosphates, especially of the alkalies (Na,K) and alkaline earths (Ca,Mg). These are mostly secondary minerals. Their solubility is the reason why surface deposits and soils generally lack alkalis and alkaline earths. Stated differently, their solubility is the reason why so much land needs fertilizing; these elements are needed by plants. Even apatite weathers readily when reduced to clay-size particles. In the Florida phosphate deposits, where the soils are ancient and weathered to a depth of 10 feet or more, the calcium phosphate (apatite) dissolves and the phosphate combines with aluminum to form hydrous aluminum phosphate (wavellite) as follows (Altschuler and others, 1956):

$$Ca_5(PO_4)_3(OH) \longrightarrow CaAl_3(PO_4)_2(OH)_5 \cdot H_2O \longrightarrow Al_3(PO_4)_2(OH)_3 \cdot 5H_2O$$
$$\text{apatite} \qquad\qquad \text{crandallite} \qquad\qquad\qquad \text{wavellite}$$

Paralleling this change, montmorillonite alters to kaolinite with a loss of silica. The lost silica, or much of it, is in the underlying Miocene limestones, where it forms nodular chert masses that replace the limestone and its contained fossils.

The primary minerals biotite and olivine are less stable than augite and hornblende. Fine-grained biotite alters readily to vermiculite. Clay-size feldspars, even calcic plagioclase, seem to be more stable than clay-size dark minerals. Clay-size sodic plagioclase is more stable than calcic plagioclase, and potash feldspar is the most stable of the feldspars. Muscovite persists longer than the feldspars, but clay-size muscovite grades to a clay mineral, illite. Finally, of the common primary minerals, the most stable is quartz. The weathering of the silicates releases silica, part of which may be removed in solution and part redeposited as secondary quartz, chalcedony, or opal.

The most stable minerals are the clays, in about the order indicated in Table 11.2, and the oxides.

The results of most studies about the stability of minerals agree with relative stabilities given here, but exceptions are numerous. One reason for some of the

TABLE 11.3

The common rock-forming silicate minerals arranged in their general order of crystallization from magmatic melts, their susceptibility to weathering, and their occurrence in various igneous rock types

Sequence of crystallization	Susceptibility to weathering	Dark minerals	Light minerals	Rock types	
				Volcanic	Intrusive
early	least resistant	Olivine $(Mg,Fe)_2SiO_4$	Calcic plagioclase (anorthite) $CaAl_2Si_2O_8$		
		Augite $Ca(Mg,Fe,Al)(Al,Si)_2O_6$	Calcic plagioclase (labradorite) with sodium	basalt	gabbro
		Hornblende $(Ca,Na,Fe,Mg,Al)_7(Al,Si)_8O_{22}(OH)_2$	Sodium plagioclase with calcium (andesine, oligoclase)	andesite	diorite
		Biotite (dark mica) $K(Mg,Fe)_3(Al,Si_3)O_{10}(OH,F)_2$	Sodium plagioclase (albite) $NaAlSi_3O_8$	latite	monzonite
			Potash feldspar (orthoclase, microcline) $KAl_2Si_3O_8$	rhyolite	granite
late	most resistant		Muscovite (white mica) $KAl_2(Al,Si_3)O_{10}(OH,F)_2$		
			Quartz, SiO_2		

Source: Adapted from Goldich (1938).

exceptions may be that the grain sizes used in the studies were not comparable. Moreover, in different environments, the processes differ. Calcite is resistant in arid regions, where soil water and ground water are scanty and likely to be alkaline, but is highly soluble and readily removed in humid regions, where water is plentiful and likely to be acid.

The feldspars and dark silicates contain more silica than do the clay minerals that develop from them. The least stable primary minerals (Table 11.3), have less silica than the more resistant ones. In anorthite and olivine the ratio of silica to alumina is about 1 to 1; in albite, orthoclase, and biotite the ratio is about 3 to 1. The proportion of sodium and potassium to aluminum also increases.

STRUCTURE OF MINERALS

The relative resistance of minerals to weathering is partly a function of their structure. The hardest minerals resist abrasion and are most likely to survive transportation by wind, water, or ice. Mineralogists use 10 minerals to form a scale of increasing hardness:

1. talc	6. feldspar
2. gypsum	7. quartz
3. calcite	8. topaz
4. fluorite	9. corundum
5. apatite	10. diamond

Except for diamond, each mineral in the series can be scratched (abraded by) the next harder mineral. One's fingernail will scratch gypsum and talc, an ordinary pin will scratch calcite, a knife will scratch apatite, and quartz will scratch a beer bottle.

Resistance of minerals to fracturing, abrasion, and weathering also is affected by a crystal structure property that controls the way minerals break, or cleave. This property is quite like jointing in rocks, but on a micro-scale.

Because of certain parallelisms in their crystal structures, minerals exhibit, to varying degrees, the property of splitting, or being cleaved easily along parallel planes, or *cleavage planes* (Fig. 11.3). If a cube of halite (sodium chloride) is broken, the resulting grains are cubes, because halite has well developed cleavage planes parallel to each face of the cube. Pyrite also crystallizes in cubes, but has poor cleavage and consequently fractures irregularly rather than breaking into cubes. The external crystal form only partly reflects the cleavage.

Minerals that fracture or cleave easily are more subject to weathering for two reasons. First, dividing a mineral into particles by abrasion or alteration greatly increases the surface area subject to weathering; second, the solutions that cause weathering gain access to the interiors of minerals along the cleavage planes.

Some minerals have one set of parallel cleavage planes; the micas and chlorite are examples. When mica weathers, the folia swell, and the edges may take on the appearance of a feather duster.

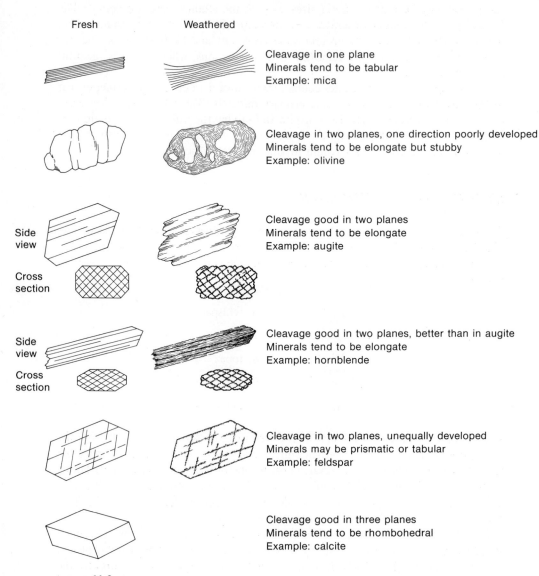

Fresh   Weathered

Cleavage in one plane
Minerals tend to be tabular
Example: mica

Cleavage in two planes, one direction poorly developed
Minerals tend to be elongate but stubby
Example: olivine

Side view

Cross section

Cleavage good in two planes
Minerals tend to be elongate
Example: augite

Side view

Cross section

Cleavage good in two planes, better than in augite
Minerals tend to be elongate
Example: hornblende

Cleavage in two planes, unequally developed
Minerals may be prismatic or tabular
Example: feldspar

Cleavage good in three planes
Minerals tend to be rhombohedral
Example: calcite

FIGURE 11.3
Examples of cleavage in minerals and how it affects weathering and alteration. Compositions of the minerals and their alteration products are discussed in the text.

Other minerals have two intersecting sets of cleavage planes. In some, like olivine and feldspar, the two cleavages are unequally developed; in general, the crystal structure reflected by these cleavages results in stubby minerals. Those that have two good cleavages, like augite and hornblende (Fig. 11.3), tend to be elongate, or *prismatic,* minerals.

Still other minerals have three cleavages at right angles. Halite, for example, has cubic cleavage; other minerals have cleavage planes that are oblique to each other and produce rhombs—the cleavage pattern of calcite and dolomite.

Hardness, density, and cleavage also affect the ease with which minerals loosened from a parent rock may be transported from the site. Heavy dense minerals (e.g., magnetite, garnet, hornblende) are likely to lag behind while the light ones (e.g., feldspar, quartz, mica) are washed away. But the size and shape of the fragment, which is partly controlled by the cleavage, may be as important as its density. Clay-size particles of dense minerals may be washed away by sluggish runoff; twice the velocity may be needed to wash away sand-size grains of the same mineral. Further, the more tabular crystals may be washed away while stubby ones remain behind.

CLAY MINERALS

The clay fraction probably is the most important component of surface deposits and soils, whether in terms of origin, agricultural or engineering use, or as a factor in the environmental processes—physical, chemical, and biological. The term "clay" can be misleading. It may refer to very fine materials, regardless of their mineralogy, or it may refer to a definite series of minerals. In this book the specific use is indicated by the term "clay minerals"; the term "clay" used alone refers only to very fine materials without implying they are necessarily the clay minerals. Because clay is cohesive, it holds soil aggregates together and controls many soil structures. Clays retain moisture, and thereby impede the movement of water and air in the ground. They also hold plant nutrients, some in forms available for plant use, others in unavailable forms. The plasticity of clay is partly a function of the particular clay mineral.

Clay mineralogy is a difficult, specialized study, because the particles are so small, mostly in the range 0.002 to about 4 microns. Even electron microscopy, which gives magnifications from 3,000 to 15,000 times, is only partly satisfactory because the clays become heated and may be altered. Also, it is difficult to obtain clay minerals freed of other clay-size particles and colloids.

The importance of clay mineralogy and of colloids in surface deposits and soils is fairly evident from their unusual properties. Some practical examples of some engineering effects may be cited. During construction of the library at The Johns Hopkins University, swelling of clays in the ground buckled a steel I-beam. In the western United States, roads crossing montmorillonitic shale formations are among the roughest in the country, because the swelling clay produces irregular hummocks in the pavement. Finally, kaolinitic clays are more permeable than illites, which in turn are more permeable than the montmorillonitic clays; calcium clays are more permeable than sodium clays (Trefethen, 1959).

Clays in general have very low permeability, but some kinds are less permeable than others. Montmorillonitic clays, especially those with absorbed sodium ions, are generally the least permeable, partly because the clay particles are small and partly because their mineral structure causes water to be held between the layers. Kaolinitic clay particles generally are larger and hold less water; they also are more permeable. Clays containing silt or rock-flour are most permeable.

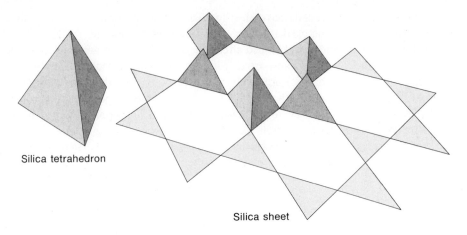

Silica tetrahedron

Silica sheet

FIGURE 11.4
Silica sheet structure. Individual tetrahedra consist of a central silicone atom (Si^{++++}) with four oxygen atoms equidistant from it, forming the corners of the tetrahedra. The tetrahedra are joined at their corners to form an open hexagonal network, as shown in the ground plan.

Clay Mineral Structures

Clay minerals are mostly hydrous aluminum silicates composed of two sheet structures, one containing silicon atoms and the other containing aluminum atoms. The silica (SiO_2) sheet is built of units composed of silicon atoms surrounded by four equally spaced oxygen atoms; the structural unit forms an equilateral triangular pyramid (tetrahedron) (Fig. 11.4). The pyramids, or tetrahedra, are arranged in an orderly pattern, with their apices all pointing in the same direction and the bases joined at the corners in 6's to form a hexagonal network of pyramids. In this structure, the oxygen atoms in the basal plane are shared by the hydroxyl $(OH)^-$ substituted for the oxygen (O^{--}). There remains in the structure an excess negative charge, and this is what joins the silica to the positively charged alumina sheet.

The alumina sheet consists of an orderly arrangement of octahedra (Fig. 11.5), which are made up of a central aluminum atom (sometimes magnesium) equidistant from 6 oxygen atoms or hydroxyls. The octahedra lie on their sides and are oriented with their long axes parallel. The oxygen and hydroxyl atoms are in 2 planes, one below and the other above the plane of the aluminum atoms, and these oxygens and hydroxyls form a hexagonal network. The hydroxyls $(OH)^-$ substitute for oxygens (O^{--}) in the basal plane and share oxygens with the apices of the tetrahedra of the silica sheet; the charge thus becomes balanced and the two sheets are tightly joined.

The simplest and most stable of the clay mineral structures is that of kaolinite $(Al_2Si_2O_5(OH)_4)$ (Fig. 11.6), referred to as 1:1 layering. Silica and alumina sheets form stacked pairs of layers that readily glide over one another. This is what accounts for its plasticity.

A second kind of clay mineral structure, illustrated by montmorillonite, is formed of 3 sheets; two sheets of silica are joined to a central alumina sheet in the planes

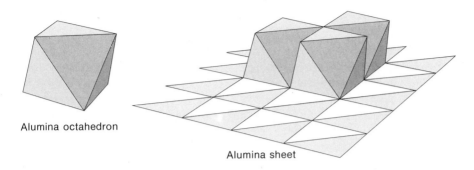

Alumina octahedron

Alumina sheet

FIGURE 11.5
Alumina sheet structure. Individual octahedrons consist of a central atom of aluminum equidistant from 6 oxygen atoms or hydroxyls. In the sheet, the octahedrons lie on one side and are all oriented the same way, which leaves triangular openings. Each octahedron is joined to its neighbors along 6 of its 12 edges.

of the apices of the silica tetrahedra (Fig. 11.7). This is referred to as $2:1$ layering. The three layers have a low net negative charge, and adjoining threesomes are loosely bonded together by a few cations, Na^+, K^+, Ca^{++}, or Mg^{++}. Because the bonding is weak, water can force the layers apart. This is what is known as an expanding clay, a major problem in engineering works.

A third kind of clay mineral structure, illustrated by illite, also has $2:1$ layering like that of montmorillonite, but in illite some Al^{+++} substitutes for Si^{+++} in the tetrahedral layers. This increases the negative charge, which becomes neutralized by cations. As a consequence of the $2:1$ layering, and the attachment of cations, these clay minerals have less definite compositions than do the $1:1$-layered clay minerals. Montmorillonite and illite have more silica and are less stable than kaolinite, and are likely to weather to it.

Two other $2:1$-layer clay minerals that are common in surface deposits and soils are chlorite and vermiculite, which readily alter to montmorillonite, illite, or kaolinite. Clays may be mixtures of more than one clay mineral; mixed clays are referred to as mixed-layer clays. Still other clays are amorphous (e.g., allophane) and have indefinite chemical compositions.

An important consequence of the crystal structure of the clay minerals is the large number of sites at which the negatively charged anions and positively charged cations are not in electrostatic balance. There are sites within each sheet within the hexagonal framework of tetrahedra and octohedra; there are sites between the sheets where they are joined; and there are sites around the ragged edges of the crystals. Clays that have a considerable excess negative charge may act like acids in the presence of ground moisture.

Stability of the Clay Minerals

The relative stability of the clay minerals may be judged by their occurrence in various weathering profiles. Chlorite is less stable than vermiculite and alters to it. Vermiculite, in turn, is less stable than montmorillonite and subject to alteration to

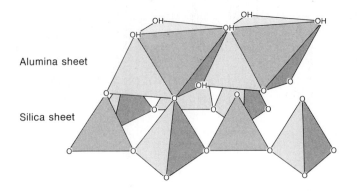

FIGURE 11.6
Kaolinite structure, the simplest and most stable clay structure, consists of two sheets, one of silica tetrahedra and the other of alumina octahedra. The basal plane consists of (OH) ions, which are dipolar, and their H^+ ions bond to the O^{--} ions of the next lower sheet. [After Grim, 1962.]

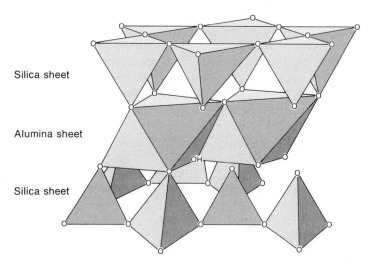

FIGURE 11.7
Montmorillonite consists of a sheet of alumina between two sheets of silica.

it by leaching of the magnesium. Illite is less stable than montmorillonite, which is less stable than kaolinite and alters to it by leaching one of the silica sheets. Mixed-layer clays are probably transitional between any of these. How these alterations occur is not at all clear, for the sequential changes involve not only rearrangement of the sheets but, for some, their resynthesis.

In Florida, montmorillonite alters to kaolinite (Altschuler, Swornik, and Kramer, 1963). In the middle west, deeply buried, unaltered, pre-Wisconsinan till contains illite and chlorite. As the following table shows (after Brophy, 1959), in the deepest weathered layer, where there has been oxidation but no leaching, the chlorite is altered to mixed chlorite-vermiculite; illite is unchanged. In the next higher layer, which is both oxidized and leached, the chlorite is altered to vermiculite, and the illite to mixed illite-montmorillonite. In higher layers, both are altered to montmorillonite. The base of the montmorillonite zone is about the upper limit of the preservation of hornblende.

1. Modern soil	montmorillonite	montmorillonite
2. Layer with decomposed silicates	montmorillonite	montmorillonite
	Approximate upper limit of hornblende	
3. Oxidized and leached	mixed illite and montmorillonite	vermiculite
4. Oxidized; not leached	illite	mixed chlorite and vermiculite
5. Unaltered	illite	chlorite

In pre-Wisconsinan soils, alteration of chlorite to vermiculite may be as deep as 16 feet; in Wisconsinan soils this alteration is only half that deep. Also the alteration of illite to mixed illite-montmorillonite is very much deeper in pre-Wisconsinan than in Wisconsinan soils. The sequence of mineral alteration is the same in both ages of soils, but the chief difference is in the hard rocks. In pre-Wisconsinan soils, hard rocks commonly are altered to clay, *argillized;* in Wisconsinan soils, they are not.

In general, the clay minerals are most extensively developed in ancient soils. Pre-Wisconsinan soils in northern Pennsylvania, for example, contain as much as 30 percent kaolinite. Saprolite on the Piedmont Province is almost entirely clay minerals and oxides. In the arid and semiarid parts of western United States the clay fraction of most pre-Wisconsinan soils is chiefly montmorillonite and mixed montmorillonite-illite; kaolinite is minor. In Holocene and late Pleistocene soils the clay fraction is little developed except in those soils on clayey parent material or on fine-grained parent materials, in both of which the parent mineral grains expose extreme amounts of surface to weathering. Most clay in soils of Wisconsinan and Holocene age is inherited from the surface deposit and parent material of the soil. Some may have been physically washed from the upper to the lower layers of the soil, but only rarely is much of the clay of secondary origin in such young soils.

Contrast between the pre-Wisconsinan and younger soils is further emphasized by some details revealed by partial chemical and particle-size analyses of soils near the border of the Wisconsinan drift in northern Pennsylvania, the Great Plains, and the Lake Bonneville Basin, Utah. In Pennsylvania (Denny and Lyford, 1963) the layer of deposition in the pre-Wisconsinan soils contains 30 percent clay, mostly kaolinite; similar layers in the younger soils contain only 10 percent clay—mostly the less stable ones, vermiculite and illite. Alkalis and alkaline earths are almost completely removed from the pre-Wisconsinan soils, less than 0.5 percent remaining, whereas the Wisconsinan and younger soils retain about 7 percent.

In the Lake Bonneville Basin the pH of the clay layer of the old soil is about neutral near the surface and increases downward to about 8.0 in the lime-enriched layer. The silt and clay content of the clay layer in these old soils, both on the Great Plains and in the Lake Bonneville Basin, generally exceeds 50 percent and in places is as high as 70 percent. Illite seems to be the dominant clay mineral on the Great Plains, but in the Lake Bonneville Basin illite is mixed with montmorillonite, which may reflect the presence of volcanic ash. In all soils tested, the dominant clay mineral in the clay layer was the same as in the lime-enriched layer. Some of the clay contains sand coarser than 20 mesh screen, and a considerable proportion of these grains are subangular and pitted. Partial chemical analyses of clay from the Lake Bonneville Basin show SiO_2, 68 percent; Al_2O_3, 15 percent; Fe_2O_3, 5 percent; MgO, 1.7 percent; CaO, 0.02 percent; TiO_2, 0.8 percent. Presumably, most of the remaining 10 percent or so that was not determined is combined water, because the slight amount of calcium and magnesium indicates considerable leaching and makes it doubtful that much sodium or potassium remain.

The origin of the clay layer in the pre-Wisconsinan soils is obscure. Part of the clay seems to have been derived from the decomposition of hard rocks in the parent material, but much of it probably formed by alteration of loess, because the clay varies little over a wide region, regardless of differences in the underlying lime zone and bedrock.

Except for ion exchange at the sites of electrostatic imbalance, clay minerals do not seem to change much when transported from one environment to another. For example, on the Piedmont Province and Coastal Plain in the humid southeastern United States, the clay mineralogy of alluvial deposits is like that of the residual deposits or other sediments that were source for the alluvium. In the arid part of the country, in Death Valley and in the Mohave Desert, the clay mineralogy on the salt pans and alkali flats is the same as on the alluvial fans from which the clays were derived.

SOLID OXIDES AND HYDROXIDES

This section considers only the solid oxides and hydroxides, especially of iron, aluminum, and titanium; the gaseous oxides, notably carbon dioxide and sulfur dioxide, are excluded. The solid oxides and hydroxides are as ubiquitous in the ground and as difficult to study as the clays; in fact, the clays probably are better understood. These oxides and hydroxides occur as colloidal films on rock and mineral particles—coatings on soil aggregates and on the walls of cracks or other openings. They are difficult both to separate and to identify. Listed below are the more common, or better known, of the minerals that occur in surface deposits and soils.

> Iron minerals
> Hematite (Fe_2O_3); goethite ($Fe_2O_3 \cdot H_2O$); limonite,
> actually a mixture of oxides ($Fe_2O_3 \cdot 3H_2O$).

> Titanium minerals
> Anatase (TiO_2); doelterite (leucoxene, $TiO_2 \cdot H_2O$ or $TiO_2 \cdot nH_2O$).

Alumina minerals
> Diaspore ($Al_2O_3 \cdot H_2O$); gibbsite ($Al_2O_3 \cdot 3H_2O$).

Silica minerals
> Chalcedony (quartz, SiO_2); opal ($SiO_2 \cdot H_2O$).

All these minerals are secondary products in the ground, and most are nearly insoluble in ordinary soil solutions. Solubilities are measured in parts per million (ppm) or millemoles per liter except in very acid or very alkaline solutions. Alumina, for example, is virtually insoluble (0.1 ppm) in solutions in the pH range 4 to about 9, but is readily soluble (10 ppm) in solutions more acid than pH 4 or more alkaline than pH 9. Silica is slightly soluble (ca 2 ppm) in solutions in the range pH 2 to about 9, but is readily soluble (10 ppm) in solutions more alkaline than pH 9.

Like the clay minerals, the oxide and hydroxide minerals are most abundant and most varied in ancient soils. In northern Pennsylvania, free iron oxides total 6 percent in the pre-Wisconsinan soils but only 1.5 percent in the younger ones, and the pre-Wisconsinan soils contain up to 3 percent gibbsite, which is lacking in the Wisconsinan and younger soils. In general, since the beginning of Wisconsinan time, little gibbsite has formed in the United States. The generalization certainly is valid, even though there may be exceptions where environmental conditions are optimal, as they may well be in the weathering of volcanic ash or at the weathering front at the base of saprolite in the Piedmont Province.

ION EXCHANGE

The addition of limestone to acid ground produces calcium-rich ground and frees carbonic acid that can be washed from it. And if ammonium sulfate be added to the calcium-rich ground, calcium sulfate precipitates from solution. In the first reaction two H^+ ions in the ground are replaced by a Ca^{++} ion; in the second, two NH_4^+ ions replace a Ca^{++} ion. The reactions leading to these changes are caused by the exchange of ions, including both anions and cations. To a considerable degree the ions are those held loosely on the surfaces and edges of clay particles and particles of living and dead organic matter.

In cation-exchange reactions, positively charged ions replace others so charged. For example, in Figure 11.8, A, a calcium ion replaces two hydrogen ions, as happens when limestone is added to acid ground. The reaction goes to the right, and the particles of clay or organic matter become less acid. But this reaction is reversible, because as plant growth continues on the calcium-rich ground, the increased supply of carbonic acid causes the hydrogen ions to replace the calcium ones and the particles of clay or organic matter again become acid. More "yard work" is needed at this point, because the reaction now goes to the left. This is simply an illustration of the law of mass action (Le Chatelier's Principle)—that the direction taken by a reaction depends on the availability of the ions. In our example, adding limestone increased the availability of the CA^{++} ion, whereas subsequent plant growth increased the availability of the H^+ ion. Similarly, the exchange between the $2NH_4^+$ and the Ca^{++} ions is reversible, as illustrated in Figure 11.8, B.

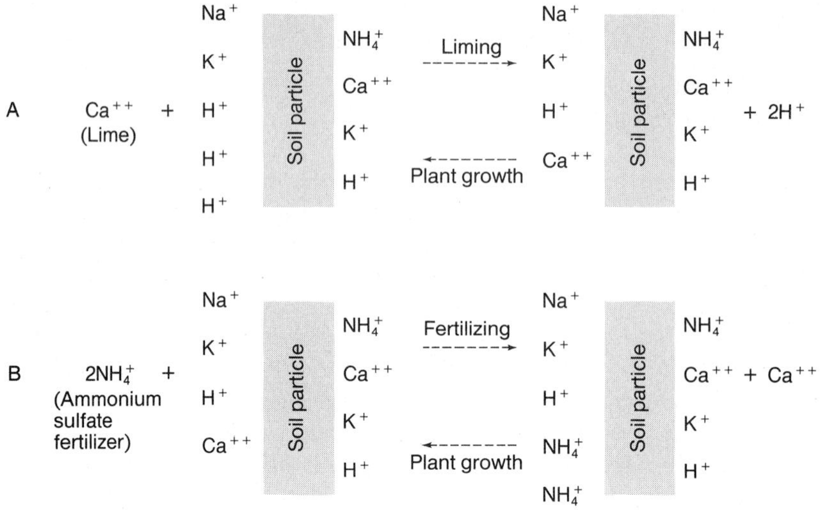

FIGURE 11.8
In cation exchange, positively charged ions replace each other. The reactions are reversible depending on the relative availabili y of the cations. In (A), liming a soil causes the reaction to progress to the right; renewed plant growth tends to slow the reaction and finally reverse it. In (B), adding ammonium sulfate fertilizer to a soil causes the NH_4^+ to free the Ca^{++}, making it available for plant growth; as the plants return the Ca^{++} to the ground, the reaction is slowed and finally reversed.

Moreover, particular ions are not equally exchangeable because some are held more tightly in their exchange positions than others. For example, ions within the hexagonal network of a crystal lattice are held tightly; those on the edges of a sheet less so (Rankama and Sahama, 1950). How tightly the ions are held on a surface is partly a function of the charge of the ion, partly a function of the charge on the surface of the clay or colloidal particle (which in turn is a function of the kind of clay mineral and position on its surface), partly a function of the radius of the ion, and partly a function of the exchange site. According to this, the divalent ions (Ca^{++}, Mg^{++}) are held more tightly than the monovalent ones (NH_4^+, Na^+, K^+, H^+), and the smaller ions are held more tightly than the larger ones. The sizes of the ions cited are roughly: Mg^{++}, about 0.75 Å; Na^+ and Ca^{++}, about 1.00 Å; K^+, about 1.33 Å; NH_4^+, about 1.43 Å. The relative ease of replacement in montmorillonite clay is, according to Ross (1943),

$$Na^+ < K^+ < Mg^{2+} < Ca^{2+}$$

Ion exchange apparently is a major factor in plant nutrition too. Rootlets have a negative charge on their surfaces, and the films surrounding them hold quantities of H^+ ions, which can be exchanged for cations in the surrounding medium and become available to the plant. The retention and release of chemical elements applied in fertilizers and lime is largely a matter of ion exchange.

Ion exchange may involve anions or cations, and the ions differ in their suscepti-bility to replacement (Hauser, 1939). Anion exchange capacity generally is less than cation exchange capacity because clay mineral surfaces have few positive sites for holding the negatively charged anions. Moreover, the relative susceptibilities are not constant, but depend on the composition and concentrations of the invading solutions. In general, though, the exchange processes cause the replacement of all anions and cations by OH^- and H^+, respectively, on the chemically active surfaces. With only a few exceptions (e.g., serpentine ground), depletion of alkalis and alkaline earths in surface deposits and soils provides a fair index of their degree of weathering.

Finally, clays differ greatly in their capacity to exchange ions. According to Grim (1962), the exchange capacity of kaolinite is least; that of vermiculite and montmoril-lonite is greatest; illite and chlorite are intermediate. Very likely there are equally great differences in the exchange capacity of the various organic colloids. In both clay minerals and organic colloids, it seems likely that exchange capacity increases with decrease in particle size because of the increase in surface area.

CHEMICAL WEATHERING PROCESSES

Some kinds of rocks and minerals readily break down by weathering in one environ-ment, but are resistant in another. The example of limestone has already been cited.

Weathering may produce a wide range of changes. Minerals can dissolve and the components be carried away in solution. Weathered ground may lose a large part of its mineral ingredients in this way, and those that remain may form *pseudomorphs,* or replacement minerals. The new mineral replaces an original mineral in such a way that the form of the latter is preserved by the former in spite of the change in composition. Thus gibbsite or kaolin may replace feldspar; goethite may replace olivine. Alternatively, the original minerals may lose enough soluble bases and perhaps silica to destroy the crystal structure and form a gel, which rearranges and recrystallizes to a clay in the same space or nearby. Or the gels and solutes may move some distance and combine with products from other minerals that have weathered or from decomposing organic matter. Again, minerals may lose some of their bases or even some of their silica and still retain their essential framework. The processes and their effectiveness vary greatly depending on the environment.

Oxygen, carbon dioxide, and moisture from the air and the biota all enter into the chemical processes that cause mineral alteration. Among the principal processes are *exchange reactions* (discussed in the preceding section) *carbonation, oxidation, hydration, solution, reduction, silication, and silicification.* The relative effects of these processes differ greatly, depending on the environment on and in the ground, espe-cially the climate.

Carbon dioxide (CO_2) is readily soluble in water (H_2O), producing carbonic acid (H_2CO_3, or $H^+ + HCO_3^-$), which dissolves such carbonates as calcite, the chief component of limy ground.

$$\text{carbonic acid} + \text{calcite} \longrightarrow \text{calcium bicarbonate (soluble)}$$
$$\underset{H_2CO_3}{} \qquad \underset{CaCO_3}{} \qquad\qquad \underset{Ca(HCO_3)_2}{}$$

Carbon dioxide in the atmosphere amounts to about 300 parts per million of air; in the soil atmosphere, the CO_2 content is 10 to several hundred times greater than in the air above ground. The amount soluble in water, by volume, is about equal to the volume of the water at atmospheric pressure.

Carbonic acid probably is the most widespread and the most important of the soil acids, but acid clays also contribute to chemical weathering of rocks and minerals. Where leaching is intense, the alkalis, alkaline earths, and other bases are removed from the system, and the clay minerals become acid by absorbing hydrogen ions in their place. In the pre-Wisconsinan tills in the middle west, the upper limit of preservation of hornblende about corresponds to the lower limit of montmorillonite clay, which suggests that the montmorillonite was derived by breakdown of the hornblende. Mineral acids may be important locally; sulfuric acid for example results from oxidation of sulfide minerals, particularly pyrite.

Oxygen also dissolves in rain and ground water and combines with various bases in the process of oxidation. The process is most evident in connection with alteration of iron-bearing minerals; the iron becomes oxidized and stains the ground brown or red. Iron carbonate breaks down to the oxides.

$$2FeCO_3 + H_2O + O \longrightarrow Fe_2O_3 \cdot H_2O + 2CO_2$$

   siderite   water + oxygen   goethite   carbon dioxide

Magnetite, the magnetic iron mineral, which is in part ferrous iron oxide ($FeO + Fe_2O_3$) oxidizes to hematite (Fe_2O_3):

$$2Fe_3O_4 + O \longrightarrow 3Fe_2O_3$$

Pyrite, the iron sulfide (FeS_2), oxidizes to ferrous sulfate ($FeSO_4$) which in turn oxidizes to geothite (or limonite) ($Fe_2O_3 \cdot H_2O$).

$$FeS_2 + 7O + H_2O \longrightarrow FeSO_4 + H_2SO_4$$
$$2FeSO_4 + H_2SO_4 + O \longrightarrow Fe_2(SO_4)_3 + H_2O$$
$$Fe_2(SO_4)_3 + 4H_2O \longrightarrow Fe_2O_3 \cdot H_2O + 3H_2SO_4$$

The sulfuric acid freed alters any calcium carbonate (limestone or calcite) to the sulfate (gypsum). Sulfuric acid is widespread and is an important factor in the weathering of surface deposits in most coalfields and metal mining districts, for in these areas the sulfide minerals are abundant in the bedrock and become freed and oxidized in the course of mining or by circulation of ground water.

Hydration is the process by which water combines with various compounds. Combined with the processes of carbonation and solution, this alters feldspars to clay minerals. A soda-calcium feldspar, for example, may alter to a clay mineral and to calcium and sodium bicarbonate:

$$CaAl_2Si_2O_8 \cdot 2NaAlSi_3O_8 + \quad 4H_2CO_3 \quad + 2(nH_2O) \longrightarrow$$

   soda-calcium feldspar   carbonic acid   water

$$2Al_2(OH)Si_4O_{10}nH_2O \quad + Ca(HCO_3)_2 + 2NaHCO_3$$

         clay   calcium and sodium bicarbonate;
                                          but probably ionized

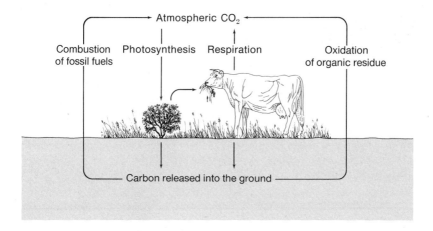

FIGURE 11.9
The carbon cycle. All living organisms contain carbon, and roughly as much organic matter is produced each year as is decomposed on and in the ground. The carbon dioxide in the atmosphere is increasing, however, because of the burning of fossil fuels.

Silica also is released and may be removed by ground water. The alteration of biotite illustrates how the dark minerals alter to residual iron oxides and clay and to silica and salts that may be removed from the soil by solution.

$$2KMg_2Fe(OH)_2AlSi_3O_{10} + O + nH_2O \longrightarrow$$
$$\text{biotite}$$
$$Fe_2O_3 \cdot H_2O + Al_2(OH)_2Si_4O_{10}nH_2O$$
$$\text{iron oxide} \qquad\qquad \text{clay}$$
$$+ 2KHCO_3 + 4Mg(HCO_3)_2 + 2SiO_2 + 5H_2O$$
$$\text{salts; probably ionized} \qquad \text{silica} \qquad \text{water}$$

Hornblende and augite alter similarly and react more readily. The silica released by this reaction may be redeposited as a cement in the pores of the surface deposit or soil, a process known as *silicification*, or it may combine with released bases to form new silicates, a process referred to as *silication*. The process of forming clay sometimes is referred to as *argillic alteration*.

ORGANIC MATTER

Organic matter on and in the ground is usually taken to mean only the dead matter, but in actual practice it is not possible to separate the dead from the living organic matter, at least the part that exists as microorganisms. Total organic matter in soil therefore includes both. The production and decomposition of organic matter is a balanced economy, like the hydrologic cycle; roughly as much organic matter is produced each year as is decomposed on and in the ground. This is the carbon cycle (Fig. 11.9).

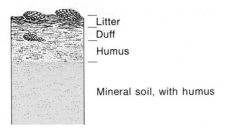

FIGURE 11.10
Organic layers at the ground surface under coniferous forest. Thickness 6 to 10 inches.

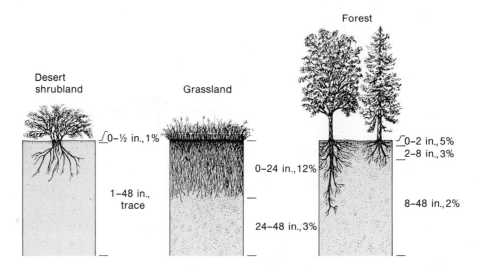

FIGURE 11.11
Differences in the content and distribution of organic matter in the upper 4 feet of a desert shrubland soil (left), grassland soil (center), forest soil (right). Grassland soils commonly contain half again as much organic matter as forest soils. Desert soils in general have a very low content of organic matter.

In forested areas, a layer of leaves, twigs, and other litter accumulates annually at the surface (Fig. 11.10). This buries the partly decomposed layers, or duff, that accumulated during the preceding year, and this in turn buries the thoroughly decomposed organic matter that accumulated still earlier. This well-decomposed organic matter is the fraction known as humus. It is an amorphous substance and its composition is variable. The layers are mixed to varying degrees depending on the fauna and on the weathering processes.

The amount of organic matter in soils can be roughly estimated by the color. Organic-rich materials tend to be gray, brown, or black whereas mineral soils with little organic matter tend to be brightly colored. A closer approximation of the organic content can be made by weighing a sample after washing it with hydrogen peroxide, but this method tends to give too low a value because the peroxide has little effect on undecomposed roots and fibers. Another method is to determine the loss of weight after ignition of a sample, but the ignition tends to give too high a value because some of the minerals, especially the carbonates and hydrates, also break down under high temperatures.

Organic matter in soil is intimately mixed with the inorganic matter in several ways. Organic matter of colloidal size may coat individual mineral grains with a film. Or the film may coat aggregates of soil particles. Indeed, aggregation probably is largely a result of the organic matter. Other colloids may occur as tiny veinlets along anastomosing cracks. Equally important, however, is the fact that the organic molecules can combine with clay minerals, especially montmorillonite or mixed-layer clays. The organic molecules can penetrate between the layers of 2:1-layered or mixed-layer clay minerals and displace the interlayer water. These organic clays may be effective weathering agents.

Clay is crystalline and composed chiefly of silicon, oxygen, aluminum, and water, whereas humus is amorphous and composed chiefly of hydrogen, carbon, nitrogen, oxygen, and some phosphorus and sulfur. Humus is not the final product of the decomposition of organic matter in the ground; it is merely an intermediate stage. The ultimate products are carbon dioxide, water, and other simple chemicals. If the supply of humus is not replenished, the ground becomes depleted in organic matter. Differences in the balance between replenishment and destruction of humus accounts for the differences in the content of organic matter in different environments (Fig. 11.11).

Production of Organic Matter

The production of organic matter depends ultimately on photosynthesis. In the presence of light, green chlorophyll in the leaves of plants creates complex organic molecules from the simple molecules carbon dioxide (CO_2) and water (H_2O). For example, six molecules of carbon dioxide become joined with six molecules of water to form one molecule of sugar and six of oxygen, as follows:

$$6CO_2 + 6H_2O \xrightarrow[\text{chlorophyll}]{\text{light}} C_6H_{12}O_6 + 6O_2$$

In turn these and more complex molecules decompose to simple salts, water, carbon dioxide, and free nitrogen.

Under forest conditions organic matter is added chiefly to the surface, though the root systems may be deep and diffuse. Under grassland, however, the root systems are shallow and the roots are crowded. As a consequence, in grassland soils, abundant organic matter is added at moderate depth as well as to the surface, and the organic-rich layer is correspondingly thicker. Grassland and forest soils that are adjacent and on similar parent material differ greatly in their content of organic matter and in its distribution, and these contrast with the desert or shrubland soils (Fig. 11.11).

Decay of Organic Matter

The breakdown of organic matter in soil is accomplished almost entirely by the microorganisms, that is, the processes are biochemical; physical chemical processes are less important. The balanced economy, therefore, depends on the maintenance of the living organisms that feed on and decompose the dead organic matter. If these are inhibited, as by too much wetting, organic matter accumulates. Peats accumulate in southern climates (e.g., Florida) where growth of vegetation is lush and where there is excess moisture, and they accumulate also in cold northern climates partly because of excessive moisture and partly because the microorganisms are inhibited by the cold and by the short growing season.

Peats vary greatly in composition, depending on the kind of plant matter accumulated and on the stage of its alteration. The alteration involves

1) loss of hydrogen and oxygen as water (H_2O),

2) loss of carbon and oxygen as carbon dioxide (CO_2),

3) loss of carbon and hydrogen as methane (CH_4).

The chemical changes may be approximated as follows:

$$2C_6H_{10}O_5 + 2O_2 \longrightarrow C_6H_{10}O_5 + H_2O + 4CO_2 + 2CH_4$$

(cellulose)   (plant   (water)   (carbon   (methane)
                              residue)   dioxide)

Plant matter containing roughly 50 percent carbon and 40 percent oxygen (ash-free basis) gradually changes to peat containing roughly 60 percent carbon and 30 percent oxygen (Clarke, 1924). The remainder is mostly hydrogen and nitrogen.

The bacteria and other microorganisms that cause decay of organic matter may be classified on the basis of their nutritional requirements, their morphology or form, or on the basis of their need for oxygen. Oxygen is needed by most living organisms, particularly by the so-called *aerobic* ones that depend on free oxygen. For example, breaking down the sugar molecule ($C_6H_{12}O_6$) releases heat and provides energy. We burn our sugar by oxidation to form carbon dioxide and water:

$$C_6H_{12}O_6 + 6O_2 \longrightarrow 6CO_2 + 6H_2O$$

Breaking down the sugar by oxidation releases 670 calories of energy, which we can use to climb mountains.

Anaerobic microorganisms, however, can live where oxygen is deficient and can break down the sugar molecule without the aid of oxygen:

$$C_6H_{12}O_6 \longrightarrow 3CO_2 + 3CH_4$$

This reaction releases only 50 calories. The decomposition or decay of a given quantity of organic matter by oxidation is far faster and yields more total energy than do the anaerobic processes. The leaves in a yard can be consumed quickly by oxidation by igniting them with a match, whereas they are consumed but slowly in a compost heap where oxygen is deficient.

Where there is an excess of organic matter, its decomposition consumes the available oxygen, and the result is a reducing environment. Sulfates lose their oxygen and become sulfides; iron sulfide, for example, may become precipitated. Such environments are common in surface deposits, in estuaries and lakes where circulation is retarded, in swamps and bogs, and in poorly drained locations on floodplains. The most important products of the decomposition of organic matter are carbon dioxide, which combines with water to form carbonic acid; marsh gas (CH_4); and ammonia (NH_4) and nitric acid (HNO_3) from the nitrogenous compounds.

In deserts, oxidation is more important than the biochemical processes in breaking down the large complex organic molecules. The annual contribution of new organic matter is slight. Fallen twigs and leaves become dried and readily blown around and fragmented to small particles. Occasional rains may wash some of this organic matter into the ground where the soil organisms can utilize and decompose it, but in the absence of rain, fine particles must simply be oxidized, as if burned. In deserts, the soils may contain only a minute fraction of a percent of organic matter, except directly beneath the shrubs.

The organic matter that accumulates on and in the ground consists of many different substances, which, like minerals, vary greatly in susceptibility to alteration and decomposition. Listed in order of decreasing susceptibility, the principal substances are

1) proteins,

2) carbohydrates (hemicellulose, cellulose, and lignin),

3) fats, resins, waxes.

The more resistant organic substances may become coated and soaked by the colloidal solutions resulting from decomposition of the least resistant ones, so that the mass of organic matter is a mixture of much altered and slightly altered organic substances. Cell walls, composed of cellulose and lignin, become stained brown with the decomposition products of more easily altered substances, and these decomposition products may be more stable than the cellulose or lignin. If so, the decomposed cell walls may be faithfully marked by the pseudomorphic, brown, early-to-form, alteration products. Such molecule-by-molecule replacement of organic compounds seems quite reasonable despite the fact that atom-by-atom replacement of carbon in organic molecules is difficult if not impossible.

The retention of organic matter in soils is affected too by the temperature. Assuming comparable replacement and aeration, the content of organic matter in soil is reduced a half or a third for each $10°$ (C) rise in ground temperature (Berger, 1965). One effect of this is the difference between the organic fraction in soils under comparable vegetation in the southern part of the United States and those in the northern part. The decay rate also is dependent on composition of the water in the ground. Carbonate waters, for example, neutralize organic acids. Given equal wetting and temperature, decomposition is faster in nearsurface layers than in deep ones because of access to air and oxygen.

NITROGEN COMPOUNDS

The nitrogenous compounds are especially important in agricultural soils as plant food. The nitrogen cycle is illustrated in Figure 11.12. Unlike carbon dioxide, little ammonia escapes from the ground to the air because it is strongly absorbed by the soil particles, especially by the humus and clays. In humid regions, unless conditions are too acid (less than about pH 6.0), this absorbed ammonia is changed first to nitrite and then to soluble nitrate by nitrifying bacteria; in arid regions, the ammonia probably escapes.

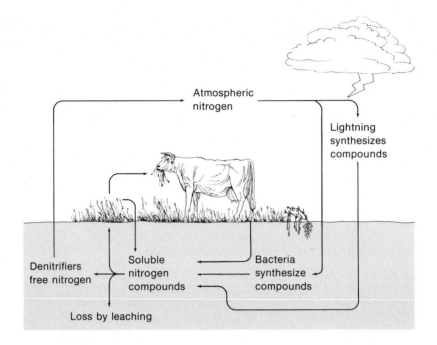

FIGURE 11.12

The nitrogen cycle. Nitrogen is an essential constituent of protein and is needed by all living organisms. Except for the little bit supplied by lightning, man has been until recently dependent upon the microorganisms in the ground that are capable of synthesizing the inert element, nitrogen, into the soluble ammonia, nitrites, or nitrates. Now it is possible to produce nitrogen compounds synthetically.

The bacteria that live in the root nodules of leguminous plants convert free nitrogen to soluble nitrates that can be absorbed by the plant (p. 289). This nitrogen is returned to the ground in the plant residue.

TRACE ELEMENTS

Chemical elements other than the eight principal ones occur in minute quantities, and are referred to as trace, or minor, elements. The term "minor" refers only to their minute quantity, because some of the minor elements are of major importance to plants. Some of the trace elements necessary for plant growth are phosphorus, sulfur, boron, chlorine, copper, manganese, molybdenum, zinc, and cobalt.

Some trace elements are used in geochemical prospecting by making systematic tests of surface deposits, soils, stream waters, or vegetation; their abnormal concentration may reflect the presence of a concealed mineral deposit. Trace elements can be important in engineering too: for example, minute quantities of phosphorus or hydrous silica are deleterious in concrete.

Copper, lead, zinc, and molybdenum are frequently used in geochemical prospecting because there are quick and easy semiquantitative field tests for them. Quantities as small as 0.002 ppm of copper or zinc can be determined readily. Geochemical prospecting has contributed importantly to the discovery of uranium and copper deposits in the western United States. Recently the U.S. Geological Survey developed semiquantitative field tests for gold, a development that may have important national and international effects in view of the world monetary situation.

The quantities of the trace elements in a particular piece of ground, of course, vary greatly. The general order of abundance that might be expected is as follows (data from Clarke, 1924; Mason, 1958; Poldervaart, 1955; Swaine, 1955; Carlisle and Cleveland, 1958):

> Commonly 0.1 percent (1000 ppm); locally > 1.0 percent
>   titanium, manganese
>
> Commonly 0.01 percent (100 ppm); locally > 0.1 percent
>   beryllium, rubidium, zirconium, vanadium, strontium, zinc, chromium, lithium
>
> Commonly 0.001 percent (10 ppm); locally > 0.01 percent
>   germanium, tin, lead, yttrium, copper, lanthanum, nickel, boron, arsenic, cobalt
>
> Commonly 0.0001 percent (1 ppm); locally > 0.001 percent
>   arsenic, uranium, barium, scandium, molybdenum
>
> Generally < 0.0001 percent
>   silver, mercury, gold

Sulfur occurs in the mineral fraction of soils and surface deposits principally as sulfates of the alkalies and alkaline earths, but also as sulfides of various metals, especially iron. The iron sulfide pyrite (FeS_2) is widespread in many surface deposits, but it readily oxidizes to soluble sulfates and consequently is not common in soils, even in young ones. Sulfur is added to the ground not only by weathering of sulfur-bearing minerals, but also by rain water, which absorbs sulfur dioxide from our increasingly polluted atmosphere.

Sulfur is an essential constituent of many proteins and is necessary for plant growth. It occurs in "excess" amounts in arid and semiarid regions, where evaporation causes sulfate salts to accumulate in the ground. Deforested hillsides leeward of some pulp mills and smelters are mute evidence of an excess of sulfur dioxide in the air, and the cause, which remains uncorrected, seems fairly obvious. Some of the difficulties caused by smelters are attributable to accumulation of metals, especially zinc, in addition to the sulfur dioxide.

Plants absorb dissolved sulfur from the ground, mostly in the form of the sulfate ion (SO_4^{--}), but also obtain it in the form of sulfur dioxide gas, which can be absorbed through the leaves and combine with the plant water to form soluble sulfates. Some plants accumulate sulfates in large amounts (e.g., arrowweed, *Pluchea sericea*).

Chlorides of the alkalies and alkaline earths are highly soluble and are taken up by all plants. In some salt-tolerant species, like pickleweed (*Allenrolfea occidentalis*), salts may accumulate to as much as a third of the dry weight of the plant. In humid regions chlorides are supplied by rainwater, and the amount of chlorine received from rains increases seaward (p. 257).

Boron is another trace element essential for plant growth, but only small amounts are needed and greater amounts can be toxic. In the soluble forms in which it is available to plants, it is readily washed from the ground, and consequently the areas of boron deficiency are typically where there is considerable leaching. Boron occurs in marine sedimentary formations and in the surface deposits and soils developed from them, but the mineralogy is uncertain. Tourmaline, a complex borosilicate, is widespread in metamorphic rocks in the Piedmont Province, but it is a stable mineral (Table 11.3) that probably has survived the weathering conditions of the Wisconsinan glaciation and the Holocene Epoch, although it has decomposed under the more intensive alteration that formed the saprolite.

Copper, another trace element required by plants, probably is contributed to the ground largely by the alteration of the sulfide and carbonate minerals, of which there are several. These minerals are not very stable, and are likely to be altered even in young soils and deposits. Where copper sulfides are concentrated in ore bodies the weathering develops layers of minerals parallel to the surface like those in a soil profile. Above an ore body, there may be a leached zone from which the copper is largely removed. Below this is a layer of copper carbonates mixed with secondary copper sulfides, and below this are the primary copper sulfide ores. These layers may be hundreds of feet thick and evidently reflect ancient weathering processes such as those that developed the deep saprolite in the Piedmont Province. At Ajo, Arizona, the configuration of the interface between the weathered zone and primary ore accords rather well with the present ridges and valleys.

Manganese, which also is essential for normal plant development, is widely distributed in several oxides. None is very stable, and even in young soils the minerals can provide the necessary metallic ion.

Other trace elements are important because they are harmful to plants. Among these are arsenic, chromium, lithium, nickel, lead, and selenium. Arsenic, lead, and lithium are problems mostly because of their use in sprays. Chromium even in small quantities can be toxic. Nickel occurs in toxic quantities near some smelters. Selenium is very poisonous to animals, but less so to plants. The Cretaceous shales in western United States are notoriously high in selenium, and in some places are suitable only for growing fiber crops, not food crops, because the plants accumulate the selenium in quantities toxic for animal use. Zinc is essential for plant growth but large quantities are toxic to many plants.

BIBLIOGRAPHY

Allen, V. T., 1948, Weathering of heavy minerals: Jour. Sed. Petrology, v. 18, pp. 38–42.
Altschuler, Z. S., Jaffe, Elizabeth B., Cuttitta, F., 1956, The aluminum phosphate zone of the Bone Valley Formation, Florida, and its uranium deposits: U.S. Geol. Survey Prof. Paper 300, pp. 495–504.
Altschuler, Z. S., Dwornik, E. J., and Kramer, H., 1963, Transformation of montmorillonite to kaolinite during weathering: Science, v. 141, pp. 148–152.
Bear, F. E., 1965, Chemistry of the soil: Reinhold, New York, 373 pp.
Brophy, J. A., 1959, Heavy mineral ratios of Sangamon weathering profiles: Illinois Geol. Survey Circular 273, 22 pp.
Cady, J. G., 1941, Soil analyses significant in forest soils investigations and methods of determination. III. Some mineralogical characteristics of Podzol and Brown Podzolic forest soil profiles: Proc. Soil Sci. Soc. (1940), v. 5, pp. 353–354.
Clarke, F. W., 1924, The data of geochemistry: U.S. Geol. Survey Bull. 770, 832 pp.
Droste, J. B., 1956, Alteration of clay minerals by weathering in Wisconsin tills: Geol. Soc. America Bull., v. 67, pp. 911–918.

————, and Tharin, J. C., 1958, Alteration of clay minerals in Illinoian till by weathering: Geol. Soc. America Bull., v. 69, pp. 61-68.

Dryden, L., and Dryden, C., 1946, Comparative rates of weathering of some heavy minerals: Jour. Sed. Petrology, v. 16, pp. 91-96.

Engel, C. G., and Sharp, R. P., 1958, Chemical data on desert varnish: Geol. Soc. America Bull., v. 69, pp. 487-518.

Goldich, S. S., 1938, A study in rock weathering: Jour. Geology, v. 46, pp. 17-58.

Graham, E. R., 1941, Acid clay—An agent in chemical weathering: Jour. Geology, v. 49, pp. 392-401.

Grant, W. H., 1964, Chemical Weathering of biotitie-plagioclase gneiss: Clays and Clay Minerals, v. 12, pp. 455-463.

Grim, R. E., 1968, Clay mineralogy (2nd ed.): McGraw-Hill, New York.

Harris, R. C., and Adams, J. A. S., 1966, Geochemical and mineralogical studies on the weathering of granite rocks: Am. Jour. Sci., v. 264, pp. 146.

Holmes, A., 1930, Petrographic methods and calculations: Thos, Murby, London, 515 pp.

Jackson, M. L., and Sherman, G. D., 1953, Chemical weathering of minerals in soils: Advan. Agronomy, v. 5, pp. 219-318.

Keller, W. D., 1955, The principles of chemical weathering: Lucas Bros., Missouri, 88 pp.

————, 1966, Geochemical weathering of rocks—source of raw materials for good living: Jour. Geol. Educ., v. 14, no. 1, pp. 17-22.

Kelley, W. P., 1948, Cation exchange in soils: Reinhold, New York.

Loughnan, F. C., 1969, Chemical weathering of the silicate minerals: Elsevier, New York, 154 pp.

Marshall, C. E., 1949, The colloid chemistry of the silicate minerals: Academic Press, New York.

Marshall, C. E., 1964, The physical chemistry and mineralogy of soils: Wiley, New York.

Mason, B., 1966, Principles of geochemistry (3rd ed.): Wiley, New York, 329 pp.

Pettijohn, F. J., 1941, Persistence of heavy minerals and geologic age: Jour. Geology, v. 49, pp. 610-625.

Rankama, K., and Sahama, Th. G., 1950, Geochemistry: Univ. Chicago Press, 912 pp.

Robinson, W. O., and Edgington, G., 1945, Minor elements in plants and some accumulator plants: Soil Sci., v. 60, pp. 15-28.

Ross, C. S., 1943, Clays and soils in relation to geologic processes: Jour. Wash. Acad. Sci., v. 33, pp. 225-235.

Seidell, Atherton, 1940, Solubilities of inorganic and metal organic compounds (3rd ed.), Van Nostrand, New York, 1698 pp.

Shand, S. J., 1952, Rocks for chemists: Pitman, New York, 146 pp.

Temple, K. L., and Koehler, W. A., 1954, Drainage from bituminous coal mines: West Virginia Univ. Eng. Expt. Sta. Research Bull. 25, 35 pp.

Trelease, S. F., 1945, Selenium in soils, plants, and animals: Soil Sci., v. 60, pp. 125-131.

Van Houten, F. B., 1953, Clay minerals in sedimentary rocks and derived soils: Am. Jour. Sci., v. 251, pp. 61-82.

van Olphen, H., 1963, An introduction to clay colloid chemistry: Interscience, New York, 301 pp.

Waksman, S. A., 1938, Humus: origin, chemical composition and importance in nature (2nd ed.): Baltimore, 523 pp.

Watson, T. L., 1902, Preliminary report on a part of the granites and gneisses of Georgia: Geol. Survey Georgia Bull. 9 A, pp. 282-356.

With so much bare ground between shrubs, deserts appear to be infertile. Actually, the reverse is true; deserts abound with mineral nutrients—but lack water. In the growth habits of desert plants there may be a lesson for us in human relations. In the desert ". . . we find no fierce competition for existence, with the strong crowding out the weak. On the contrary, the available possessions—space, light, water and food—are shared alike by all. If there is not enough for all to grow tall and strong, then all remain smaller. This factual picture is very different from the time-honored notion that nature's way is cutthroat competition among individuals" (Frits Went, Scientific American, *April, 1955, p. 74*). [*Photograph by John Stacy, U.S.G.S.*]

12 / Ground Fertility and Erosion

MAINTENANCE PROBLEM

Fertility, productivity; plant nutrients, the important chemical elements, losses caused by farming; overcoming deficiencies, nitrogen, phosphorus, potassium, other elements; toxic elements; erosion control and soil conservation, crop rotation, contour plowing, and stripcropping, improving farmed ground to retard erosion, buffers, erosion at mines, quarries, and construction sites; suitability of land for agriculture; maintaining food supplies for the exploding population; bibliography.

The term *fertility* is used to refer to the nutrient content of soil. The term *productivity* is used to refer to the capacity of soil to support the growth of vegetation—crops for food and fiber, ornamental plants for the garden, or forests. In order for soil to be productive, it must of course be fertile. It must also have a texture and a structure that renders it easily tillable. Tilling gives soil the porosity needed for aeration, allows the infiltration and retention of water needed by plants, and enables plant roots to spread easily. Some ground is actually too permeable to be productive because it drains too quickly. Such ground may be good only for growing grapes for winemaking. Other ground is impermeable. Both of these conditions can be remedied, but only at a price.

Availability of moisture is the major factor in ground productivity. Even if all other conditions are met, ground is unproductive without water. Most persons probably regard the bare or sparsely vegetated ground in deserts as infertile, since there is a tendency to equate absence of vegetation with infertility. Actually, deserts include some of the potentially most fertile ground on earth. What is lacking is water.

Water can be delivered to desert regions, but here again price is usually the controlling economic factor. Tremendous volumes of water are lost through evaporation. To produce crops on desert ground requires consuming a resource that is already in short supply—water.

There is no standard for either fertility or productivity because both depend upon the crop to be produced. Ground that is productive for potatoes may be too acid to be productive for sugar beets. Ground productive for onions may be too mucky for tomatoes.

PLANT NUTRIENTS

Of the 90 or so chemical elements that form the earth's crust, 16 are known to be essential for plant growth. Seven needed in quantity are:

> hydrogen, oxygen, nitrogen, and carbon from air and water,
> and phosphorus, potassium, and calcium from mineral particles in the soil.

The other 9 elements, needed only in small amounts and mostly only in traces, are:

> magnesium, sulfur, boron, copper, iron, manganese, zinc, molybdenum, and chlorine.

Curiously, sodium, a common constituent of organic matter, is not known to be essential to plants, although it increases the yields of some products, notably the sugar in sugar beets.

In the natural habitat, plant matter is returned to the ground through decay, and unless there is excessive leaching the inorganic nutrients become available again for new growth. Many farmed crops, however, are removed from the ground in toto, and the supply of essential elements thereby becomes depleted.

An acre of fertile soil might produce a ton of corn or wheat containing 10 to 50 pounds of nitrogen, 4 to 8 pounds of phosphorus, 10 to 30 pounds of potassium, a pound or so of calcium, and 3 or 4 pounds of magnesium. A ton of soybeans produced on another acre might contain 75 or 100 pounds of nitrogen, but these are legumes and produce much of their own nitrogen from the symbiotic bacteria living in the nodules on their roots. Still another acre might produce 3 tons of alfalfa hay (also a legume) containing 150 pounds of nitrogen, 15 pounds of phosphorus, 100 pounds each of potassium and calcium, and 25 pounds of magnesium. Manures or fertilizers are needed to replace these nutrients after they have been removed from the ground by crops. Unless this is done, crops will suffer from deficiencies in the essential elements. Deficiencies become reflected in the growth of the crop in several ways. Some cause a reduction in yield as a result of poor plant growth. Other deficiencies cause plants to mature late, a factor that may be vital to crop yield in northern latitudes, where the growing season is short. The symptoms of mineral deficiency may be dwarfed, spotted, distorted, curled, or wilted leaves, or rotting at the center of fruits. Animals need phosphate for bone, and they need nitrogen for protein; feed crops deficient in these elements pass their deficiency along to livestock. One can properly turn up his nose at livestock manure as a fertilizer, despite its long use; it does improve soil tilth, but is low in the important nutrients phosphate, nitrogen, and potassium.

OVERCOMING DEFICIENCIES

The essential elements can be supplied by fertilizers, which may be applied as a top dressing or be plowed into the ground. Elements needed in only minute quantities may be applied by spraying the leaves or even by injection.

Acidity (or alkalinity) is perhaps the most important single chemical property of the ground that affects fertility. Acid ground is readily leached of mineral nutrients, but this can be corrected by adding slaked lime or pulverized limestone. Large quantities may be needed. The degree to which the acidity must be reduced depends upon the crop. Some indicated limits of acidity for certain crops are:

Red clover, barley, sugar beets	pH limit about 5.5
Wheat	pH limit about 5.0
Potatoes, oats, rye	pH limit about 4.0

Highly acidic soils, such as the red and yellow soils and the acid forest soils, generally require liming. For some plants, such as ornamentals, trees grown in nurseries, or potatoes, it may be desirable to increase the acidity. The acidity can be increased by applying fertilizers containing sulfates, which react in the ground to form sulfuric acid. Some nitrogen fertilizers also increase acidity.

Nitrogen

Figure 12.1 shows the average number of pounds of nitrogen per acre in various soil regions of the United States east of the Rocky Mountains. Four to eight pounds of nitrogen per acre are added each year by rain and snow (p. 257), and although this suffices to maintain the supply of natural vegetation, it is not enough to maintain the supply on farmed land.

High temperatures increase the rate of organic chemical reactions (p. 88) and as a consequence the nitrogen content of the ground is less in southern than in northern latitudes. For each 18°F (10°C) increase in annual temperature the nitrogen content decreases two- to threefold (Fig. 12.2). For this reason it is difficult to maintain the nitrogen content of ground in the South by plowing green crops back into the ground (green manuring); the organic matter decays too fast. Green manuring is feasible, however, in the North, and it helps minimize nitrogen deficiency, especially if the green crop is a legume. This might maintain the regional base level of nitrogen content in the ground, but would not increase it.

Nitrogen deficiencies may be corrected by using any of several kinds of fertilizers, chiefly organic ammoniates $(NH_4)^+$ and nitrates (NO_3). The organic ammoniates are

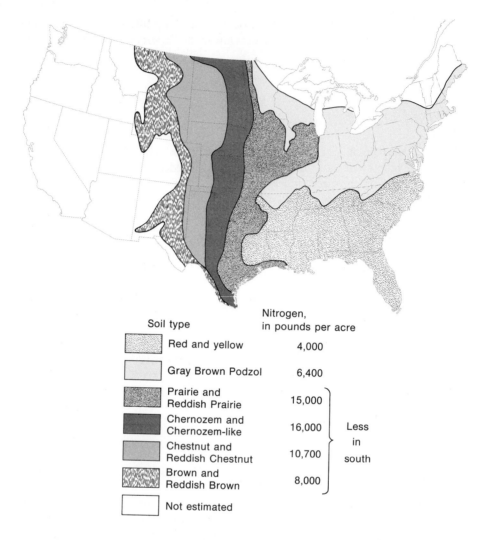

FIGURE 12.1
Outline map showing average number of pounds of nitrogen per acre in various soil regions of the United States to a depth of 40 inches. [After U.S.D.A., 1938.]

derived from nitrogen-bearing compounds, such as protein and related decomposition products contained in stable manures, fish scraps, dried guano, meat or garbage waste products, and sewage sludge. Before about 1900, organic ammoniates were a major source of fertilizers, but increasingly they are being replaced by chemical fertilizers. One wonders, though, if sewage sludge might not be used more than it is to help relieve pollution around our cities; its nitrogen content is 5 to 8 percent, which compares favorably with most other organic ammoniates.

Ammoniacal nitrogen may be obtained in several ways. In the manufacture of coke for the steel industry, ammonia gas (NH_3) that is driven off is treated with sulfuric acid to produce ammonium sulfate. The ammonia also can be obtained synthetically.

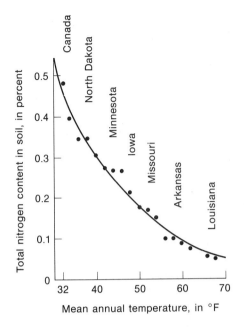

FIGURE 12.2
Decline of nitrogen content in soils with rise in mean annual temperature in the semi-humid region. [After U.S.D.A., 1938.]

Ammonium sulfate, which is a white or gray crystalline powder, contains 20.5 percent nitrogen. It also contains small amounts (less than 0.025 percent) of free sulfuric acid.

Ammonium sulfate plowed into the ground becomes absorbed by the clay and organic matter. The ammonia displaces calcium ions attached to the clay particles, and the calcium is removed as calcium sulfate. The ammonia then becomes nitrified and the nitrate ions join with sodium, potassium, or calcium ions, to form soluble nitrates, which can be absorbed by the plants. Because of their high solubility, however, these nitrates are readily leached from the ground. Consequently, ammonium sulfate should not be applied until a crop is sufficiently mature to utilize the nitrates; it may be applied in the spring seed bed or as top dressing after the crop begins to appear at the surface.

Ammoniacal nitrogen fertilizers also are produced from ammonium nitrate (NH_4NO_3), which contains 33.5 percent nitrogen, and from anhydrous liquid ammonia, which contains 82 percent nitrogen. As can be seen from the chemical formula of ammonium nitrate, half of its nitrogen is in the form of ammonia and half in the form of nitrate; if it is mixed with granulated limestone, the nitrogen content is reduced but calcium is added.

The sodium and potassium nitrates were originally obtained from caliche deposits in Chile, but today they are synthesized. These salts are white, coarsely crystalline, very soluble, and deliquescent—that is, they absorb moisture. Because of their high solubility, the nitrogen they contain is immediately available to plants; the nitrates thus act somewhat more rapidly than ammonium sulfate. The nitrate-bearing fertilizers generally are applied as top dressing.

Still another nitrogen compound used as a fertilizer is calcium cyanamide ($CaCN_2$). This compound undergoes a series of changes in the ground, eventually forming ammonia and finally nitrates. First, though, it changes to a very toxic compound that has been used as a weed killer and which is of some value in composting. When used as a fertilizer, calcium cyanamide must be applied some weeks before a crop is planted.

Nitrogen deficiencies can also be met by rotating crops. For example, wheat may be grown successfully on fertile land without fertilizers if it follows a good stand of clover or beans.

It is a curious fact of nature that plants and animals must obtain their nitrogen by such devious means when they live in and are surrounded by an atmosphere containing 78 percent of nitrogen! The natural cycling of nitrogen is illustrated in Figure 11.12.

Phosphorus

Phosphorus is another of the essential elements needed in quantity by plants. According to the Department of Agriculture, the amount of phosphorus in the first 6 or 7 inches of the ground in farmed land in the United States commonly is between 500 and 1,500 pounds per acre (roughly 0.025 to 0.075 percent of the plowed layer, which is considered to weigh 2 million pounds per acre). The amount of phosphorus in this part of the ground is about half the amount of nitrogen. Phosphorus is most available to plants when it is combined with organic matter or with calcium and magnesium. Phosphorus also occurs in combination with iron and aluminum and is present in certain rock-forming minerals, such as apatite; in these forms the phosphorus is not available to plants. The minerals must be broken down.

The mineral apatite can be broken down and changed to soluble compounds by treatment with acid or heat. Treatment with orthophosphoric acid (H_3PO_4) produces triple superphosphate with about 50 percent soluble P_2O_5. The raw rock can be used in acid soils if it is ground very finely because the soil acids will, in time, break down the apatite. Release of the phosphorus, however, is slow. Rock phosphate is not suitable on alkaline soils.

The resistance of apatite to weathering may be illustrated by its properties in another acid environment, the mouth. Fluoridation of water supplies and toothpaste contributes to converting tooth enamel to fluorapatite, which helps retard tooth decay.

Some of the principal phosphatic fertilizers and their approximate phosphorus content, expressed as the water-soluble phosphoric acid (P_2O_5), are listed below (data from U.S. Department of Agriculture Yearbook, 1938, p. 502):

Superphosphate	13.5–22.0 percent available P_2O_5
Double (triple) superphosphate	40.0–50.0 percent available P_2O_5
Sewage sludge	2.0–3.6 percent available P_2O_5
Ground phosphate rock	12.0–35.0 percent total P_2O_5
Basic slag	5.0–20.0 percent available P_2O_5

The principal source of phosphate is the calcium phosphate mineral apatite, $Ca_2(F,Cl)(PO_4)_3$, which occurs in various kinds of mineral deposits. The most productive ones are bedded deposits, such as those in the Bone Valley Formation (Pliocene) in Florida. Phosphate rock (or phosphorite) is rock containing 20 percent or more of P_2O_5. Most phosphorite deposits are of marine origin. They are thought to have formed by precipitation from phosphate-rich, deep cold water containing 0.3 parts per million PO_4 upwelling into warm surface water containing only 0.01 part per million or less. The deposits in Florida subsequently were further concentrated by weathering. United States production of phosphate rock in 1966 was almost 40 million tons, half of which was used for fertilizer.

Plants take up phosphorus chiefly as inorganic phosphate ions, PO_4^{---}, HPO_4^{--}, or $H_2PO_4^{-}$. The availability of these ions depends upon the acidity of the ground. They are most available in ground that is nearly neutral, and they become nearly insoluble in strongly acid or strongly alkaline soils, and plants grown in such soils exhibit symptoms of phosphate deficiency. Because it is comparatively insoluble, phosphorus is concentrated by the organic matter in the surface layers of the ground; not much is removed by drainage. Probably most of the phosphorus in the ground is relict—a product of ancient weathering. Because apatite is a moderately resistant mineral (Table 11.2), it seems reasonable to say that the apatite present in surface deposits or soils that are of Wisconsinan age or younger has probably undergone little alteration unless very fine grained.

Although superphosphate provides phosphate in a soluble form, not much of it is washed from the soil, because it changes gradually to insoluble forms. Consequently, superphosphate is best added in small amounts for each crop rather than attempting to build up reserves in the ground. It may be applied at the rate of a few hundred pounds per acre. In general, greater amounts are needed as acidity and rainfall increase.

Basic slag, a by-product of the manufacture of steel from pig iron, can be used as a source of fertilizer where the iron ore is rich in phosphorus, as it is at the steel furnaces at Birmingham, Alabama. The phosphorus is removed from the iron ore by causing it to combine with lime. This slag can be poured off, ground, and used directly as phosphate fertilizer. In the United States this kind of fertilizer contains about 8 percent or more total phosphoric acid. Fertilizers made from slags also contain calcium and some manganese. A dressing that would provide about 100 pounds P_2O_5 per acre requires applying about 1,000 pounds per acre of slag containing 10 percent P_2O_5.

Phosphate is metabolized by animals and concentrates in their bones; before the introduction of chemical fertilizers, animal bone (bone meal) was an important source of phosphatic fertilizer. The American Indians are said to have buried fish in the hillocks of corn plants; this may have been done as a ceremonial act, but the decaying fish would have supplied phosphorus and nitrogen to the plants.

Potassium

The principal mineral sources of potassium fertilizers today are:

> Potassium chloride (KCl, chiefly sylvite, muriate of potash), which, as fertilizer, contains 47 to 61 percent potassium.

Mixed minerals containing potassium sulfate (K_2SO_4).

Potassium nitrate (KNO_3; niter, saltpeter), which contains 38.7 percent potassium (K) or 46.6 percent potash (K_2O).

In natural deposits the potassium in the chloride, sulfate, or nitrate may be diluted by sodium. The potassium content of nonmineral sources is comparatively low. For example, the ashes of hardwood trees and farmyard dung, the original sources of potassium, contain less than 8 percent potash—literally, "pot ash," a concentrate obtained by evaporation of potassium carbonate leached from wood ashes. Potassium is present in nearly all animal tissue but not in the high concentrations that nitrogen is present in protein or that phosphate is present in bone. Most of the potassium metabolized by animals is excreted as soluble salts in urine. Such salts in barnyard manures were long an important source of potash fertilizer.

Potassium from all these sources is highly soluble, which means that it is readily available to plants but also susceptible to removal from the open-system soils in humid regions.

Potassium salts, applied in a mixed fertilizer to a soil, go into solution; not all of the potassium content, however, is immediately available to plants. Part of it unites with colloidal matter or becomes attached to clay particles, and is not available to plants until it is freed from the colloid or clay by ion exchange. This fraction constitutes a reservoir that becomes available when the fraction in the ground moisture is depleted. The quantity in the reservoir depends on the content of clay and colloid in the ground; ground without much clay or colloids develops no such reserve. It all moves out of the system either via plant metabolism or downward percolating water. Potash is stored in the ground more readily than nitrogen but less readily than phosphorus.

The amount of potassium in the surface layers of the ground—say, within 7 inches of the surface—is much greater than the amount of either nitrogen or phosphorus. A given piece of ground might contain 0.2 percent each of nitrogen and phosphates and 10 times that amount of potassium (about 40,000 pounds per acre), supplied largely by weathering of feldspars and micas. Much of the potassium, however, is held in unavailable form by the clays or colloids, although that reserve does become available in time. The amount of potash removed from the ground depends partly on the kind of crop. The depletion may amount to 20 to 25 pounds per acre for grains (corn, wheat, oats, barley), 50 to 100 pounds per acre for alfalfa or clover, and 125 pounds per acre for beets. Repeated losses such as these must be replaced by fertilizers in order to maintain production.

Other Elements

The alkaline earths, magnesium and calcium, are necessary both for plant structure and plant metabolism. Magnesium is a component of chlorophyll and contributes to carbohydrate synthesis. Calcium is an essential component of roots and cell walls, and affects the activity of several enzymes.

Sulfur is a component of some amino acids, vitamins, and enzymes, and is thought to contribute to formation of chlorophyll, although it is not part of the chlorophyll molecule. Boron is essential for healthy plant growth, but just what role it plays in plant structure or metabolism is not known.

Copper, iron, zinc, manganese, and molybdenum are important chiefly because they are components of essential enzymes. Iron, zinc, and copper are components of enzymes that facilitate oxidation, especially of carbohydrates; molybdenum and manganese are components of enzymes necessary for nitrogen metabolism.

Toxic Elements

Ground that has an overabundance of particular elements in soluble form may be as unsatisfactory for plant growth as ground that has deficiencies. Even the elements necessary for plant growth can be present in excess and be accumulated in sufficient quantities to be toxic. Ordinary sodium chloride and the alkalis are familiar examples; as already noted (Chapter 11), flowering plants do not grow where soil water contains more than 6 percent of sodium chloride and other salts. The elements, of course, are not equally toxic, and the various species of plants differ in their susceptibility to toxic elements. Ground may have to be treated for excesses as well as for deficiencies.

As already noted, some kinds of plants find acid ground unfavorable for growth, and other plants find alkaline ground unfavorable. Acid soil may be unfavorable not only because of the direct effects of the acidity, but also because the mineral salts are readily soluble in the acid solutions and may become available to the vegetation in toxic concentrations, or in concentrations that interfere with absorption of other necessary elements. Both aluminum and iron may damage plants growing in acid soils (Janick and others, 1969, p. 575).

EROSION CONTROL—SOIL CONSERVATION

Soil conservation has many purposes besides minimizing erosion. It seeks to help reduce deposition of sediment where it is not wanted, to help maintain proper soil density, to minimize exhaustion of plant nutrients through overleaching, to conserve water by improving infiltration of drainage, and to minimize waterlogging and accumulation of salts or alkalis.

Erosion of farmed land removes the topsoil; unless it is replenished, the ground surface may be lowered to the clayey enriched layer, which is more difficult to farm than the original surface layers. Soil losses increase as slopes steepen. Ground with a slope of 1 percent might lose 5 tons of soil (about 100 cubic feet) per acre annually, whereas ground with a slope of 10 percent loses 50 tons and ground with a slope of 20 percent loses 100 tons.

Erosion on cropland can be and has been reduced by various conservation practices. One of these is crop rotation, which consists in planting a succession of different crops on the same piece of ground. It is based on the principle that cultivated crops expose the ground to maximum erosion, that small grains cause less exposure, and that grasses or mixed grasses and legumes effectively protect the ground against erosion. A common rotation plan therefore includes a cultivated crop followed by a small grain, and this followed by grasses. Crop rotation not only saves soil, but increases production. Examples of some combinations of rotated crops are (Bennett, 1955):

1. In the northeastern United States: oats, red clover, potatoes.

2. In the southeastern United States: corn with cowpeas, small grain, lespedeza, cotton.

3. In the Ozark Plateau: corn with cowpeas, cotton, small grain, cowpeas.

4. On the Great Plains: sorghum, oats and sweet clover, sweet clover, wheat for 2 years.

Another common practice is contour plowing and the cultivation of row crops along the contour. Runoff is checked by the furrows. Crop rotation and contour plowing may be combined, and either or both of those practices may be combined with strip-cropping—the practice of alternating row crops with closely growing ones.

There are ways of improving farmed ground in order to retard erosion. Adding vegetable matter to the soil increases its aeration and the spread of roots. Sloping ground can be terraced to minimize runoff and prevent gullying. Natural channels can be controlled artificially by dams, and the broad shallow ones may be stabilized with stands of grass or other vegetation. Ground that is subject to washing by runoff from upslope can be protected with diversion channels; the runoff might even be saved by collecting it in ponds.

Farmed land can also be protected against erosion by using part of the land for planting permanent buffers in the form of strips of perennial grass or shrubs. Steep slopes generally should not be used for farming; they can be used for growing perennial vegetation.

The foregoing examples perhaps suffice to illustrate some of the principles about the need for and use of fertilizers and/or crop rotation to maintain fertility in farmed land. Applying fertilizer is not a matter of one element being more essential than another; just as an automobile engine needs a fueling, firing, and cooling system, so do plants need proper amounts of *all* the essential elements. A small quantity of one may be just as vital as a large quantity of another.

Today, most fertilizers supply mixtures of nitrogen, phosphorus and potassium in various forms. The mixture formula is given by numbers on the container; the numbers 5-10-10 mean that the fertilizer contains 5 percent N, 10 percent P_2O_5, and 10 percent K_2O. This bag may be less expensive than one having the number 10-20-20, but it also has only half as much plant food.

Other erosion caused by Man's use of the land is caused by quarries, sand and gravel pits, strip mines, and the waste piles at mines and other excavations. The materials obtained from these operations are needed and used, as is the produce from the croplands, but there is no need for the excavations or waste piles to continue muddying streams and scarring the landscape. Highway engineers excavate through hills and fill valleys and contribute greatly to erosion and muddying of streams in the process, but when the construction is completed the scars are covered and healed so well that geologists protest the greenery that covers what had been outcrop. Abandoned quarries, sand and gravel pits, and strip mines could similarly be healed.

SUITABILITY OF LAND FOR AGRICULTURE

Land is classified by the U.S. Department of Agriculture in categories of excellence according to its suitability for agriculture. The categories range from lands that are regarded as suitable for virtually all-purpose use to those regarded as wastelands.

In the best category are the lands on which few restrictions are imposed as to their use. Besides being nearly level, they have ground that is easily worked, favors root penetration, provides nutrients and retains moisture for plant growth, is little subject to erosion either by wind or washing, is well drained (that is, not subject to flooding or waterlogging), and is not subject to accumulation of salts. As the restrictions increase, the kinds of use become reduced, and need for special farming methods increases (such as strip-cropping). Some land may be suited only for particular crops or may require costly management practices (for example, terracing, drainage systems, levees). In arid and semiarid regions, salinity may be an important limiting factor. For agricultural use, ground is considered free of salt if the salt content is less than about 0.2 percent. Ground containing more than 0.65 percent of salts is considered strongly affected and in need of treatment to be suitable for agricultural use (U.S.D.A. *Soil Survey Manual,* 1951, p. 360).

Lesser categories of land are not suitable for cultivation. The best of these may support grasses and trees, but not be suitable for cultivation because the growing season is short or because they are subject to flooding. The limitations on use become increasingly severe where the topography is rough, the surface deposits shallow, stony, alkaline, salty, excessively wet or dry, and the climate too dry, too hot, or too cold. Lands in the lowest category, called "wasteland," are not suited for pasture or range, but commonly are valuable for scenery, recreation, mineral resources, or wildlife.

In this classification, which is based on suitability of land for agriculture, kind of ground is less important than topography and climate. Kind of soil is important in the better categories of land; kind of surface deposit is important in the lesser ones. In the foregoing chapters, we looked at the ground for more than merely its agricultural uses. Our mountain wildernesses, canyons, salt and sand deserts, and the urban areas in which most of us live are agricultural wastelands, but they do have other uses.

Moreover, land uses change. Increased multiple use of forested lands has greatly increased the need for information about the ground—for roads, construction, sanitation, minimizing erosion, improving forest and grazing productivity, and protecting the water supplies. But the basic surveys have lagged. There have been only a few soil surveys because the lands are not agricultural, and there have been few surveys of the surficial geology because the surface deposits generally have not been of interest to mining companies and mineral prospectors. We would be much better equipped to plan intelligently for the future use of our public lands—which amount to one-third of the United States—if the basic surveys of the surficial geology and soils were available.

The Atlantic Coastal Plain is being changed because of the pressures of the exploding population there and the increased demand for land. Swamps and marshes have been opened for land development by ground drainage made possible by the use of heavy machinery and use of fertilizers. The shores of estuaries, once the breeding grounds for all kinds of wildlife, are being changed by the construction of bulkheads and boat docks. And we have no satisfactory inventory of that ground; the basic surficial geology and soil surveys have not been made.

MAINTAINING FOOD SUPPLIES FOR THE EXPLODING POPULATION

Man has demonstrated that he can reach the moon, but he has failed to demonstrate an ability to provide food for those millions of human beings on this planet who exist on inadequate diets—millions in our own country, and vastly more millions abroad. The simple fact is that we have spent too much on war and prospective war and not enough to relieve poverty and malnutrition and to provide better health services and better education.

In part the problem consists in maintaining or increasing the productivity of the ground around us. To achieve this requires more basic and systematic information about what is left of this thin layer. It also requires the application of "biological engineering," a greater understanding and use of ecological principles, and the development of biological strains that can increase the nutrient value of particular crops. Plant breeding is not new, but unexplored possibilities still remain for improving the food value of plants and for increasing yields.

With intelligent use of the ground, intelligent biological engineering, and intelligent distribution of our efforts and resources, we can provide for those who are hungry as well as for those who are affluent. The price? Change our national priorities.

BIBLIOGRAPHY

Bear, F. E., 1953, Soils and fertilizers (4th ed.): Wiley, New York.

———, 1965, Soils in relation to crop growth: Reinhold, New York, 297 pp.

Bennett, H. H., 1955, Elements of soil conservation: McGraw-Hill, New York, 358 pp.

Blakely, B. D., Coyle, J. J., and Steele, J. G., 1957, Erosion on cultivated land: *in* Soil, U.S. Dept. Agriculture Yearbook, 1957, pp. 290–307.

Copeland, O. L., Jr., 1965, Land use and ecological factors in relation to sediment yields, *in* Proceedings of the Federal Interagency Sedimentation Conference 1963: Agric. Res. Serv. Misc. Publ. No. 970, pp. 72–84.

Craddock, G. E., and Pearse, C. K., 1938, Surface runoff and erosion on granitic mountain soils of Idaho as influenced by range cover, soil disturbance, slope, and precipitation intensity: U.S. Dept. Agriculture Tech. Circ. 482, 24 pp.

Francis, C. J., 1957, How to control a gully: *in* Soil, U. S. Dept. Agriculture Yearbook, 1957, pp. 315–320.

Higbee, Edw., 1958, American agriculture: Wiley, New York, 399 pp.

Hodge, C., and Duisberg, P. C. (editors), 1963, Aridity and Man: Am. Assoc. Adv. Sci. Publ. 74, 584 pp.

Janick, J., Schery, R. W., Woods, F. W., and Ruttan, V. W., 1969, Plant Science: W. H. Freeman and Company, San Francisco, 629 pp.

Jenny, H., and Ayers, A. D., 1939, The influence of the degree of saturation of soil colloids on the nutrient intake by roots: Soil Sci., v. 48, pp. 443–459.

Kelley, V. P., 1937, The reclamation of alkali soils: Calif. Agricultural Exp. Sta. Bull. 617.

Kraenzel, C. F., 1955, The Great Plains in transition: Univ. Okla. Press, Norman, Okla., 428 pp.

Larson, G. A., 1939, Reclamation of saline (alkali) soil in Yakima Valley, Washington: Wash. Agricultural Exper. Sta. Bull. 376.

Leach, H. R., 1942, Soil erosion: *in* O. E. Meinzer (editor), Hydrology: Dover, New York, pp. 606–613.

Lowdermilk, W. C., 1953, Conquest of the land through seven thousand years: U.S. Dept. Agriculture Inf. Bull. 99, 30 pp.

Muehlbeier, J., 1958, Land-use problems in the Great Plains: *in* Land, U.S. Dept. Agriculture Yearbook, 1958, pp. 161–166.

Nelson, L. B., 1965, Advances in fertilizers: Advances in Agronomy, v. 17, pp. 1–84.

Powell, J. W., 1879, Report on lands of the arid region: U.S. Geog. and Geol. Survey, Rocky Mtn. Region, Washington, 191 pp.

Renner, H. T., 1941, Conditions influencing erosion on the Boise River watershed: U.S. Dept. Agriculture Tech. Bull. 528, 32 pp.

Russell, E. J., 1956, Soil conditions and plant growth: Longmans, Green and Co., London-New York. Revised Ed., 635 pp.

Sears, P. B., 1947, Deserts on the march: Univ. Okla. Press, Norman, Okla.

Stallings, J. H., 1957, Soil conservation: Prentice-Hall, Englewood Cliffs, N.J., 575 pp.

Talbot, M. W., 1937, Indicators of southwestern range conditions: U.S. Dept. Agriculture Farmers Bull. 1782, 34 pp.

Tisdale, S. L., and Nelson, W. L., 1966, Soil fertility and fertilizers: Macmillan, New York, 694 pp.

United States Salinity Laboratory Staff, 1954, Diagnosis and improvement of saline and alkali soils: U.S. Dept. Agriculture Handbook 60.

Landslide in the Berkeley Hills, between Oakland and Orinda, California. Before the mid-1940's, an old road curved around the toe of an old slide of clayey material—montmorillonite, which is highly absorbent and swells when wet. When the old road was widened and straightened to make the four-lane undivided freeway seen here, the toe of the old slide was excavated. (Notice the steep slope of the road cut). This removed part of the ground that had been supporting the upper part of the old slide. In December of 1950, a year of higher than average rainfall, the sliding was renewed when the slide plane became lubricated by seepage. Trucks in the foreground and trees carried with the slid mass give an impression of the amount of material, estimated at about a million tons. [Photograph by Bill Young, courtesy of the San Francisco Chronicle.]

13 / Some Engineering Aspects of Ground Conditions

Engineering geology; considerations for planning; topographic and geologic maps; examples of different kinds of ground; importance of geologic history; construction materials—sand and gravel, stone, limestone, clay; ground stability—slope, creep, landsliding, avalanching; effect of texture and composition of ground, organic matter, peat; settlement of ground; ground drainage, moisture holding properties, water tables, controlling moisture; permafrost; sewage disposal, pollution; other ground problems—erosion, placers, earthquakes, creep at faults; testing ground; bibliography.

Engineering works may succeed or fail according to how well they fit their geologic environment and how well we understand the geologic processes that might affect those works. Engineering geology, as the name implies, involves both engineering and geology. The geologist's task is to determine the ground conditions at the site and their interrelationship with the land around it, for the geology of the surrounding area may be as critical as that at the site. The engineer's task is to plan, design, and construct the works at maximum safety and least cost. Both geological and engineering knowledge are needed, especially during the planning stages.

CONSIDERATIONS FOR PLANNING

Most of the ground conditions that have been described, especially the topography, geology of the bedrock and surface deposits, and the hydrology, are factors to be considered in planning, designing, and constructing engineering structures. Mining and civil engineers are as concerned with the ground as are farmers and geologists.

In general, mining engineers are concerned chiefly with the bedrock, although some open-pit mining, quarrying, or placer operations involve surface deposits; civil engineers generally are concerned chiefly with surface deposits, although they must often excavate bedrock. This chapter considers some of those aspects of surface deposits that are of special engineering interest:

1. Construction materials obtained from the ground; such as gravel and sand, rock, clay for adobe or sealing ground, limestone for cement, gypsum for plaster.

2. Ground stability and its suitability as foundation for various structures—dams, bridge abutments, pipelines, aqueducts, residential homes, high-rise apartments, or shopping centers.

3. Ease and cost of excavating different kinds of ground.

4. Effects of different kinds of ground on the quantity and quality of water supplies and their susceptibility to pollution.

5. Susceptibility of various kinds of ground to erosion.

6. Suitability of certain ground for solid-waste disposal.

Such matters also should be considered in city and county planning.

For *preliminary* planning, good topographic maps and aerial photography are available for most of the country. Where good geologic maps are not available for showing the surface deposits—state and federal geological surveys have lagged in this—much can be learned about the ground by interpreting aerial photographs. If the area under study is agricultural land, there will be available detailed surveys showing the modern soils, which also can be used to help identify the surface deposits and determine the ground conditions that may pose particular problems.

After preliminary planning and the selection of particular sites for consideration, ground studies of the sites can be made jointly by geologists and engineers in as much detail as is needed. Figure 13.1 is a topographic map of a hypothetical area; Figure 13.2 shows its hypothetical geology. These maps will be used in this chapter to show some engineering aspects of the particular geologic conditions. Although the maps are hypothetical, the problems discussed are real enough, and can be duplicated in the Great Basin area of Nevada and western Utah.

The old fan deposits (Qfg in Fig. 13.2) that underlie the lake deposits are aquifers that can be reached by drilling. Judging by the slope on the surface of the old fans (Fig. 13.1), the aquifer would be less than 50 feet deep in the southern part of the area. Water of good quality probably also can be obtained from the base of the deltaic gravel (Qbg in Fig. 13.2) at a depth of about 40 feet. The seeps around the edges of the young fan (QR) on which Jonesville is built are discharge from the shallow ground water perched under the fan. Seeps having that setting almost certainly are polluted.

The clay ground (Qbc) has poor surface drainage and no subsurface drainage. The ground is readily flooded and is difficult to drain, but there would be little or no seepage losses from reservoirs or irrigation canals. Fill for highway subgrades and road metal would have to be hauled to that ground and would require installation of subdrains.

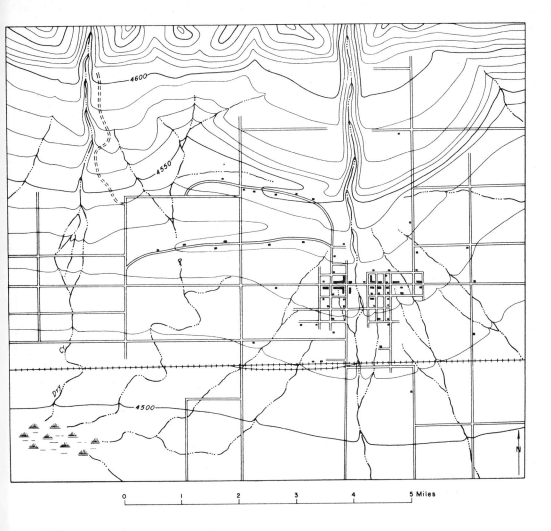

FIGURE 13.1
Topographic map. Topographic maps show quantitatively the configuration of the land surface by contours that represent level lines on the earth's surface. Irregularities in the contour lines indicate the shape of land forms; the spacing between the contours gives a measure of the amount of slope. In addition, topographic maps show the works of man, such as roads, railroads, and buildings, and drainage features, such as perennial streams, intermittent streams, springs, and marshes.

The young fans (QR) are subject to flash floods out of the canyons, but their ground is moderately permeable with satisfactory runoff except perhaps during severe local storms. The ground is stable enough but basement excavations would have to be shallow to avoid intersecting the perched water table.

The sandy ground (Qbs) is protected against flash floods, is permeable, and has good surface and subsurface drainage. The ground is stable, but there is a perched water table at the base of the sand and basements would have to be shallow to avoid intersecting it. Roads would have good subdrainage but would require clay to bind the sand.

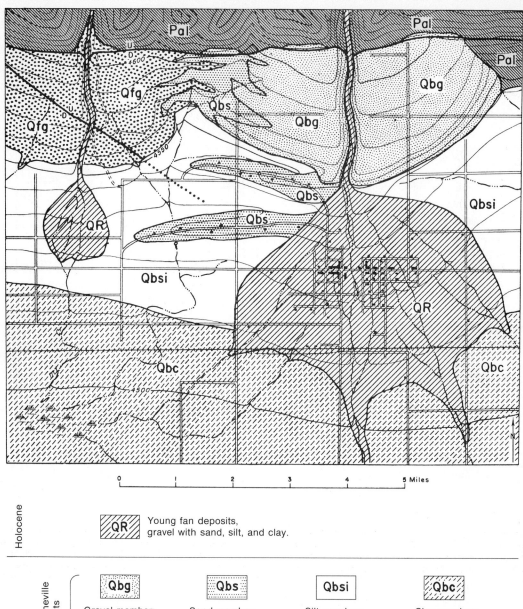

FIGURE 13.2
Geologic map. The surface deposits are of Holocene and Pleistocene age.

The deltaic deposit (Qbg) has excellent subsurface drainage and provides good stable ground for foundation for roads. The ground is not subject to flooding, but because it is highly permeable, reservoirs and canals require sealing to avoid excessive losses by seepage.

IMPORTANCE OF GEOLOGIC HISTORY

Determining the geology of an area for engineering or other purposes involves far more than merely locating and indicating the kind of surface deposits and rock formations. The relationships between these mappable units must be determined in order to learn their geologic history, because understanding that history is basic to predicting what happens to particular formations where they are buried. Such prediction focuses upon thicknesses of the formations and how they change—that is, how the formations change composition and texture laterally or with depth, and whether they grade into, underlie, or overlie neighboring formations. In our example (Fig. 13.2) the geology records 6 major events that have significance in engineering problems.

The first event was the folding and upward faulting of the ancient limestone and quartzite formations (Pal in Fig. 13.2) to form the mountains. The topographic map shows this to be a steep mountain front, and if engineering structures are placed near it, consideration must be given to the stability of the mountainside—its susceptibility to landsliding, snow or rock avalanches, and flash floods. An assessment should be made of the possibility of earthquake hazard due to renewed movement along the frontal fault. Many cities in Utah, including Salt Lake, invite disaster by zoning such ground for residential use. The view is fine, but prudence dictates that such land should be zoned for parks; more people could enjoy the view, and there would be no hazard to homes and other property.

The second recorded event in the geological history was the building of the old fan of bouldery gravel (Qfg), shown in the northwest corner of the map. These old gravels are weathered, and include a high percentage of crumbly stones. The gravels are poorly sorted and contain a high percentage of fines; they also contain considerable caliche, which is detrimental for some purposes. Not all sand and gravel deposits are equally usable.

The third event in the geologic history partly answers the question about potential earthquake hazard along the frontal fault. The bouldery fan is broken by a fault that branches southeasterly from the frontal fault, and the topographic map indicates 5 to 10 feet of displacement. This branch fault, however, is old enough so that it does not displace the younger deposits around the foot of the fan; the broken line indicates the part of the fault that is buried.

The fourth event is recorded by the Pleistocene lake deposits. This lake was formed by meltwaters from Pleistocene glaciers in the mountains. The lake deposits include gravel and sand in deltas (Qbg), sand (Qbs) that was deposited in bars across the front of the deltas, silty deposits (Qbsi) laid down on the lake bottom near the shores, and clay (Qbc) deposited on the quiet offshore bottoms. These lake beds are younger than the old fan deposits and overlie them. The difference in slope of the old fan and the silty lake beds suggests that the old bouldery gravels can be expected at a depth of about 10 feet midway between the south edge of the fan and the railroad. Deltaic sands and gravels are likely to be better sorted and better suited for engineering purposes than the fan gravels. Since the deltaic deposits are younger, they are likely to be less weathered and contain less caliche.

The fifth event is the renewal of movement on the frontal fault, shown by 20 feet of displacement of the deltaic beds near the northeast corner of the map. This helps answer the question about potential earthquake hazards. The geologic record shows that there has been repeated recent movement on the fault system, hence precautions should be taken against earthquake shocks and against avalanches that might be triggered by earth movements.

The sixth event recorded in the geologic history occurred after the Pleistocene lake dried up. There has since been 10 to 30 feet of downcutting by the streams that cross the deltaic deposits and old fans; as a result of this erosion, young fans (QR) have been built on the lake bottom sediments. These young fans are rather permeable ground, but they rest on comparatively impermeable lake bottom silt and clay. They would have a shallow, perched water table highly vulnerable to pollution.

In brief, the succession of several geologic events illustrated by this geology has considerable significance in various kinds of engineering problems. The succession of deposits provides a clue to determining what lies below the surface at given places, and at what depths. The deposits are of different kinds and have different engineering potentials because of the changes in climate and sedimentation that affected the area. The history of the mountain uplift clearly indicates that precautions should be taken against several kinds of geologic hazards. Other, more specific engineering aspects of the geology are described below.

CONSTRUCTION MATERIALS

The principal construction materials—sand and gravel, lime for cement, and stone (other than quality dimension stone)—are bulky, heavy, and used in large quantities. As a consequence, the economics of their use differs substantially from that of other mineral commodities. A considerable part of the cost of construction materials is for haulage, and for large projects sources must be sought near the construction sites.

Sand and gravel deposits are almost ubiquitous and are produced in large amounts. In dollar value, sand and gravel production ranks among the top four mineral commodities produced in three-quarters of the states. The total value of national production of sand and gravel in recent years has exceeded 1 billion dollars annually. Stone (including dimension stone) and lime for cement have each totaled more than $1\frac{1}{4}$ billion dollars annually in recent years. Together, these mundane construction materials have exceeded the total worth of our production of the more glittering metallic minerals. Their value also exceeds that of our coal production and is exceeded only by the value of crude petroleum and natural gas.

Most sand and gravel is excavated from open-pit quarries. In the past we have permitted operators to abandon their pits when production was halted. Fortunately we have been spared some of the evils of our lack of planning because large numbers of such pits have been near cities, and spreading urbanization has caused the pits to be used as excavations for foundations of large buildings. The need for construction materials should be considered in planning for urbanization, and the possible uses of the open pits should also be considered by zoning regulations that anticipate the changes in use of the land.

The geology illustrated in Figure 13.2 shows a number of potential sources of construction materials and some differences in their potential suitabilities. The old rocks (Pal) could be a source of limestone for cement and of limestone or quartzite for building stone and riprap. The old gravel (Qfg) is angular but silty; it is poorly sorted and, as already noted, weathering has affected a high proportion of the stones. In addition, there is considerable secondary lime carbonate (caliche). The materials may not be suitable for concrete aggregate, and there appear to be better nearby sources of road metal. Probably this ground should be zoned as residential.

The deltaic gravel (Qbg) is well rounded, well sorted, and the stones are firm. It should be an excellent source of road metal, but because the gravel contains considerable secondary lime, better sources should be sought for concrete aggregate. The youngest gravel (QR) is poorly sorted; the fragments are in part rounded and in part angular. The deposit has little secondary lime and may be the best source for concrete aggregate.

The deposit of lake bottom clay (Qbc) contains lime and other water-soluble salts. It could be used for brick or seal clay, but probably is not suited for high-grade ceramic purposes.

GROUND STABILITY

Ground stability, or instability, is dependent on numerous variables, one of which is slope. In general, simply because they slide easily because of gravity, steep slopes are less stable than gentle slopes, but there are exceptions.

Where surface deposits on steep slopes are thin, structures may be anchored to bedrock foundations. The stability of such foundations depends partly on the composition of the rock and partly on whether its bedding and other fractures dip down the slope or into the hill. Rock in which bedding planes or fractures dip in the same direction as the slope of the hill is dangerously subject to landsliding, particularly if water can enter the fractures. Water can do more than lubricate the fracture; in winter it may freeze, lift plates of rock, and cause them to move downhill when the ground thaws.

Hillside surface deposits that are chiefly colluvial may be subject to creep. Excavations may accelerate the movement; vibrations and jars can start slope failures. The most troublesome situations are those in which geology and topography combine to form seeps or springs. Relationships similar to those on hillsides may be created artifically at the sides of excavations in ground that is nearly level. Excavations for highways, canals, or buildings may enable water to seep into colluvial ground and activate sliding.

Landslides are common where slopes are steep, and are especially conspicuous and troublesome where urbanization has led to building on such ground. One sees pictures almost yearly of California homes sliding into the sea; the sites were chosen for development because of the magnificent ocean view, and the homes literally join the sea. Again, such land should be zoned for parks or open space, to be enjoyed by large numbers rather than be developed for residential use when risks of property damage are so great. In many hilly areas construction of residences and streets increases the hazards of landsliding by adding to the load on potential slide blocks and by modifying the natural drainage.

Landslides have been a considerable problem along the shores of the Franklin D. Roosevelt Lake, an artificial lake impounded behind the Grand Coulee Dam on the Columbia River in Washington. Where the ground consists of Pleistocene or Holocene stream or lake deposits, it is unstable and landslides develop wherever slopes are steep, as at the edges of many valleys. The newly formed lakes have raised the water table many tens of feet and thereby lubricated the clay and wettened cracks in the unstable ground. In places the landsliding has widened the lake 2,000 feet. Catastrophic slides create large waves (*seiches*) on the lake, waves very much like the wave a child can make by sliding into a bathtub, and with even more disastrous results. Some slides on Roosevelt Lake have created waves that washed up 65 feet high on the opposite shore. The hazard caused by landsliding is not restricted to the unstable ground!

Other kinds of slides and conditions that favor sliding have been described in the chapter on surface deposits.

Slides and avalanches that move across highways, railroads, or into towns commonly are removed by excavation, but this practice can add to an already bad situation, because the toe of the slide may serve as a buttress that prevents further movement. To stabilize such ground may require building retaining structures, using piles, draining the slide and slide plane, and/or grouting.

"Fallen rock zone" is a familiar sign along highways that pass below cliffs that have been oversteepened. Usually it is not feasible to prevent blocks from being loosened and falling to the pavement or railroad tracks; a common practice in such ground is to contain the fall with steel nets.

Effect of texture and composition of the ground

In excavations, safe slopes are a function of the cohesiveness of the ground—that is, its content of fine-grained binder. Moderately cohesive sandstone or shale may be stable on cuts sloping 1 to 1, where gravel with sand would not be stable on slopes greater than 2 to 1, and cuts in sand may have to be as flat as $2\frac{1}{2}$ to 1. Loess, on the other hand, has the unusual property of being stable in nearly vertical cliffs.

Some kinds of ground are unstable for engineering purposes, even on moderate slopes, because they lack sufficient fine-grained material (clay) to bind the mass together; to strengthen and stabilize such ground, binding materials must be added. In some circumstances it is feasible to haul in clay and mix it into the sand that needs to be compacted. In other circumstances cement, bituminous materials, or sodium silicate may be added. In most situations only small proportions of these binders are needed, usually no more than a very few percent by weight of the aggregate. Contrariwise, clayey ground may need to have coarse aggregate added.

Gravel and sand well mixed with clay may support loads of 75 or even 100 pounds per square inch, whereas poorly mixed clays or sands may support loads only half or a quarter as much. Loose sand and some soft clays are not capable of supporting more than 10 pounds per square inch; peat and topsoils are not suitable for supporting loads greater than about 5 pounds per square inch and the load limit commonly is no more than a pound or two. As can be seen in Table 13.1, the density (shown in the 3rd row) of a mixture is a fairly satisfactory index of its suitability as a foundation. Materials denser than 120 pounds per cubic foot (sp. g. more than 1.9) generally make a satisfactory foundation. If the density is less than about 100 pounds per cubic foot (sp. g. less than 1.6) the materials generally are not suitable.

Colluvial clay is especially subject to sliding, sand much less so. Cracks in the clay open as excavation progresses and relieve the confining pressure. These cracks collect water, which may soften the clay and cause it to swell, or the water may freeze and cause heaving. Moreover, inhomogeneities in the clay, such as layers of sand or silt, collect water and guide its underground movement. In wet weather these become pressure points in the hydrostatic system, and when cut by an excavation can become live seeps that will lead to slope failure. Dry sand containing much silt or clay is reasonably stable unless it is a sand-stratum acquifer containing water under pressure.

The kind of clay minerals in the ground is another very important factor in ground stability. Ground containing the swelling clay montmorillonite is notoriously bad. Even the best highways on such ground develop a dangerously wavy surface; telephone poles become tilted. The clayey ground on the Cretaceous shale formations in the western states is a notable example. The damage can be minimized by keeping the ground nearly dry. Kaolin clays, on the other hand, seem to cause little trouble.

The density and stability of a given piece of ground can be increased by compaction to expel the air and water and cause the particles to interlock more closely. Not all ground, though, responds to compaction. It is difficult to compact highly micaceous ground, such as the residual deposits over mica schists, or very silty ground, or ground containing 5 percent or more of organic matter.

For foundation purposes organic matter is undesirable and generally must be removed, for it lacks strength and greatly compresses under load. Moreover, acids generated by the decay of the organic matter may have a corrosive effect on cementing substances that serve to stabilize ground.

Peat is a special foundation problem. It is wholly unsuited for highway subgrade or other foundations because it has little strength (load limit commonly less than 3 pounds per square inch) and its volume may change in reaction to changes in the moisture content or load. Where roads must be built across such ground, it is generally most economical to remove the peat if it is less than about 10 feet deep. This removal can be done by excavating or by adding enough fill on the unstable peat to displace it laterally. The displacement method may be aided by jarring the ground with explosive charges beneath the fill. Where the peat or other unstable ground is more than 10 feet deep, the new load may be reduced by increasing the area of the foundation or by removing the unstable base to a depth equal to the load of the structure to be placed on it.

TABLE 13.1
Textural classes of surface deposits and soils and their suitability for highway and other construction

General classification	Granular materials 35% or less passes No. 200			Silt-clay materials more than 35% passes No. 200				Peat, muck
Group classification	A-1	A-2	A-3	A-4	A-5	A-6	A-7	A-8
Sieve analysis; percent passing: No. 10 (2 mm)	50 max							
No. 40 (0.42 mm)	25 max		51 min					
No. 200 (0.074 mm)		35 max	10 min	36 min	36 min	36 min	36 min	
Texture (percent) sand	70-85	55-80	90	55 max	55 max	55 max	55 max	55 max
silt	10-20	0-45	0-10	High	Medium	Medium	Medium	Not significant
clay	5-10	0-45	0-10	Low	Low	30 min	30 min	
Maximum dry weight when compacted, pounds per cubic foot	130 min	120-130	120-130	110-120	80-100	80-100	80-100	90 max
Shrinkage and expansion	Very slight	Slight	Slight	Variable	Variable to high	Detrimental		Detrimental
Drainage characteristics	Excellent	Fair to practically impervious	Good	Fair to poor		Practically impervious		Fair to poor
Suitability as subgrade or subbase under flexible pavement	Good to excellent	Good	Good to excellent	Fair to poor		Mostly poor	Poor to very poor	Not suitable
Value as base course directly under bituminous pavement	Poor to good			Not suitable				
Value for embankments	Stable	Fairly stable	Reasonably stable	Fair to poor stability				Not suitable
Value for fill: Less than 50 ft high	Excellent	Good		Good to poor	Fair to very poor	Very poor	Fair to poor	Not suitable
More than 50 ft high	Good	Good to fair		Fair to poor	Fair to very poor	Very poor	Fair to poor	Not suitable

Source: Generalized from U.S. Bur. Public Roads.

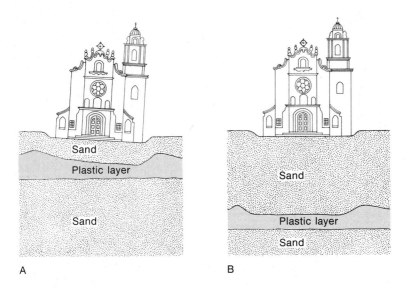

FIGURE 13.3
(A) If the plastic layer that yields is shallow, settlement is irregular and governed by the distribution of weight in the structure. (B) If the plastic layer is deep, settlement is likely to be even, and the amount of settlement is governed by the total load. This is the situation at the Washington Monument, which has settled about six inches because of compaction of a clay bed about a hundred feet below the base of the monument. In situations where the plastic layer is irregularly distributed, settlement may be irregular, even under a uniform load.

SETTLEMENT OF GROUND

Settlement of ground because of an imposed load may be due to any of several causes, some of which already have been mentioned. Ground may settle because of collapse into cavernous ground, as has happened in those parts of Florida underlain by cavernous limestone and on the Great Plains and other places where there has been underground solution of gypsum and other salts. Mining operations, particularly in the coalfields, have caused settlement of ground; in the Missouri Plateau burning of coal beds in situ (by spontaneous combustion) causes ground to settle.

Ground may settle because of withdrawals of petroleum or of water. Parts of the coast near Los Angeles, California, have settled below sea level and have had to be diked because of settling attributed to withdrawal of petroleum at nearby oil fields. Similar settling of ground has occurred as a result of withdrawals of ground water in the Central Valley of California and in the basin at Las Vegas, Nevada.

Buildings and other structures may settle because of flowage of a plastic layer or because of compaction of the ground by the squeezing of air or water from it. According to Krynine (1947), if the layer that yields is shallow, settlement is irregular and determined by the distribution of weight in the surface structure (Fig. 13.3, A). If the deformed layer that yields is deep, settlement is likely to be even and determined by the gross weight of the surface structure (Fig. 13.3, B). A uniformly distributed load at the surface may settle unevenly if the ground is stable under one part and unstable under another, as at the Leaning Tower of Pisa. Newspaper accounts report that the clock tower containing Big Ben in London has tilted 15 inches from the vertical, and continues to tilt at a rate of one inch every century.

In some situations foundations may be strengthened by using piles. Piles may be driven through unstable layers into underlying stable ones, these being so-called end piles because the load is carried by the ends of the piles. A building on a foundation like that illustrated in Figure 13.3A, could have been supported on piles driven through the plastic layer to the nonplastic one beneath it. On the other hand, equally deep piles in a foundation like that shown in Figure 13.3B, would not have prevented settlement because the piles, the containing sand, and the structure above them would have settled together into the deep plastic layer. A third situation occurs where a structure must be built on a thick plastic layer. Piles driven into such a layer distribute the load by friction between the walls of the piles and the clay, and settlement is minimized.

Still a fourth situation involving thick unstable ground may be met by excavating the foundation to a depth such that the material excavated is about equal to the weight of the structure to be erected. The load represented by the new building is no greater than the former load on the unstable ground and, consequently, there is little or no settlement.

All such relationships affecting ground stability are in turn greatly affected by the water content of the ground, and the access of water to the ground. Additional examples of stable and unstable ground will be cited after considering some of the engineering aspects of ground drainage.

GROUND DRAINAGE, MOISTURE-HOLDING PROPERTIES

In most engineering works, there is the problem of keeping water out of the ground that is being excavated or built upon. The excavation or foundation may extend below the water table or intersect an aquifer in which water is perched on an impermeable stratum. The problems are most difficult where the lands are low; in New Orleans, even the cemetaries are on stilts. Water problems correspond in degree with the permeabilities and usefulness of the surface deposits as sources for ground water.

Some perched water tables can be drained by drilling through the impermeable stratum and letting the water drain into permeable substrata (Fig. 13.4, A). Conversely, an excavation or bore hole can cause flooding should it penetrate an impermeable layer that confines an aquifer under a hydraulic head (Fig. 13.4, B). In homogeneous ground it would be possible to measure the permeability and compute the amount of water that could enter or leave the ground in a given time, but ground rarely is homogeneous, and the inhomogeneities are all important in guiding the movement of water through the ground.

Other moisture-holding properties considered by engineers include: (1) the *flocculation limit*, which is the amount of water below which the material ceases to behave as a liquid and behaves as a solid; (2) the *liquid limit*, the amount of moisture at which the material ceases to behave as a viscous liquid; (3) the *plastic limit*, the moisture content at which the material ceases to act as a plastic solid; and (4) the

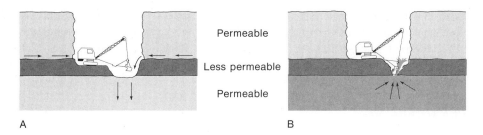

Permeable

Less permeable

Permeable

A   B

FIGURE 13.4
Drainage problems in excavations. Perched ground water can be drained into lower ground (A) if that
ground is permeable and dry. If ground contains confined water under hydraulic head (B), that water
can rise into an excavation if the impermeable confining layer is broken.

shrinkage limit, the moisture content below which further drying produces no
shrinkage. The tests for these properties are empirical. For example, the liquid limit,
is tested by applying 25 measured blows to cause flowage into a groove of specified
size. The plastic limit is tested by alternately rolling a thread of the material and
balling it until the material breaks because of water loss.

These determinations test the amount of moisture in the capillary and subcapillary
interstices, a matter of important engineering significance because the interstices
create a suction (capillarity) that tends to pull additional moisture into that ground.
Consequently, a highway sub-base course that is drier than the subgrade upon which
it rests will draw moisture into itself from the subgrade if the sub-base and subgrade
are texturally alike. Enough moisture may be introduced into the sub-base to weaken
the foundation for the pavement. Total moisture in the two layers is only one of
the variable factors; another depends upon the size and shape of the particles because
the suction force (capillarity) is greatest where the radius of curvature of the meniscus (Fig. 10.10) is least; the smaller openings—that is, those around the smallest particles—have the smaller radius of meniscus and greatest pull. A clayey sub-base can
draw moisture from a drier sandy subgrade. It is this property that justifies the
sprinkling of land fill to aid in compacting it. The capillary force may amount to
hundreds of pounds per square inch, greater than many rollers can exert.

These moisture-holding properties of the ground depend in large part on the
amount and type of clay. For example, ground with 20 percent clay may reach its
plastic limit at 15 percent moisture and its liquid limit at 25 percent moisture;
neighboring ground with 40 percent clay may reach its plastic limit at 25 percent
moisture; and still another part of the ground with 60 percent clay may reach its
plastic limit at 30 percent moisture and its liquid limit at 75 percent moisture. Organic
matter tends to increase the liquid limit.

Of two pieces of ground that have the same liquid limits, the one with the greater
plastic limit is less permeable and stronger. Of two pieces of ground that have equal
plastic limits, the one with the greater liquid limit will be the weakest and most
permeable.

For ground to be stable, its moisture must be controlled. Granular ground becomes
unstable if it is too dry; clayey ground is unstable if it is too wet. Ground may be
waterproofed by adding a layer of bituminous materials $\frac{1}{2}$ to 1 inch thick at the surface.

The moisture content also can be controlled by spraying or mixing hygroscopic substances, especially calcium chloride, into the surface layers of the ground. Some of these varied relationships may be illustrated by referring again to Figures 13.1 and 13.2.

PERMAFROST

Permanently frozen ground, permafrost, poses special kinds of ground problems, some of which were discussed in Chapter 5. The difficulties caused by permafrost relate particularly to inhomogeneities in the ground; these cause the active layer to freeze and/or thaw irregularly because of the way the water and ice are distributed. Water becomes trapped between layers that freeze and that act like a hydraulic press.

Under a heated pipeline or building, the active layer may remain melted while the surrounding ground freezes, and the structures may sink into the wet, unstable ground or become flooded and encased in ice if water is squeezed from the bordering ground that is partly frozen. The ice may form a mound large enough to survive the following summer.

A measure of the rigorous freezing and thawing that occurs every year in ground with permafrost is provided by the extensive, well-developed patterned ground that characterizes that kind of country. The modifications of the freezing and thawing process caused by man-made structures are not easy to anticipate.

SEWAGE DISPOSAL, POLLUTION

Some kinds of ground are suitable for cesspools or septic tanks, and others are not. Suitability is largely a function of ground permeability, as measured by rates of infiltration. Ground with low permeability and a low rate of infiltration drains too slowly to be suitable. On the other hand, permeability can be excessive and cause pollution. Moreover, a given acreage may be suitable for a few cesspools or septic tanks and not be capable of accomodating many.

Disposing of sewage and avoiding pollution involves not only the ground permeability but also the courses of ground water movement. Long Island's history of groundwater withdrawals and pollution by sewage and salt water are examples. Long Island consists of glacial drift overlying sand and shale formations of the Coastal Plain, which dip southeastward into the Atlantic Ocean. The early homes and farms were widely spaced, and each could have its well and cesspool dug in the glacial drift. As the population increased, groundwater in the drift became polluted, and water wells then were drilled into the underlying formations. As the cities grew, withdrawals from the drilled wells became so great that a new form of pollution developed: salt water began encroaching into the sands that had formerly contained fresh water. Some wells had to be abandoned. The recharge into the ground became less than the withdrawals, partly because natural recharge was reduced by runoff from the system of streets and buildings. As the fresh water supply became threatened by the encroachment of salt water, steps were taken to reduce withdrawals by requiring that water used for cooling systems be recycled. In brief, the original water supply from the drift had to be abandoned because of pollution by sewage, and the second supply is threatened by contamination with sea water because of the heavy withdrawals and because streets and buildings have interfered with the recharge. As our population grows and urban areas spread, other communities can expect similar successions of crises.

OTHER GROUND PROBLEMS— EROSION, PLACERS, EARTHQUAKES

Other engineering and economic geology problems include the measures needed to stabilize drifting sand, protect shores and river banks against erosion, protect water supplies, prevent floods, avoid pollution in rivers, in estuaries, and on beaches, and to facilitate recovery of placer deposits.

Drifting sand, like drifting snow, can be controlled by lath-type fences; the drift sand accumulates on both lee and windward sides. Dunes, especially coastal ones, can also be controlled by planting; along highway right of ways they may be stabilized with oil or by extending the pavement.

Protection of shores and river banks against erosion commonly require one of three kinds of protective measures. Walls built along riverbanks (including the modern kind built of abandoned automobiles) provide protection against bank erosion; similar walls along beaches built parallel to the shore and well back of high tide protect the land against storm waves. Massive structures are known as sea walls; lighter ones are bulkheads or revetments. Most commonly, along beaches, groynes or jetties built at a high angle to the shore retard the erosion by the shore drift. Wave action may be reduced by breakwaters.

Shore erosion along the Atlantic and Gulf coasts is especially severe during hurricanes. The most damaging are those that occur at high tide. Long-range planning should also take into account that the level of the sea is rising; in the next hundred years at places on the Atlantic coast we can expect storm waves to reach 6 inches higher than they do now. More than a few communities will be threatened. Moreover, we have no reason to suppose that the hurricanes we have experienced have been of maximum severity. Shores that are so exposed should be used for public recreation and not for expensive buildings.

Some beaches contain valuable mineral deposits, placer deposits of heavy minerals or metals concentrated by the waves washing away the lighter mineral fraction. At Nome, Alaska, a few million dollars worth of gold has been produced from a strip about 200 feet wide along the beach (Fig. 13.5). Gold placers also occur along the California and Oregon coasts; the Oregon beaches also have chromite placers. Some parts of Florida beaches have placers of heavy minerals containing titanium.

Alluvial ground is a more important source of placer deposits than are the beaches. The requirements for a placer deposit, whether along a beach or stream, are simply that the metal or mineral be hard or tough in order to withstand abrasion during transport, that it be inert in order to resist decomposition, and that it be dense enough to be concentrated while the water washes away lighter minerals. Gold is the principal metal produced from placers, but a large part of the world's platinum and tin are

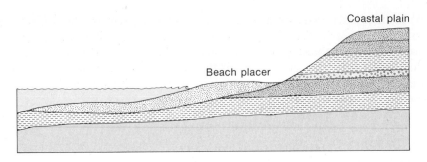

FIGURE 13.5
Beach placer at Nome, Alaska. [After U.S.G.S.]

also so obtained. In Africa, titanium, some heavy rare-earth minerals (monazite, columbite), and diamonds are obtained from both beach and alluvial placers.

The metals or minerals in a placer are generally derived from nearby sources. They are freed from the parent rock by weathering, washed to a valley bottom, and concentrated in streams, where the lighter minerals are washed away. The coarseness of gold particles is measured by the number required to be worth one cent. Particles averaging fewer than 100 to a cent are coarse. In some placers the particles are very fine, more than a thousand to a cent; these are referred to as flour gold and are very difficult to recover; the surface tension of the water suffices to float the heavy metal. Much of the gold along Glen Canyon of the Colorado River is flour gold; a prospector who owned a bank of gold bearing gravel there once said, "I have a million dollars in gold in my bank; the only trouble is there's just too much sand and gravel mixed with it". His million dollars in gold would cost twice that to extract.

Of considerable importance in some regions is the way in which earthquake waves affect the ground, as that term has been used in this book. The problem can be summarized by analogy. Shot resting on a table top can be made to jump by tapping the end of the table with a hammer. The tap sends a wave, like an earthquake wave, through the table top; the wave imparts a shock to the loose shot causing them to jump far more than the amplitude of the wave. Similarly, surface deposits jump violently when earthquake waves pass through the underlying bedrock.

Much of the damage caused by the Anchorage, Alaska, earthquake of 1964 was due to the violent shaking of the unconsolidated deposits that form the ground there. A disaster may be in the making at San Francisco, where apartments and other large buildings have been erected on unconsolidated fill at the edge of the bay. The view into the bay is magnificent, but a strong earthquake can cause such ground to shake violently, to settle, and even to slide. In areas subject to earthquakes, buildings must be designed to withstand the shaking, but there are some kinds of ground where buildings should not be allowed, no matter how sturdily they are constructed.

In earthquake-prone areas not all earth movement is violent. Along some faults the movement may occur slowly; one side of the fault might move horizontally past the other at rates as great as several tenths of an inch per year. Buildings, pipelines, highways, railroad tracks, water and sewer lines, dams or other structures that extend across such active faults require constant repair. It may not always be possible to avoid crossing active faults, but at least the maintenance problem should be recognized beforehand.

Not just because of earthquakes and earth creep, but for many reasons implicit in much of this book, city planning and other land development need to give adequate consideration to the ground around us.

TESTING

Testing ground for its materials, stability, or drainage characteristics can be a considerable task. There is little difficulty obtaining samples, and these can be tested in a laboratory, but the tests, however refined, are useful only to the degree that the ground is homogeneous. It is technically possible but generally not economically feasible to test drill sufficiently to learn all the variation in occurrence and extent of the different kinds of materials in heterogenous surface deposits. In some kinds of ground—for example, bouldery till—even drilling may not be satisfactory for determining depth to bedrock because the drill hole may drive into buried boulders. In such ground, seismic and electrical conductivity methods have been used to determine the depth of the deposit and drill holes later used to check the depth at critical locations. Knowledge of the geology, though, including knowledge of the geologic history, can help minimize unnecessary drilling and maximize the information obtained from holes that are drilled. In seismically active areas, each fault and all of its branches should be investigated and the history determined. At least the ground that poses the greatest hazards can be pinpointed.

BIBLIOGRAPHY

American Association of State Highway Officials, 1961, Standard specifications for highway materials and testing (8th ed.): The Association, 2 v., 401 and 617 pp.

Belcher, D. J., Gregg, L. E., and Woods, K. B., 1943, The formation, distribution, and engineering characteristics of soils: Purdue Univ. Res. Ser. no. 87, Highway Res. Bull. 10, 389 pp.

Bollen, R. E., 1945, Characteristics and uses of loess in highway construction: Am. Jour. Sci., v. 243, no. 5, pp. 283-293.

Capper, P. L., and Cassie, W. F., 1953, The mechanics of engineering soils (2nd ed.): McGraw-Hill, New York.

Casagrande, A., 1948, Classification and identification of soils: Trans. Am. Soc. C.E., v. 113.

Davidson, D. T., 1949, Large organic cations as soil stabilizing agents: Bull. 168, Iowa Eng. Exper. Sta., Ames, Iowa.

Edminster, T. W., and Reeve, R. C., 1957, Drainage problems and methods: *in* Soil, U.S. Dept. Agri. Yearbook, 1957, pp. 379-385.

Gilluly, J., and Grant, U.S., 1949, Subsidence in the Long Beach Harbor area, California: Geol. Soc. America Bull., v. 60, no. 3, pp. 461-529.

Guy, H. P., 1965, Residential construction and sedimentation in Kensington, Md., *in* Proceedings of the Federal Interagency Sedimentation Conference 1963: Agric. Res. Serv. Misc. Publ. 970, pp. 30-37.

Highway Research Board, 1950, Soil exploration and mapping: Bull. 28.

Hough, B. K., 1957, Basic soils engineering: Ronald Press, New York.

Jacoby, H. S., and Davis, R. P., 1941, Foundations of bridges and buildings: McGraw-Hill, New York.

Johnson, A. L., and Davison, D. T., 1947, Clay technology and its application to soil stabilization: Proc. Highway Research Board, v. 27.

Kohnke, H., 1968, Soil physics: McGraw-Hill, New York, 224 pp.

Krynine, D. P., 1947, Soil mechanics: McGraw-Hill, New York, 511 pp.

Lamar, G. E., Grim, R. E., Grogan, R. M., 1938, Gumbotil as a potential source of rotary drilling mud, bonding clay, and bleaching clay: Ill. Geol. Survey Inf. Circ. 39, pp. 1-23.

Leggett, R. F., 1938, Geology and engineering: McGraw-Hill, New York.

Peck, R. B., Hanson, W. E., and Thornburn, T. H., 1953, Foundation engineering: Wiley, New York, 410 pp.

Public Roads Adm. Staff, 1943, Principles of highway construction as applied to airports, flight strips and other landing areas for aircraft: Federal Works Agency, 514 pp.

Rhoades, R., and Mielenz, R. C., 1948, Petrographic and mineralogic characteristics of aggregates: Am. Soc. Test. Materials Spec. Tech. Publ. 83, pp. 20-48.

Schultz, J. R., and Cleaves, A. B., 1955, Geology in engineering: Wiley, New York, 592 pp.

Spangler, M. G., 1951, Soil engineering: International Textbook Co., Scranton, Pa., 458 pp.

Stoesz, A. D., and Brown, R. L., 1957, Stabilizing sand dunes: *in* Soil, U.S. Dept. Agriculture Yearbook, 1957, pp. 321-326.

Stokstad, O. L., and Humbert, R. P., 1949, Interpretive soil classification: engineering properties: Soil Sci., v. 67, no. 2, pp. 159-168.

Sweet, H. S., 1948, Chert as a deleterious constituent in Indiana aggregates: Proc. Highway Res. Board, v. 20, p. 599.

Taylor, D. W., 1948, Fundamentals of soil mechanics: Wiley, New York, 700 pp.

Terzaghi, K., and Peck, R. B., 1967, Soil mechanics in engineering practice: Wiley, New York, 729 pp.

Trefethen, J. M., 1959, Geology for engineers: Van Nostrand, Princeton, 632 pp.

Trask, P. D., and Kiersch, C. A., 1957-1969, Engineering geology case histories: Geol. Soc. America, 7 vols., 519 pp.

Waterways Experiment Station, Corps of Engineers, 1953, The unified soil classification system: Technical Memo 3-357, 48 p.

Wu, Tien Hsing, 1966, Soil mechanics: Allyn and Bacon, Boston, 429 pp.

Yong, R. N., and Warkentin, B. P., 1966, Introduction to soil behavior: Macmillan, New York, 451 pp.

Checklist

A SUMMARY OF IMPORTANT PROPERTIES
OF SURFACE DEPOSITS AND SOILS

SURFACE DEPOSITS

A. *General three-dimensional form of the unit.* Kind of deposit, e.g., dune, alluvial, morainal, soil; dimensions, including variations in thickness and extent; lenticular, persistent; relation to overlying and underlying formations

B. *Color.* Color of unit as a whole, wet or dry; pattern of colors; color of beds, layers, or spots; color of particles

C. *Bedding or layering*
   1. How manifested: sharp, by partings, by differences in texture, color, or original structure; transitional; shaly
   2. Shape of the surfaces of the beds or layers: planar, undulating, ripple marked; if irregular, form and dimensions of the irregularities
   3. Thickness of beds or layers: comparative thicknesses; different orders of thickness and their relationships, e.g., rhythmic, random; relation between thickness and composition
   4. Attitude and direction of bedding or layering surfaces: horizontal, inclined, curved; relation to each other, parallel, intersecting, tangential; angles between them; relation of composition to different kinds of beds or layers
   5. Markings on bedding surfaces: mud cracks, rain prints, bubble impressions, ice or salt crystal impressions; trails, tracks
   6. Disturbances of bedding or layers: edgewise or intraformational conglomerates, folding or crumpling of individual beds or layers

D. *Composition*
  1. Inorganic constituents
     a. Mineralogy or lithology of principal constituents; approximate percentages; vertical and lateral variations
     b. Size: prevailing size if fairly uniform; range in sizes if not; proportions of sizes; distribution of sizes with relation to other features; vertical and lateral variations in sizes
     c. Shape: angular, subangular, subround, round, crystal; relation of shape to size, material, position in bed or layer
     d. Surfaces: glossy, smooth, mat, pitted, chatter marked
     e. Orientation: direction of greater dimensions with respect to contacts and to each other
     f. Chemical and physical condition: fresh, weathered, decomposed, cracked
     g. Packing: closeness, manner
     h. Pore space
     i. Cement: presence or absence; proportion; composition; variations in composition vertically and laterally and in relation to other features; disposition with respect to bedding, fractures
     j. Color: wet or dry; inherent or a stain on constituents or in the cement; variations and their relations to composition, porosity, bedding, fracturing, fossils
  2. Organic constituents
     a. Kinds: relative abundance
     b. Size: distribution by sizes and possible effects of transport by running water at time of deposition, or by vadose water subsequently
     c. Condition: entire, fragmented, wholly or partly dissolved; alteration and replacement
     d. Distribution: with respect to kinds of beds or layers, kinds of organisms; evidence of burrowing
     e. Orientation: with respect to bedding, life habits, possible manner of death, e.g. life associations versus death associations

E. *Concretions*
  1. Form, size, color, composition and their variations
  2. Internal structure: central nucleus, organic or inorganic; central hollow; homogeneous; banded horizontally, concentrically; radial, compact, vesicular
  3. Contact with surrounding deposit; sharp, transitional
  4. Relation to bedding: bedding extends through, deflected above or below or both; thinned above, below
  5. Distribution: random, regular; if regular, intervals between groups or layers vertically and horizontally; differences between characteristics of concretions in different groups; relation of distribution to bedding, layering, and/or variations in physical and chemical composition of the formation, including internal structure; relation to topography and to present or past water tables

F. *Hardness and structure:* firmness, friability, jointing, case hardening

G. *Landform produced:* cuesta, hogback, valley, rounded hill, fan, pediment, cliff, talus; amount of slope and its regularity; relation to topographic unconformities

H. *Stream development:* stream pattern; spacing of drainage lines; channel conditions, straight, meandering, anastomosing, distributary; water falls

I. *Weathering effects on underlying bedrock*
   1. Kind of weathering and weathering products; physical disintegration versus chemical decay; frost action; solution
   2. Depth of weathering and relationship to the bedrock profile
   3. Changes wrought by weathering on the various rock constituents and rock properties; alteration of mineral particles; effect on hardness; redistribution of soluble constituents
   4. Description of weathering products; color, texture, structure, organic content, variations with depth and composition of parent rock

J. *Fossils, archeological remains*

SOILS

A. *Physiographic setting of the locality*
   1. Topography: landform, e.g., mountain top, hillside, valley bottom, plain; altitude; relief; surrounding terrain; accessibility
   2. Slope of ground surface: nearly flat, gently sloping, steep; runoff features, e.g., sheds water or collects water from neighboring terrain; susceptibility to erosion
   3. Microtopography: terracettes, patterned ground, tussock mounds
   4. Climate, vegetation, land-use

B. *General features of the ground*
   1. Parent material: kind and age of the deposit on which the soil has developed; its mineralogy, texture, and structure, such as bedding
   2. Stoniness, looseness, compactness, or other special textural feature
   3. Acidity, salts, alkali; e.g., carbonate layers, content of iron sulfide, gypsum
   4. Drainage, moisture conditions; depth to water table; runoff or infiltration of surface water; layering or bedding that affects infiltration
   5. Recent erosion or sedimentation on the surface
   6. Human artifacts in the ground; fossils

C. *Profile of the modern soil*
   1. General aspect: depth, layering, outstanding features
   2. The individual layers
      a. Thickness; depth of top and bottom; range in thickness
      b. Color, moist and dry; overall color, color of individual particles; mottling, abundance and size, sharpness of boundaries, contrast between mottles
      c. Texture: proportion of different grain sizes; shapes of the larger grains; aggregates of mineral particles; stoniness
      d. Structure and fabric: strongly or weakly developed; form and shape; sizes

e. Consistence: when dry—loose, soft (can be crumbled), hard; when moist—loose, friable, firm; when wet—stickiness, plasticity

f. Coatings; pressure faces, slickensides: location with reference to mineral grains, surfaces of mineral aggregates, structural faces; continuity or patchiness; thickness of coatings; composition, whether of clay, iron or manganese oxides and hydroxides, organic matter, soluble salts, silica

g. Cementation, pans: uncemented pans representing compact layers that may be hard when dry such as clay pans, fragipans; pans cemented by silica, iron oxides, carbonates; continuity of the pan and its structure, massive, vesicular, nodular, platey

h. Permeability

i. Acidity, salinity, alkalinity; accumulation of salts or alkalis

j. Biotic influences, roots, burrows; amount of intrusive organic matter and signs of resulting mixing

Annotated Index
and Glossary

Terms for which page references are given are discussed and explained in the text. Terms listed without page references are briefly explained without being fully defined, with the hope that the simplified explanations will facilitate and thereby encourage more reading in the several related fields.

a-axis: direction of movement or transport, as down the slope of cross beds

A horizon (upper layer of mineral soil in which leaching is greatest), 167. *See also* leached layer, Fig. 8.2

ablation zone: on a glacier, the zone below the snowline, where ice is reduced by melting and meltwater runoff

a-b plane: plane of the *a* and *b* axes—for example, the plane of a stratum

absolute age: an age given in years before the present, as determined by radiometric measurements, dendrochronology, varves, or other methods

absolute zero: −273°C or −461°F

accelerated erosion: erosion faster than normal erosion, as at a construction site where ground has been scraped bare of vegetation

accessory minerals: uncommon minerals in a surface deposit or soil that are not peculiarly characteristic of the deposit or soil

accretion gley: (gleyed ground on which sediment is deposited and added to the gleyed layer), 170

accumulator plants: (plants that accumulate particular chemical elements), 120

acetate: a salt of acetic acid (CH₃COOH); solubility, 259

Acheulian Culture (prehistoric culture in Europe and Mediterranean lands), Table 2.4

acid igneous rock: one that contains more than about 65 percent SiO₂

acid soil: one whose *p*H is less than 6.5; causes, 43, 98, 167, 170, 184, 215; degree of acidity, 255–256; discontinuities suggest unconformity, 214; distribution of acid soils, 97, 116, 182, 185–189, 191, 193, 198, 207; Figs. 6.10, 8.2, 8.7—8.10, 8.12; Table 8.1; favors leaching, 289; kinds of dissolved acids, 39–40; *See also* clay

acidophilous: plants that favor acid ground

acre foot: amount of water necessary to cover an acre to a depth of 1 foot, i.e., 43,560 cubic feet

actinomycetes (microorganism intermediate between bacteria and fungi), 103, 104; Fig. 6.1

active layer (layer above permafrost and subject to annual thawing), 244; Fig. 10.15; on badland slopes: layer of finely divided shale fragments subject to downhill creep

adfreezing strength: strength of adherence of frozen ground to objects

Adirondack Mountains, 58, 63, 64, 128; Fig. 4.2

adit: horizontal or gently inclined entry into a hillside; unlike a tunnel it does not go through the hill

Admiralty drift (early Wisconsinan drift in the Pacific Northwest), Table 7.1

adobe: clayey material with which sun-dried and sun-baked bricks are made

adventitious: plant or animal that moves into a new environment

aeolian. *See* eolian

Aeolian Buttes till (pre-Wisconsinan till at east foot of the Sierra Nevada), 150; Table 7.2

aerated, aerobic (ground condition with oxygen present), 43, 86, 104, 107, 233, 280; Fig. 6.3

aerosol (matter suspended in atmospheric vapor), 160, 257

aforestation: establishing forest where there had been none. *Cf.* reforestation

Aftonian (interglaciation following the Nebraskan glaciation), 31; Fig. 9.8; Tables 2.2, 2.5, 7.3

agate: banded chalcedony

age, of soils, 182, 187, 189, 199, 214; Table 8.1

agglomerate: volcanic breccia; may occur as a flow or as a fill in a vent aggraded: ground where surface deposits have accumulated

aggregate; construction material: coarse inert fraction added to concrete, of soil (mass or cluster of soil particles stuck together), 267, 276 bitumen, or soil mixes

agric (depositional soil layer—B horizon—of clay and humus formed by cultivation), 180

agriculture, crops, farming, 49, 98, 117, 120, 141, 165, 233–235, 295, 296; Fig. 5.5

agronomy: pertains to farming

air, 85, 86, 275; composition, 257, 259

albedo (lightness or darkness of ground), 91; Fig. 5.5

albic (sandy or silty soil layer having little clay or oxides coating the particles), 180

albite (sodium plagioclase feldspar), 265; Tables 11.2, 11.3

alcove, Fig. 3.10

alder, Fig. 2.4

alfisol, 181

algae (simplest form of green plants), 103, 104, 117, 148, 160, 216

alkali: elements of the first group in the periodic table forming monovalent ions—lithium, sodium, potassium, rubidium, cesium; in pre-Wisconsinan soils, 153, 260, 271; readily soluble, 263; soils, 167, 174, 176, 185, 188, 191–193; in surface deposits, 40; tolerance of plants, 117

alkaline: opposite of acid; *p*H above 7.0. *Cf.* saline; compared to acidity, 255–256; effect on plants, Figs. 6.8, 6.10; pre-Wisconsinan soils, 43, 97; versus acid soils, 116, 169, 184, 185, 188, 193, 208; water, 168

alkaline earth (elements of the second group in the periodic table forming divalent ions—beryllium, magnesium, calcium, strontium, and barium), 40, 153, 167, 175, 185, 187, 191, 192, 193, 260, 263, 289; in pre-Wisconsinan soils, 271

allophane (amorphous clay mineral), 269

alluvial, alluvium (sediment deposited by a stream, on a floodplain), 76, 112, 128, 130, 141–144, 156, 160, 187–189, 191, 208, 215, 216; Figs. 4.9, 4.19, 4.20, 5.11, 6.4, 6.12, 7.1, 7.10, 9.3; alluvial fans, 3, 43, 47; Figs. 1.2, 6.3, 6.6, 8.4, 8.12, 9.5, 9.8, 9.12; Tables 7.5, 9.1

alpine: mountain summit above timberline

Alpine Meadow Soil (dark-colored soils of usually wet meadows at high altitudes), 188, Table. 8.2

altalf: degraded (leached) Chernozem Soil

altithermal (early part of the Holocene), 144

alumina (Al$_2$O$_3$, oxide of the element aluminum), 152, 153, 155, 167, 185, 191, 192, 254

ammonia (NH$_3$, produced by decomposition of organic matter in soil), 280, 290

ammonification: process by which ammonia forms from nitrogenous compounds

amorphous: noncrystalline

amphibole (group of dark aluminum silicate minerals containing iron, magnesium, and/or calcium), 109, Table 11.2

amphibolite: metamorphic rock composed largely of amphibole

anaerobic (condition in which oxygen is deficient), 43, 104, 230

anatase (mineral composed of titanium oxide), 272, Table 11.2

anchor ice: ice formed at the bottom of stream channels

Ancylus Lake (fresh-water lake stage in Holocene history of the Baltic Sea; named after a gastropod), 22

andalusite: aluminum silicate mineral

Andept, 181

andesine (soda-lime plagioclase feldspar), Tables 11.2, 11.3

andesite (igneous rock composed of soda-lime feldspar with biotite, augite, or hornblende), Table 11.3

angle of repose: maximum angle of slope at which loose material remains at rest, generally around 30 to 25 degrees

angularity, of particles in surface deposits and soils, 227; Figs. 10.2, 10.4

anhedral: crystal in igneous rock that developed in a confined space and failed to develop its own crystal faces

anhydrite (anhydrous calcium sulfate), Table 11.2

animal, effects on soil development, 170, 175; Fig. 8.2

anion: negative charged ion. *See* ion exchange

anorthoclase (lime plagioclase feldspar), Tables 11.2, 11.3

ants, ant hills, 109; Fig. 6.3. *See also* termite

antecedent: stream that has maintained its course by eroding into a mountainous barrier uplifted across that course

anthropic (surface soil layers modified by farming that has involved addition of organic matter), 180

anticline (upfold of rock formations; the formations dip away from the axis of the fold), 57; Figs. 4.1, 4.15, 4.16, 4.18

anticyclone: mass of high pressure air rotating clockwise

apatite (phosphatic mineral of calcium with fluorine or chlorine), 265, 292; Table 11.2

aphanitic: rock or soil texture too fine-grained to be visible except by microscopy; grain sizes less than about 0.01 mm

Appalachian Highlands, Plateau, 63, 71, 107, 110, 111, 145, 146, 153, 203; Figs. 4.3, 4.13, 6.5

Aquafalfa: ground with high water table

Aquent, 181

Aquept, 181

Aquert, 181

aquiclude: impermeable bed that perches ground water

aquifer (permeable water bearing stratum), 168

arachnid (mite), 103

archeology, usefulness for dating surface deposits and soils, 25, 30, 45, 99, 130, 136, 141, 143, 144, 145, 149, 156, 160, 204, 206, 208; Figs. 2.6, 3.10, 5.11, 7.9, 7.19, 7.20, 9.1, 9.7; Tables 2.4, 7.4

Archisdiskodon meridionalis (mastodon regarded as diagnostic of earliest Pleistocene), 19

Arctic soils, vegetation, 42, 145, 175, 182, 185; Figs. 6.3, 8.2; Table 8.1

arenaceous, arenite: sandy, sandstone

arête: sharp crested ridge dividing two cirques

argillaceous, argillite: clayey, clay stone

argillic alteration (alteration to clay minerals), 198, 199, 276

argon, 86

Aridosol, 181

arkosic: sand or sandstone containing much feldspar

arroyo, in Southwest (steep-sided dry valley usually inset in alluvium), 47, 144; Figs. 3.8, 7.10

arrowweed, (*Pluchea sericea;* desert plant that accumulates sulfates), Fig. 6.11

artesian water: ground water under hydrostatic pressure

ash, volcanic (explosively erupted, fine-grained fragmental material ranging in size from 0.25 to 4.0 mm in diameter but generally includes some coarser fragments), 36, 40, 45, 49, 128, 138, 148, 149, 150, 273; Figs. 3.1, 7.1; Tables 7.3, 10.1

association plant: plant stand dominated by characteristic species; soil: one or more types that occur together in a characteristic pattern in a given area, providing a repetitive appearance to the ground

Astian (Pliocene formation in Italy), Fig. 2.3

Atchison Formation, Table 7.3

atmosphere, 1, 259; Table 1.1

Atterberg scale (scale of grain sizes in soils), Table 10.1

aufeis: surface ice formed by freezing of successive sheets of water discharged by seeps or springs

auger: instrument for drilling and collecting samples of soils and surface deposits

augite (most common of the pyroxene minerals), 263, 266; Figs. 3.1, 11.3; Tables 11.2, 11.3

Aurignacian Culture (late Pleistocene culture in Europe), Table 2.4

authigenic: mineral grain that formed in the surface deposit or soil that contains it

autotrophic (microorganism capable of deriving carbon from carbon dioxide in soil air or soil moisture) 104. *Cf.* heterotroph

avalanche (catastrophic slides of snow or of debris), 146, 147

Azilian Culture (early Holocene culture in Europe), Table 2.4

avulsion: sudden change of stream course

Azonal Soil (those without well-developed profiles), 173, 174, 175, 181

B horizon (depositional layer in a soil profile; underlies the leached layer), 167; Fig. 8.2

backfill: refilling of an excavation

back slope: side of a hogback or cuesta that slopes in the direction of the dip of the beds

bacteria, soil, 103, 104, 117, 281; Fig. 6.1

badlands (arid or semiarid landform developed on easily eroded formations, especially shale), 47, 56, 74

bajada (Southwestern United States): coalescing gravel fans sloping from mountains in the Basin and Range Province

bald rock surfaces and exfoliation, 42

ballast: unscreened sand and/or gravel

balley: mountain; local in northern California

Baltic Sea, Quaternary history, 22

Baltimore area, saprolite, 203, 215; Figs. 9.3, 9.5

banco: cut-off meander; local in Southwest

bank storage: ground water from flood stage stored in alluvial banks of a floodplain and returned to the stream as the flood ebbs

bar: unit of atmospheric pressure, about 29.5 inches of mercury; bayhead bar: bar built around the head of a bay; baymouth bar: bar crossing or partly crossing the mouth of a bay; beach bar: bar built along a shore; point bar: stream bar built along the inside of a river bend; stream bar: ridge of gravel and sand along a stream

barbed drainage: drainage system in which tributaries, flowing in directions opposite that of the main stream, join it with the acute angle on the downstream side of the junction

barchan (crescent-shaped sand dune concave toward leeward side), 139

barranca, Southwest: gully, gorge, or canyon having rock side walls

barrier beach (sand beach separated from the mainland by an estuary or lagoon), 116, 148; Figs. 4.5, 4.6, 6.3, 6.10

barrier island: segment of a barrier beach isolated from the rest by an inlet

basal area, of tree: area of cross section of a tree at breast height (4½ feet); used for estimating lumber as total basal area per acre

basal conglomerate: conglomerate at base of a stratigraphic section

basalt (dark, fine-grained igneous rock occurring as lava or small intrusion), 40, 151, 180, 193; composition, Tables 11.1, 11.3

base in chemistry: compound containing the hydroxyl (OH⁻) ion; in engineering: that part of the structure resting on the ground or on the sub-base

base course: first or lowest layer supporting a pavement or other structure

base exchange. *See* ion exchange

base level: lower limit to which a stream can erode its channel; along coasts, base level is sea level; tributary streams have base level determined by position of the master stream; local base level may also be determined by a resistant ledge of rock

basic rock: an igneous rock that contain less than about 55 percent SiO_2

Basin and Range Province, 58, 76, 135, 138, 144, 148, 156, 174, 188; Figs. 4.3, 7.19

basin listing: method of tillage that creates small basins by damming furrows at regular intervals to increase moisture penetration

basophilous: alkaline tolerant plant

bauxite (hydrous aluminum oxide, common in ancient residual deposits in southerly latitudes, 153, 155, 203; Figs. 7.16, 7.17

b-axis: axis at right angles to the *a*-axis and in the plane of the bedding or other foliation

bayou, Gulf Coast: inlet with sluggish drainage

BC soil: soil without an A or leached layer; presumably removed by erosion

beach, 40, 114, 128, 148; Figs. 6.9, 6.10, 7.1

beachrock: beach sand cemented with calcium carbonate, in tropics and subtropics

beaded drainage: series of small pools connected by short water courses, common in areas having permafrost

bed, bedding (layers and layering of sedimentary rocks and surface deposits), 41, 232, 319; Fig. 3.10

bed load: sediment moved by rolling along the bottom of a stream

bedrock (solid rock underlying surface deposits), 36, 51, 55, 58, 85, 125, 128, 129, 136, 146, 167, 215, 216, 263; Figs. 1.3, 6.4, 6.5, 6.6, 7.3, 7.15, 9.3; affects water in overlying ground, 238, Figs. 10.11, 10.12; for foundation, 307

beech, tree, 110, Fig. 5.4

behead: capture of one stream by another

bench gravel: terrace gravel along the side of a stream

bentonite: volcanic ash altered to clay

bergschrund: crevasse between glacial ice and head wall of a cirque, opens during warm season

berm, along a beach: a bench eroded by storm waves or high tide

bicarbonate: acid carbonate

bight: rounded bank along the shore of a river, bay, or coast

Bignell Loess (late Wisconsinan loess deposit), Table 7.3

binder (material used for stabilizing ground), 308

Binghamton Drift (early Wisconsinan drift in southern New York), Table 7.1

biofacies: fauna or flora characteristic of a particular environment. *Cf.* facies

biogeography, biome: total fauna and flora in a large natural region

biological oxygen demand (BOD): nutrients introduced into a body of water increase the organic activity and need for oxygen (BOD), which becomes depleted

biostrome: an organic layer or bed

biota (fauna and flora), Chapter 6. *See also* ecosystem

biotite (dark mica), 263, 265, 277; Tables 11.2, 11.3

biotope: small scale plant and animal community, as a prairie dog colony

Bishop Tuff (Pleistocene volcanic ash deposit at east foot of Sierra Nevada), 150

crystal tuff: volcanic ash containing high percentage of mineral crystals

cuesta (landform having escarpment facing up the dip of the formation and long slope—*see* backslope—in the direction of the dip; formed by outcrop of gently dipping resistant formation), 60, 63, 70; Figs. 4.4, 4.7, 4.13, 4.15

cutan (film or coating on walls of an opening in the ground), 232

cyclone: mass of low pressure air rotating counterclockwise. *Cf.* anticyclone

D horizon: unweathered parent material of a soil

dacite: andesite containing quartz

Danger Cave (in Great Salt Lake desert), 136; Fig. 7.6

Darcy's law: equation concerning rate of flow of water through the ground, notably sandy ground; quantity passing depends on the kind of sand, area of filtration, and hydraulic head

Dark Gray Gleysolic Soil: poorly drained forest soil grading laterally to Podzolic Soil; Canada

David City Sand and Gravel, Table 7.3

dealkalinization: removal of alkali from ground by leaching

Death Valley, 57, 85, 135, 137, 272; Figs. 4.19, 7.7

decalcification: removal of calcium carbonate from ground by leaching

deferred grazing: method of controlling grazing by waiting until growth has reached a satisfactory height; contrasted with rotation grazing, continuous grazing

deficiencies, need for fertilizer, 289–294

deflation (wind erosion, usually of a depression), 49

deflocculate: breaking down clay aggregates

deglaciation: melting back of glaciers uncovering the ground

degradation: change of an alkaline soil (Solonetz) to a Soloti; change of a soil to one that is more highly leached (e.g., a Degraded Chernozem becomes a podzolized Chernozem where forest advances onto grassland), 175, 187; lowering by stream erosion; permafrost that has thawed because of engineering structures or shifts in drainage

delta, 65, 135, 137, 141, 214; Figs. 4.9, 7.1

dendritic drainage (drainage system with branches giving a pattern resembling tree growth), 58; Fig. 4.2

dendrochronology (determining age by tree rings), 29; Fig. 2.7

denitrification: reduction (by soil organisms) of nitrates, nitrites, or ammonia to free nitrogen

density (weight of material expressed as a multiple of the weight of an equal volume of water; dry density is weight after drying at 105°C; Table 13.1; modern soil, 234, 309; peat, 161, 234; saprolite, 152

Denver, Colorado, soils, 208–212; Figs. 9.1, 9.10; Table 9.1

Depression Podzol: modified Podzolic Soil in a depression, as result of poor drainage; Canada

desalination: removal of salts from ground by leaching

Desert holly (*Atriplex hymenelytra*, most xeric shrubs in the Southwest), 238; Figs. 6.11, 6.12

desert pavement (closely set pebbles and cobbles on gravelly surfaces in deserts), 160, 230; Figs. 6.12, 6.13, 10.6

Desert Soil, 175, 182, 188, 287; Figs. 8.7, 8.10, 8.11; Table 8.1. *See also* Polar Desert

desert varnish (stain of iron and manganese oxide on rock surfaces), 45, 46, 133, 158, 159, 198; Figs. 7.12, 7.20, 8.7, 8.10, 8.11, 9.2

desilication: removal of silica from the ground by leaching

detrital: mineral grains transported as sediment and deposited in the surface deposit in which they are contained. *Cf.* authigenic

deuterium: isotope of hydrogen having atomic weight of 2 (*Cf.* tritium); deuterium oxide (D_2O) is heavy water

devitrification: weathering of volcanic glass to crystalline substances

Devonian (fourth period of the Paleozoic Era), Table 2.1

dew, 93, 95, 104, 137, 160

diabase (like basalt but more completely crystallized), 63, Fig. 4.7

diagenesis: compaction and solidification of sediment to form rock, e.g., mud becomes shale, sand becomes sandstone

diamond, 265

diaspore (hydrous aluminum oxide), 272; Table 11.2

diastem: minor unconformity in a succession of sedimentary beds

diatom: microscopic plant related to algae that secretes silica in a skeleton

diatomaceous earth: soft, porous deposits containing high percentage of diatoms

differential compaction: variation in compaction of ground because of differences, usually in texture and water content

diluvium: flood deposits; originally applied to glacial deposits and attributed to the great flood of Noah

dimension stone: building stone suitable for shaping blocks

diorite (intrusive igneous rock with intermediate silica content), Table 11.3

dip (angle of slope of a stratum), 307

discharge, of streams: volume of water that flows in a certain period of time; commonly expressed as cubic feet per second, gallons per minute, or acre feet per year

discordant bedding: lack of parallelism between strata due to irregular deposition

dismembered stream: coastal streams originally in one system but now isolated by flooding because of rise of sea level; e.g., valleys draining east and west into Chesapeake Bay

distributary: opposite of tributary; branches separate downstream, as on Mississippi River delta

doelterite (leucoxene; hydrous titanate), 272; Table 11.2

doline: depression in limestone; a sink

dolomite (related to limestone but contains calcium magnesium carbonate; gradational to limestone), 38, 267; Table 11.2

Donau (earliest Pleistocene glaciation in Europe), Tables 2.3, 2.4

drainage: engineering problems, 312–314; Fig. 3.14; ground, 43, 167, 174, 185, 187, 321; Figs. 6.6, 8.6; irrigation, 240–244; river systems, 58, 64, 65, 67; Figs. 4.2, 4.10, 4.14, 4.18

dreikanter (stone having faces smoothly shaped by wind), Fig. 3.9

drift (glacial deposit, whether by the ice or its meltwaters), 3, 129, 130, 141, 153, 185, 191, 198, 201; Figs. 2.2, 7.1, 7.2, 7.3, 9.2, 9.3

dripstone: stalactites or other calcium carbonate deposits attributable to drip from the roof in limestone caverns

drizzle: refers to small size of droplets

drowned coast, drowned valley (submergence due to rise of sea level or downwarping of the land; creates dismembered valleys), 61; Fig. 4.6

drumlin (hill of glacial drift, oval in plan with long axis parallel to direction of ice advance), 65

dry farming: farming without irrigation in semiarid country

field crop: grain, hay, etc; contrasts with truck crop (vegetables) and fruit crop

fill, materials for, Table 13.1

fine earth: soil fraction having particles less than 2 mm in diameter

Finger Lakes, New York, 65, 67, Fig. 7.3

Fini-glacial pause, Table 2.4

fir, Fig. 2.4; Table 2.6

fire, forest, grass, 121, 122

fire clay: clay that can be molded and that has properties suitable for use in commercial refractory products

firn, on glaciers: snow compacted into granules because of alternate melting and freezing

fissure (fracture in bedrock or in compact ground but without noticeable displacement), 38, 40, 150; Figs. 3.2, 3.3,F, 6.13

flagstone (flat rock suitable for pavement), 40

flake (angular or subangular rocks $\frac{1}{4}$ to $2\frac{1}{2}$ inches in diameter), 229. *Cf.* slab, pebble

flint: dense rock with conchoidal fracture composed of chalcedony and opal. *Cf.* chert

flocculate: aggregation of clay and colloidal particles into small groups or masses

flocculation limit (amount of water below which material ceases to behave as a liquid and behaves as a solid), 312

flood, 121, 138; Figs. 6.14, 6.15

floodplain (alluvial surface, usually little higher than a river and subject to flooding at high water stages), 47, 57, 109, 128, 137, 141, 142; Figs. 3.8, 4.10, 6.4, 7.1

flora, Chapter 6. *See also* vegetation

fluorine, fluorite in bone for dating, 29; hardness, 265

flow bog: peat bog that rises or falls with changes in water supply

flow structure: alignment of stones or minerals because of flowage of the mass, particularly in mudflows and lavas

flow till: till that moves like a mudflow or slurry off the surface of a glacier

flowage: movement of a surface deposit by flow because of excess water; e.g. mudflow

flowstone: calcium carbonate deposits in a limestone cavern due to evaporation of water from a film or sheet moving over the surface. *Cf.* dripstone

fluvial, fluviatile: pertaining to streams

fluvial glacial (pertaining to streams of glacial origin), 17

fog: visibility less than 1 kilometer. *Cf.* haze

foliation in bedrock, especially metamorphic rocks: parallelism of minerals in a plane (plane of the a, b axes) along which the rock tends to cleave; in soils (parallelism of soil aggregates in a plane), 40, 232

forest: affects ground temperature, wind, 93, 97; Figs. 5.5, 5.10; litter, 51, 107, 111; kinds, distribution, 109, 110; Fig. 6.2; reflect ground conditions, moisture supply, 109-113; Figs. 6.4-6.10; tree throw, 111, 144, 146; Fig. 6.3

forest floor: layers of litter, duff, and humus on ground under forest, 107, 111; organic matter, Figs. 11.10, 11.11

fosse: linear depression separating two therminal moraines or separating an outwash plain from terminal moraine; attributed to sudden retreat of an ice front

fossil: generally, the petrified remains of animals or plants; also refers to ancient soils. *See also* paleontology and particular species

foundation, engineering problems, moisture, 234, 307

fragipan: acid hardpan cemented by clay skins around grains of silt; platey fracture, generally at layer of deposition; consistence hard, resists root penetration, only slightly permeable

fragmental ejecta: violently erupted and fragmented volcanic rocks; *see* pyroclastic, volcanic ash

freeze, frost, 41, 42, 44, 89, 93, 94, 144, 147, 153, 170, 175, 182, 187; Figs. 3.4, 5.7, 6.3, 8.2; Table 8.1; develops fabric, 232; instability, 307; patterned ground, Fig. 10.7

frost boil: under highway pavement; unstable, thawed, saturated layer that retains water because underlying part of the subgrade is still frozen

fruticose: shrubby

Fullerton Silt, Clay, Table 7.3

fungi (soil organism lacking chlorophyll), 103, 104, 117; Fig. 6.1

fungicide: poison for killing fungus. *Cf.* herbicide, insecticide, pesticide.

fusion, heat of, Fig. 5.6

gabbro (coarse-grained, basic igneous rock with chemical and mineral composition like that of basalt), 64, Table 11.3

garnet (complex silicate mineral characteristic of metamorphic rocks), 267

gastropoda. *See* snail

gel, semisolid colloidal system, 275

geode: hollow concretion

geographic cycle: evolution of a landscape

geolifluction: mass wasting associated with permafrost

geologic maps, for engineering planning, 302-306; Fig. 13.2

geologic time, Chapter 2; Table 1.1

geomorphic cycle: same as geographic cycle

Gettysburg, Pennsylvania, 63

gibbsite (hydrous aluminum clay mineral), 272; Table 11.2

gilgai: patterned ground developed on swelling clays, mostly unsorted nets but, on hillsides, elongated with slope

glaciations, 17, 45, 98, 99, 128-133, 136, 153, 160; Tables 2.2, 2.3, 2.4, 7.1-7.3

glaciers, glacial deposits: cause drainage changes, 45; central U.S., 67, 70, 138; Figs. 4.10,D, 4.11, 7.8; Table 7.3; degree of weathering, 45, 128, 198, 205-206; northeastern U.S., 64, 65, 129, 153; Figs. 4.7, 6.6, 7.3; occurrence in Pleistocene, kinds of deposits, 17, 160; Figs. 4.9, 4.10, 9.2; Table 7.1; western U.S., 68, 74, 88, 132, 136, 149, 150; Fig. 4.12; Table 7.2

glass, volcanic (lava chilled quickly enough to prevent crystallization of minerals; obsidian is an example), 36, 150; Fig. 3.1

glauconite (hydrous silicate of iron and potassium), 120

glazed frost: freezing, wet weather condition that causes ice to coat branches, wires, and other surfaces

gleization: processes developing gley soil

Glenns Ferry Formation (Pliocene and Pleistocene formations on Snake River Plain), Table 7.5

gley (mottled or dark gray, sticky, organic rich layer in ground that is frequently or continuously saturated, resulting in a reducing condition that preserves the organic matter), 43, 169, 170, 174, 175, 185, 216; Figs. 7.5, 8.4; Table 8.1

gneiss (metamorphic rock, banded, commonly with quartz, feldspar, micas, or other minerals in separate layers), 38, 152; Figs. 3.1, 4.7

goethite (hydrous iron oxide), 272, 276; Table 11.2

graded, graded bedding: textural term referring to gradational sizes of particles; a common sequence is a layer with coarse sediment at the base grading upward to fine-grained sediment at the top

grain size, 223-225; Table 10.1; field identification, 225; Table 10.2

gram mole. *See* mole

Grand Canyon, 56, 74, 78, 150

Manneto Gravel, Table 7.1

manure (organic matter used to fertilize ground; may be animal excrement—stable or barnyard manure; may be green manure; near coasts may be seaweed), 288, 289

maple, Figs. 6.4, 6.6

marble (metamorphosed limestone or dolomite), 30, 109, 254; Figs. 3.1, 9.5

Marella Drift (late Wisconsinan drift in western Pennsylvania), Table 7.1

marine (pertains to the sea), 129

marl (clayey limestone or calcareous clay), 174

marsh (wet ground with more or less open water, may be saltmarsh at sea shore or fresh-water marsh in lake country), 109, 113, 114, 128, 135, 148, 160, 161, 174, 182, 215; Figs. 6.3, 6.6, 6.7, 6.9, 6.10, 7.1, 7.21, 8.2, 8.12, 9.12; classes, Table 10.4; extent, 241; Fig. 10.14

marsh gas (methane, CH_4, product of decomposition of organic matter), 107, 257, 279, 280

mass action (in reversible chemical reactions, the direction taken by the reaction depends upon availability of the ions), 273

mass wasting (slow flowage or creep downhill of colluvial material because of intermittent saturation with water, freeze and thaw, rain drop splash, tree throw, or surface washing), 144, 153, 188, 191, 216; patterned ground, Fig. 10.7

massive, soil structure, 231

mastodon: Pleistocene elephant. *See also* elephant

Matterhorn: sharply peaked mountain between cirques, resembling Swiss mountain of that name

mature: soil with well developed profile; topography with broad open valleys between rounded hills

Meade Formation, Table 7.3

meadow: grassland in woods or forest; also, field where grasses are grown

Meadow Soil: acid Holocene soils with dark brown organic rich layer grading downward at 6 to 30 inches into gray soil; develops under grasses and sedges in humid and subhumid climates

mean: in statistics, the average; short for arithmetic mean; obtained by adding the measures and dividing by the number of them. *Cf.* mode, median

meander (one of a series of winding bends along a stream course; may be intrenched between high valley walls; may be cut off by the stream eroding laterally), 135, 137. *See also* incised and ingrown meander

mechanical analysis: separation of soil or surface deposit into different sizes of particles

median: in statistics, the second quartile or 50th percentile; half the measures are above and half below the median. *Cf.* mode, mean

mellow soil: one that is easily worked

Melon Gravel (late Pleistocene deposit on the Snake River Plain, related to overflow of Lake Bonneville), Table 7.5

meniscus, Fig. 10.10

mesa, Southwest (flat topped butte), 56

mesophyte: plant adapted to medium conditions of atmosphere and ground moisture

Mesozoic (third of the four great eras of geologic time), 49, 56, 74, 197, 217; Tables 1.1, 2.1

mesquite (desert tree of genus *Prosopsis*), 117, Fig. 6.11

meta: prefix meaning altered by metamorphism, as metasedimentary or metavolcanic rock

metamorphic, metamorphism (refers to igneous or sedimentary rocks that have been altered and reconstituted by crystallization of new minerals under great pressure and/or heat; alteration by weathering excluded), 4, 38, 40, 64, 109, 174; Figs. 1.1, 3.1, 7.1, 8.12. *Cf.* katamorphism

metapedogenesis: change of soil type because of such changes as those caused by irrigation, drainage, plowing and mixing; burrowing animals, including earthworms, can cause metapedogenesis

metazoa (animals more complex than the protozoa), 2

methane. *See* marsh gas

mica (any of several platey minerals), 265, 267; Fig. 11.3. *See also* biotite and muscovite

microclimate (climate on and near the ground), Chapter 5, 92

microcline (potash feldspar), Tables 11.2, 11.3

microcrystalline: texture in which grains are visible only under a microscope, mostly in range 0.0001 to 0.01 mm

microfauna, microflora, 103. *See also* microorganisms

microorganism (small plants or animals that require a microscope to be seen), 103, 104, 107, 160

Micropodzol Soil: Podzol Soil with distinct layers but no more than about 3 inches thick

microrelief (small scale topography, usually measured in inches or at most a few feet), 55, 107; Fig. 6.3

milliped, 107

Mindel (third glaciation in the Alps), Tables 2.3, 2.4

mineral: in mineralogy, inorganic matter characterized by definite chemical composition and definite crystal structure and form; in sense of mineral deposit, includes numerous non-minerals such as coal, oil, sand, gravel; relative stability, 261–265; Tables 11.2, 11.3; scale of hardness, 265

Minford Silt: lake deposits of Nebraskan or Kansan age, in Ohio

minor elements. *See* trace elements. *Cf.* essential elements

Miocene (fourth epoch of the Tertiary Period), 16, 204; Fig. 9.6; Table 2.2

Mississippi River valley, 138; Fig. 8.12

Missouri Plateau, 129

Missouri River valley, 58, 68, 70; Figs. 4.12, 4.14

mite: arachnid; part of the soil fauna

mode, modal: Petrology, minerals actually present in rocks, surface deposits, or soils. *Cf.* norm; Statistics, the measure or group of measures that occur most frequently; a given distribution having two points of maximum frequency is bi-modal. *Cf.* mean, median

moisture. *See* water

moisture holding capacity. *See* field capacity

mold, fungi, 104

mole: molecular weight of a substance in grams

molecular weight: total of the atomic weights of elements composing a molecule

Mollic: applied to thick, dark, organic rich surface layers of Prairie, Chernozem, and Chestnut Soils; layers are rich in calcium and magnesium

Mollisol, 181

molybdenum, 288, 295

monadnock: isolated hill or mountain rising above a plain or plateau surface, an erosional feature of resistant rock and not a constructional feature; i.e., not a volcano

monazite (phosphate mineral containing rare earths), Tables 11.2, 11.3

monocline, Fig. 4.18

Mono Craters, 150

monolith, Fig. 3.10

montmorillonite (clay mineral, a swelling clay), 263, 267, 268, 269, 270, 271, 272, 276; Table 11.2; structure, Fig. 11.7

monument: erosion remnant, a column or pillar of rock

monzonite (intrusive igneous rock of intermediate composition), Table 11.3

moor: marshy, peaty land

mor: humus without much admixed mineral matter; highly acid

in different kinds of soils, Figs. 11.10, 11.11; Table 8.1

organic matter: decay of, 279; estimating, 278; production, 279; in soils, 36, 46, 85, 107, 109, 119, 158, 160, 167–170, 185, 193, 201, 245, 277–281; Figs. 3.6, 6.3, 8.2

organisms: first, 3; Table 1.1; on and in the ground, 51, 85, 86, 103, 104, 107, 160; Figs. 6.1, 6.3

Orterde: hardpan weakly aggregated with organic matter. *Cf.* Ortstein

Orthid, 181

orthoclase (potash feldspar), 265; Tables 11.2, 11.3

orthogneiss: gneiss derived by metamorphism of an igneous rock. *Cf.* paragneiss

orthophosphoric acid (H_3PO_4), 292

Ortstein: hardpan strongly cemented with organic matter. *Cf.* Orterde

osmosis: passage of liquid through permeable membrane; osmotic pressure draws less saline liquid into the more saline side of the membrane

Ouachita Mountains, 67, 70; Figs. 4.3, 7.16, 8.12

oued: arroyo (Mediterranean)

outcrop: usually refers to exposure of bedrock, as in Figs. 1.2, 1.3, but also applicable to exposures of surface deposits and soils

outwash (deposits by glacial meltwaters issuing from ice), 136; Figs. 4.9, 9.3

outwash plain (plain built of glacial outwash), 65; Fig. 4.9

overgrazing. *See* land use

overlap: deposition of a young deposit beyond the limits of an older one

overturn, in lakes: there is likely to be a spring and fall overturn and consequent mixing of the epilimnion and hypolimnion each spring and fall

ox-bow lake (crescentic lake occupying a cutoff meander on a floodplain), 137

oxic: soil layer from which silica has been leached by alteration of the silicate minerals

oxidation, -oxidize (addition of oxygen to an element or compound, opposite of reduction), 43, 107, 167, 169, 191, 198, 205, 208, 272–273, 275; Figs. 3.6, 9.9; of organic matter, 279–280

oxidation-reduction potential: symbol E_h; capacity to give up electrons (oxidation) and to accept them (reduction). *Cf.* redox potential

oxide, 191, 192, 216, 222

Oxisol, 181

oxygen, 43, 86, 107, 160, 257, 259, 280; in rocks, 254

oxyhypopedon, 182

Ozark Plateau, 67, 71, 153; Figs. 4.3, 7.16, 8.12

ozone: molecule with three atoms of oxygen, O_3; mostly in upper atmosphere where it absorbs lethal ultra-violet rays; amount would make a layer only 3 mm thick

paleobotany (paleontology of plants), Figs. 2.4, 2.5; Table 2.6

Paleocene (earliest epoch of the Tertiary Period), 155; Fig. 7.16; Table 3.2

Paleolithic, in Europe (Magdelenian and older cultures), 27; Table 2.4

paleontology (study of ancient life forms and the environments they indicate; preserved as fossils), 15, 18, 20, 22, 27, 54, 143, 155, 203, 206, 208, 210, 211; Figs. 2.1, 2.3, 2.4, 9.6; Tables 2.3, 2.5, 2.6, 7.4

Paleozoic (second of the four areas of geologic time), 56, 64, 70, 72, 128, 129, 217; Figs. 4.4, 4.8, 4.15, 8.12

Palisades, of the Hudson, 63

Palouse Soil, 141

paludal: marshy

palynology: study of pollen

Pampas: treeless plain, Argentina

pan. *See* hardpan, salt pan

paragenesis: sequence in which minerals develop, in rock or soil; original (primary) minerals become replaced by younger (secondary) ones because of weathering or metamorphism

paragneiss: gneiss formed by metamorphism of sedimentary rock. *Cf.* orthogneiss

parasite, 104

parent material (rock from which a surface deposit was derived, or surface deposit from which a soil was derived), 167, 169, 174, 185, 187, 191, 193, 198, 199, 201, 203, 204, 211, 216; Figs. 7.1, 8.2, 8.9, 8.12, 9.9; Table 9.1; may control fabric, 232

pathogen (microorganism causing infection or disease), 104

patina: satiny weathered surface of prehistoric artifacts, or similar surfaces on early Holocene and older rocks

patterned ground (ground having more or less repetetive patterns of polygons, circles, nets, stripes, terracettes; characteristic of surface deposits and soils over permafrost or in saline ground), 232; Fig. 10.7

Patuxent Formation (Cretaceous deposit on Coastal Plain; overlaps saprolite), 155

pavement: course of roadway above the subgrade; may be flexible or rigid. *See also* desert pavement

Pearlette Ash (bed, or beds?, of Pleistocene volcanic ash on Great Plains and in Rocky Mountains; best known one is of Yarmouth interglaciation), Table 7.3

peat (plant matter accumulated under reducing conditions in marshes and only slightly decomposed; may be woody if composed of stemmed shrubs or trees; or fibrous if composed of mosses, sedges, rush, or grass), 4, 87, 98, 114, 115, 160, 161, 182, 185, 244, 279, 281; Figs. 6.3, 7.1, 7.21, 8.2, 10.15; Table 8.1; instability, 309; Table 13.1

pebble (round or subround stones $\frac{1}{4}$ to $2\frac{1}{2}$ inches in diameter), Fig. 10.2; Table 10.1. *Cf.* cobble, flake; pebble counts, 228

peccary, Table 2.5

ped: individual soil aggregate or other soil structure

pedalfer (acid soil in which iron and alumina have been removed from leached layer and redeposited in depositional layer; open system soil from which alkalis and alkaline earths are removed), 167. *Cf.* pedocal

pedestal rock, Fig. 3.10

pediment (eroded, planed surface sloping from base of hills or mountains; stream eroded; may or may not be covered by alluvium or gravel), 47, 189

pedocal: alkaline soil of arid or semiarid regions characterized by depositional layer containing calcium carbonate and other salts; closed system soil that retains its alkalis and alkali earths. *Cf.* pedalfer

pedogenic: pertaining to a soil origin

pedology: soil science

pedon: plant community dependent on lake or river bottom; smallest unit that can be called soil; width may be 1 to 10 meters; depth down to parent material

pelite: deposit composed of fine particles as clay, shale

peneplain: literally, almost a plain, and as such a useful word; has unfortunate genetic connotation that a peneplain is developed by stream erosion and reduced nearly to sea level

Pennsylvanian (sixth period of the Paleozoic Era), Table 2.1

Pensauken Formation (deeply weathered pre-Wisconsinan formation along the Delaware River), 203; Fig. 9.4

Peorian (mid-Wisconsinan loess), Table 7.3

percentile in statistics: frequency distribution expressed in percent